Chemical Containment of Waste in the Geosphere

It is recommended that reference to all or part of this book should be made in one of the following ways:

METCALFE, R. & ROCHELLE, C. A. (eds) 1999. *Chemical Containment of Waste in the Geosphere.* Geological Society, London, Special Publications, **157.**

SAVAGE, D., ARTHUR, R. C. & SAITO, S. 1999.Geochemical factors in the selection and assessment of sites for the deep disposal of radioactive wastes. *In:* METCALFE, R. & ROCHELLE, C. A. (eds) *Chemical Containment of Waste in the Geosphere.* Geological Society, London, Special Publications, **157,** 27–45.

GEOLOGICAL SOCIETY SPECIAL PUBLICATION NO. 157

Chemical Containment of Waste in the Geosphere

EDITED BY

R. METCALFE & C. A. ROCHELLE

British Geological Survey,
Kingsley Dunham Centre,
Keyworth, Nottingham, UK

1999

Published by

The Geological Society

London

THE GEOLOGICAL SOCIETY

The Geological Society of London was founded in 1807 and is the oldest geological society in the world. It received its Royal Charter in 1825 for the purpose of 'investigating the mineral structure of the Earth' and is now Britain's national society for geology.

Both a learned society and a professional body, the Geological Society is recognized by the Department of Trade and Industry (DTI) as the chartering authority for geoscience, able to award Chartered Geologist status upon appropriately qualified Fellows. The Society has a membership of 8600, of whom about 1500 live outside the UK.

Fellowship of the Society is open to those holding a recognized honours degree in geology or cognate subject and who have at least two years' relevant postgraduate experience, or who have not less than six years' relevant experience in geology or a cognate subject. A Fellow with a minimum of five years' relevant postgraduate experience in the practice of geology may apply for chartered status. Successful applicants are entitled to use the designatory postnominal CGeol (Chartered Geologist). Fellows of the Society may use the letters FGS. Other grades of membership are available to members not yet qualifying for Fellowship.

The Society has its own publishing house based in Bath, UK. It produces the Society's international journals, books and maps, and is the European distributor for publications of the American Association of Petroleum Geologists, (AAPG), the Society for Sedimentary Geology (SEPM) and the Geological Society of America (GSA). Members of the Society can buy books at considerable discounts. The publishing House has an online bookshop (*http://bookshop.geolsoc.org.uk*).

Further information on Society membership may be obtained from the Membership Services Manager, The Geological Society, Burlington House, Piccadilly, London W1V 0JU, UK. (Email: *enquiries@geolsoc.org.uk*: tel: +44 (0)171 434 9944).

The Society's Web Site can be found at *http://www.geolsoc.org.uk/*. The Society is a Registered Charity, number 210161.

Published by The Geological Society from:
The Geological Society Publishing House
Unit 7, Brassmill Enterprise Centre
Brassmill Lane
Bath BA1 3JN, UK

(*Orders*: Tel. +44 (0)1225 445046
        Fax +44 (0)1225 442836)
Online bookshop: *http://bookshop.geolsoc.org.uk*

First published 1999

The publishers make no representation, express or implied, with regard to the accuracy of the information contained in this book and cannot accept any legal responsibility for any errors or omissions that may be made.

British Library Cataloguing in Publication Data
A catalogue record for this book is available from the British Library.

ISBN 1-86239-040-1
ISSN 0305-8719

Typeset by Aarontype Ltd, Bristol, UK

Printed by Cambridge University Press

Distributors

USA
    AAPG Bookstore
    PO Box 979
    Tulsa
    OK 74101-0979
    USA
Orders: Tel. +1 918 584-2555
        Fax +1 918 560-2652
        Email *bookstore@aapg.org*

Australia
    Australian Mineral Foundation Bookshop
    63 Conyngham Street
    Glenside
    South Australia 5065
    Australia
Orders: Tel. +61 88 379-0444
        Fax +61 88 379-4634
        Email *bookshop@amf.com.au*

India
    Affiliated East-West Press PVT Ltd
    G-1/16 Ansari Road, Daryaganj,
    New Delhi 110 002
    India
Orders: Tel. +91 11 327-9113
        Fax +91 11 326-0538

Japan
    Kanda Book Trading Co.
    Cityhouse Tama 204
    Tsurumaki 1-3-10
    Tama-shi
    Tokyo 206-0034
    Japan
Orders: Tel. +81 (0)423 57-7650
        Fax +81 (0)423 57-7651

Contents

Containment of metalliferous wastes

Investigative methods

Chemical containment of waste in the geosphere

RICHARD METCALFE & CHRISTOPHER A. ROCHELLE

British Geological Survey, Kingsley Dunham Centre, Keyworth, Nottingham N12 5GG, UK

Abstract: The aim of this introductory paper is to highlight those underlying chemical principles that are common to all forms of waste management by geological means, and that rely to some extent upon chemical containment. Until recently, chemical processes were usually considered mainly because they can affect the physical performance of engineered containment systems. However, in recent years, many researchers have recognized that chemical processes themselves can offer containment to wastes. Thus, it is no longer possible to view physical and chemical containment processes separately. The containment system can be optimized only if both the engineered and natural barriers are considered together, and if the engineered barrier system is designed taking the features of the geosphere into account. However, there has been relatively little reliance upon the geosphere itself as a chemical barrier. It is concluded that the potential for chemical containment should be considered in all forms of geological waste management. Even if the chemical barrier function of the geosphere is not relied upon to meet safety targets, the confidence of regulators and public alike will be enhanced if it can be demonstrated that the geosphere at the site functions as a chemical barrier.

Geological disposal is an option that is taken for many different kinds of waste, ranging from domestic refuse (e.g. Christenson *et al.* 1994; Department of the Environment 1990, 1995) to radioactive wastes (e.g. Savage 1995; Kass *et al.* 1997; Anon 1999). These diverse materials are usually contained using some anthropogenic, or 'engineered' barrier system that is constructed within the geosphere. The wastes are contained by a combination of both physical and chemical processes operating within both artificial and natural parts of this system.

It is important to recognize the coupling between the chemical and physical containment processes when designing any particular waste containment strategy. It is also necessary to evaluate the interactions between processes operating within any engineered barrier system and processes operating within the geosphere beyond. The lack of this will lower the confidence in evaluating the temporal variation in risk arising from any waste. Consequently, at best, possible advantageous contributions to containment from chemical processes within engineered barriers and/or the geosphere may be ignored. Conversely, at worst, there may be unforeseen increases in risk at some time in the future, owing to changing chemical conditions. For example, both chemical and physical processes may be important in containing waste early in the life of a barrier. However, later in the life of the barrier, chemical conditions might change and cause the pollutants that have accumulated within the barrier to be released. The efficacy of the barrier will then depend mainly upon the physical processes restricting contaminant transport and rather less upon chemical processes. It is critical to the development of safe containment strategies that such coupling is understood. Such an understanding is essential if we are to prevent the chemical processes governing containment early in the life of a barrier from causing the later release of much greater concentrations of pollutants than otherwise would have been the case.

However, until relatively recently, most workers concerned with waste management have tended to consider chemical processes primarily because they may affect the physical containment properties of engineered barrier systems. Several texts have examined these physical aspects of containment in considerable detail (e.g. Bentley 1996). Implicitly, there has been a tendency to view 'chemical containment' as an aspect of physical containment. For example, any collapse of expandable clay minerals, such as may be caused by interactions involving polar organic molecules, will affect the physical integrity of clay barriers (e.g. Bowders & Daniele 1987; Hettirachi *et al.* 1988). However, this view of containment is simplistic. In reality, chemical and physical processes must be considered holistically. For example, where clay is used to confine a waste, it should be considered as a physico-chemical barrier to contaminant migration (Horseman *et al.* 1996).

Furthermore, the term 'containment' means different things to different people. Many members of the public require reassurance that any waste will be rendered completely immobile within the disposal environment. However, the earth scientist may view 'containment' as 'highly restricted leakage' because, over a sufficiently long time, some contaminants will leak from any containment system around any waste form. Regulators tend to be influenced by both schools of thought and the particular regulatory framework usually depends upon the timescale for which the waste will be hazardous. For example, the management of domestic wastes typically considers timescales of the order of tens of years (Department of the Environment 1995). Over such time intervals the retention of waste within a tightly defined area is often seen as a technically feasible option. Coupled with an increasing political and social aversion to the 'dilute and disperse' or 'dilute and attenuate' approaches to waste management (Stief 1989), this view has caused renewed emphasis to be placed upon active containment technologies, rather than passive containment strategies. By definition, chemical processes are important aspects of such active containment strategies. However, active containment implicitly focuses attention upon chemical processes within engineered barriers and rather less attention is given to chemical processes within the surrounding geosphere.

At the other end of the spectrum is radioactive waste management, which considers time spans greater than 10^6 years (e.g. Nagra 1994; Kass *et al.* 1997; Nirex 1997). While many members of the public would certainly like to see these wastes contained indefinitely, regulators recognize that containment in the strict sense of the word is impossible over these timescales. Thus, relatively more attention is given to the migration of pollutants through the geosphere.

These issues present several important questions to researchers, notably:

- What are the most important chemical processes that can be relied upon to impart chemical containment in the management of different wastes?
- Can we predict the contribution of chemical processes to containment?
- If so, over what timescales can we make such predictions?
- How do we acquire the relevant data with which to evaluate the efficacy of chemical containment?
- To what extent should engineered barrier systems be relied upon to provide chemical

containment and to what extent can the geosphere itself be used as a chemical containment barrier?
- What generic lessons about chemical containment processes can be learnt by comparing the results from different waste management disciplines?

For all these reasons, it is now timely to evaluate what is known about chemical containment processes in different geological environments and to review how this knowledge is put to use by different waste management industries. Therefore, the present volume attempts to provide an overview of the topic of chemical containment, with the hope that important lessons can be drawn from the experiences of different industries.

The principles of chemical containment

Fundamental issues

The overriding aim of this volume is to assess the extent to which chemical processes might contribute to the confinement of a particular waste within the geosphere. In any type of waste containment, it is not necessary to ensure that no waste will ever leak from the disposal site. Rather, it is required to show that for as long as the waste remains hazardous it poses an acceptably low risk to the biosphere. Thus, this volume is really concerned with the degree to which chemical processes will restrict the movement of contaminants through the geosphere.

Potentially, both natural and 'engineered' barriers can be used to provide chemical containment at or around waste disposal sites. As a general rule, waste managers concerned with short-term waste containment, typically over timescales of tens of years, tend to rely upon engineered barriers. Conversely, those waste managers dealing with long-term containment, even up to millions of years, tend to give a greater emphasis to deep, natural geological barriers. This general division is logical since it is to be expected that many deep, natural barriers will be more stable, even over geologically significant periods of time, than will anthropogenic barriers. On the other hand, natural geological barriers are likely to be less suitable than engineered barriers in near-surface environments, where conditions are more oxidizing, groundwater flow rates are generally greater and the potential for human intrusion is higher. However, it is necessary to consider to what extent natural barriers can, or should, be relied upon in near-surface disposal situations and to what degree these barriers can provide long-term containment of the waste.

There are three distinct but interlinked aspects to chemical containment

- chemical transformation of contaminants
- retardation
- attenuation

Yong deals these basic aspects and their implications for regulators. He points out that the ultimate fate of contaminants in the longer term will depend to a large degree on the redox conditions in the barrier system and upon the oxidation/reduction reactions that occur. However, when predicting the transport of contaminants it is essential to consider not only their chemical transformations within the barrier system, but also contaminant retardation and contaminant retention mechanisms. Such predictions require an understanding of the chemical mechanisms controlling the partitioning of pollutants between fluid phases (mobile) and solid phases (generally immobile). While many of the underlying processes are well known individually, the complex interactions between them are not yet fully understood. Therefore, it is currently impossible to incorporate them into fully quantitative transport models.

The contribution by **Yong** develops its themes by reference to the interactions between contaminants and soil fractions that are responsible for chemical partitioning. However, the key concepts are equally applicable to geological management of many wastes in many diverse environments. Much of the work reported in the other contributions of this volume may be viewed as dealing with the mechanisms by which the geosphere may retard and/or attenuate various actual and potential pollutants.

Chemical containment in engineered barriers

Chemical containment in engineered barriers may be divided into two main types:

- permeable active containments, which rely upon the chemical properties of the containment to remove the pollutant, but the barrier does not restrict significantly the flow of groundwater;
- low-permeability active containments, which rely on low-permeability barriers to retard the movement of both pollutants and groundwater.

The first type of containment is particularly relevant where large volumes of water must be treated. Such containment is dealt with by

Bowell *et al.* who review of the role of chemical containment in the management of mine wastes. In broad terms, three main management strategies are used by this industry:

- controlling the source of contaminants from the solid phases, e.g. by removing sulphide minerals from the waste
- restricting the interaction between solid phases and environmental agents, e.g. preventing reaction with air or water
- controlling the overall release of contaminants into the environment, e.g. by collecting and treating contaminated water

However, they point out that mining is usually conducted so as to return a profit to investors, and the nature of waste containment is likely to be determined by cost implications as well as the types of methodology that are suitable. In general, therefore, if waste containment becomes too expensive then the mine will become uneconomic and will either close or never open in the first place.

At the other extreme, active containment by low-permeability barriers is commonly relied upon by the radioactive waste management industry. Both **Alexander & McKinley** and **Frank** consider the role of chemical containment in the engineered barrier system proposed for a high-level radioactive waste repository in Switzerland. The former authors point out that, in this case, the initial design of the waste form was based solely on engineering criteria. However, several chemical phenomena of relevance to containment would occur as a result of the engineering criteria that had to be met. For example, the engineered system included: redox buffers (steel canisters), pH buffers (bentonite backfill), a capability to filter colloids (owing to the bentonite backfill), low permeability causing mass transport to be by means of diffusion (which can be viewed essentially a chemical process) and a high sorptive capacity (once again owing to the backfill).

The degree to which low-permeability active containment is applicable to other types of wastes is currently an area of some debate. Until relatively recently, chemical processes were considered only to assess the extent to which they might cause an engineered barrier system to degrade. This has been a particular concern for managers of landfills in many parts of the world. Degradation of the waste will produce leachate, which may in turn contaminate groundwater resources. Lining the landfills with a low-permeability barrier will help to attenuate the migration of this leachate. Although a variety of different materials are used as liners for landfills, one of the more common materials used, either on its own or in combination with other materials, is sodium bentonite. Despite its widespread use, there is uncertainty about its ability to attenuate landfill leachate in the long term. **Spooner & Giusti** review the properties and reactions of bentonite with leachates from municipal solid wastes. The high surface area and interlayer chemistry of sodium bentonite suggest that it has the potential to be an effective low-permeability barrier. However, these properties also make this material susceptible to damage. For example, leachates having high ionic strengths may cause bentonite shrinkage, and hence an increase in permeability.

For this reason, sodium bentonite might not be the most appropriate barrier material. Furthermore, other materials may be available more readily and might be more cost-effective to use. Therefore, much research has been directed at other low-permeability materials that might be used for active containment. **Thornton et al.** present results from a study of a high-attenuation liner material constructed from low-cost natural and industrial waste available in the UK. Laboratory column experiments were used to assess the performance of this and other potential liner materials to contain mobile organic and inorganic components by chemical means. Results of the experiments are discussed and linked to important landfill liner design issues. For example, additives could be included within the main liner material, or a separate liner could be placed beneath the main liner to act as a secondary chemical containment system.

Glacial clays are widely available in many countries from the northern hemisphere and also have some potential to be used as active containment media in landfills. **Hurst & Holmes** present results from a study using column and tank experiments to investigate the potential for these clays to attenuate heavy metals. When typical landfill leachate was passed through the columns, retardation of heavy metals was observed (relative to columns flushed with de-ionized water). This occurred primarily at the inlet end of the columns where a black, metal-rich 'sludge' formed. Tank experiments confirmed the presence of this sludge. Possible origins of this sludge are suggested, including; metal reduction and precipitation as sulphides as a result of either inorganic or microbial reactions, and adsorption of metals onto colloidal or particular material which settles out on the top of the clay surface.

All these laboratory studies provide an important means by which the suitability of active containment of low-permeability liners might be

assessed. However, **Spooner & Giusti** note that although laboratory studies can provide valuable insights into the specific processes causing these effects, it is often difficult to extrapolate such relatively simple studies to more complex real systems. Such complexity means that long term predictions are currently uncertain. This is an area where additional research would be beneficial.

Cementitious barriers have also been used to contain a wide variety of wastes. These barriers may cover a range of permeabilities, but chemical processes are always prominent controls on their containment properties. To date, most research into these barriers has been undertaken by the radioactive waste management industry. However, it is recognized that there is considerable potential to use such barriers to manage other wastes (Jefferis 1981; Evans 1985, 1997).

Two important features of cementitious materials are that they have internal chemical environments conducive to containing many types of waste and they may also favourably modify the surrounding chemical environment within the geosphere. In particular, these barriers generate highly alkaline conditions and can retard the mobility of radionuclides. **Hodgkinson & Hughes** present a study of reaction products from laboratory experiments, and from archaeological samples. The nature of secondary phases formed by the interaction of cement pore waters with the cement aggregate or surrounding rocks were investigated using Analytical Transmission Electron Microscopy. If protected from carbonation (which would occur if in contact with atmospheric carbon dioxide), many of the reaction products are stable for prolonged timescales even though they are highly disordered, calcium silicate hydrate (CSH)-type phases. It is also noteworthy that many of the phases studied were relatively aluminous which makes such systems difficult to model using predictive software packages because of the general lack of basic thermodynamic data for such phases. For example, many modelling studies predict the precipitation of CSH phases (as a sink for Ca and Si) and zeolites (as a sink for Al and Si). Highly disordered aluminous phases are not predicted because data for them are not generally available. This highlights the need for new basic data to help improve the predictive capability of geochemical computer codes in dealing with cementitious systems. A similar finding is presented by **Bateman et al.**, who also emphasize the fact that available thermodynamic and kinetic data are inadequate to predict accurately the mineralogical variations that occur in experimental simulations of cement–porewater interactions. However, general temporal variations in the Ca : Si ratios of the solid phases can be modelled successfully. In contrast, the predicted variations in porosity were less amenable to modelling. Therefore, these results demonstrate that the coupling between chemical and physical containment processes cannot be simulated fully at present, at least partly due to the lack of basic thermodynamic and kinetic data.

Chemical containment in natural barriers

The role of the geosphere itself as a chemical containment barrier is relatively difficult to evaluate. This is partly because the geosphere is complex and heterogeneous, and partly because it is difficult to acquire information with which to quantify relevant chemical processes. This latter problem reflects the fact that the natural conditions of the geosphere are inevitably perturbed by any procedure that might be used to investigate them. For these reasons, there has been a tendency to focus attention on the chemical containment properties of engineered barriers, rather than natural barriers. Thus, **Alexander & McKinley** point out that the most recent Swiss safety assessment for a high-level radioactive waste repository assumed that the main role of the geosphere is to provide a suitable environment in which the near-field barriers can function properly over the required time intervals. In this case, it is currently envisaged that high-level waste will be isolated by a multi-barrier system consisting of the waste-form itself, a leach-resistant glass matrix surrounded by steel containers, surrounded by a steel overpack, surrounded in turn by bentonite backfill. However, chemical containment in the geosphere potentially may still contribute to the safety of any disposal system. For example, **Frank** points out that at Wellenberg, a potential site in Switzerland for a repository of low- and intermediate-level radioactive wastes, chemical containment of the radionuclides is a feature of both engineered and natural barriers.

The most important geochemical parameters governing waste isolation in the geosphere are highlighted by **Savage et al.**, groundwater pH, redox, partial pressure of carbon dioxide (p_{CO_2}), ionic strength, colloid concentration, organic and inorganic ligand concentrations, and the mineralogical composition and structure of groundwater flow paths. However, they also emphasize that it is important to determine the chemical buffering capacity of the geosphere when evaluating the future performance of a barrier system. It is insufficient simply to show that the present chemical conditions are suitable for waste containment. It must also be demonstrated that the barrier system will resist chemical changes that might otherwise be caused by transient phenomena such as climate change (including glaciation) and variations in heat flow. They also emphasize the need to evaluate whether the chemical conditions of the natural geosphere are suitable to maintain an adequate performance of the engineered barrier system.

Quantifying the stability of chemical conditions and the chemical buffering capacity is particularly challenging to researchers. One reason for this is that moving groundwater will normally be in contact with only a small proportion of the total rock mass. Identifying which portions of the rock mass can react with a particular groundwater, characterizing the reactions that occur, and quantifying their effects is difficult. However, it is easier to quantify the chemical containment properties of the geosphere where the waste being considered is chemically simple and has well-defined physical and chemical properties. Carbon dioxide (CO_2) is one such waste. However, even then, a large research effort is required to estimate the containment capacity of the geosphere. Such research is described by **Rochelle et al.**, who deal with problems resulting from anthropogenic emissions of CO_2 to the atmosphere, and the potential for reducing these by its storage as a supercritical fluid within the deep geosphere. Their work indicates that over relatively short timescales, physical processes act to 'trap' the CO_2, but over longer timescales it will dissolve into the formation porewater to form a variety of dissolved species, or precipitate as secondary carbonate minerals such as calcite. The containment of CO_2 by such chemical reactions depends upon the type of rock involved. For example, sandstones appear preferable to carbonates as potential host rocks, primarily because of their greater capacity to buffer pH and hence greater potential for storage.

In the cases of complex wastes, such as radioactive wastes, it is even more difficult to quantify the role of the geosphere adequately. An approach to tackling these issues is given by **Iwatsuki & Yoshida**, who present approaches for characterizing the containment properties of the deep geosphere in the Tono area of central Honshu, Japan. Their work shows that the chemical containment properties of structurally distinct portions of the same fractured rock mass may differ. They found that different fractures within a particular rock formation contain chemically distinct groundwaters, each with a differing potential to mobilize radionuclides. Notably, the redox states of groundwaters in different types of fracture varied significantly. This phenomenon probably reflects the differing degrees of interconnectivity shown by different fractures and demonstrates an important coupling between chemical and physical containment processes within the geosphere.

It is particularly challenging to evaluate the chemical containment properties of natural low-permeability lithologies. These difficulties arise because it is very difficult to obtain reliable water or mineral chemical data at the spatial scales required for evaluating containment. These difficulties are tackled by **De Windt et al.**, who investigated the chemical containment properties of indurated fractured shales at the Tournemire research site in southern France. They used detailed petrographical observations to identify naturally open fractures. Constraints were placed on the chemistry of porewaters by analysing leachates. Analyses of the rock matrix and fracture-filling minerals were used as evidence for the past immobility of U, Cs, Ce (an analogue for Th), and Ba (an analogue for Ra). They concluded that, as far as these elements are concerned, the shales had effectively remained a closed system throughout diagenesis.

One particularly heterogeneous chemical containment process in the geosphere is sorption. This is highly dependent upon the distribution of a relatively small number of minerals, notably clays and iron oxides, within the rock mass. These minerals may be distributed only in certain parts of the rock mass, such as in fractures. The capacity of the rocks to sorb radionuclides will depend upon whether or not these parts of the rock mass are hydraulically active. **Berry et al.** investigated the contribution that would be made by sorption to the possible containment of radionuclides by Ordovician volcanic rocks from Sellafield, northwest England. They adopted a conservative experimental approach to measure sorption onto rock samples (the measurements of sorption are considered to be underestimates). They concluded that the geosphere, rather than engineered barriers, is the most important component of containment in the cases of radionuclides such as ^{79}Se, ^{135}Cs and ^{242}Pu. Sorption is an important component of this containment, particularly for nuclides such as those from the ^{238}U decay series that will not be contained completely within the engineered barriers.

The reduction/oxidation (redox) state of the geosphere is a very important control on its containment properties. Unfortunately, the redox state is one of the most difficult parameters to determine quantitatively. This difficulty is partly due to the fact that any investigative procedure will tend to perturb redox conditions. Furthermore, in any case, groundwater/rock systems are characteristically in a state of redox disequilibrium. However, it is possible to build confidence that redox conditions will lie within specified ranges by identifying key chemical mechanisms that control redox states. The results of one such investigation, at the Äspö Hard Rock Laboratory in Sweden, are presented here by **Banwart et al.** who show that, although inorganic redox reactions can be relatively slow at near-surface temperatures, they are much faster when enhanced by microbial processes. They investigated a fracture zone that was intercepted during the driving of a tunnel at about 70 m depth. The tunnel enhanced fluid flow in this fracture zone, providing a route for relatively oxidizing surface water into more reducing parts of the fracture zone. However, even though the region has exposed bedrock and almost no soil, the surface water also contains significant dissolved organic carbon. This has reacted with the dissolved oxygen whilst the groundwater was still in the fracture zone, thereby maintaining reducing conditions in the tunnel.

Evaluating chemical containment

An important question when evaluating the extent to which chemical containment occurs is 'How well do we need to demonstrate the adequacy of chemical containment?' This is a question that taxes both regulatory authorities and organizations responsible for waste management alike. It can often be stated that containment has been demonstrated adequately when the general public feels comfortable with the particular management option being evaluated. However, there is clearly a complicated coupling between the public perception of safety, objectively determined health risks, and regulatory requirements. For example, usually, the public will not be satisfied that a management option is safe unless the relevant regulatory limit is met. Conversely, the regulatory limit may be set at some suitably low level that is considered likely to be accepted by the public. This coupling can be well illustrated by the approach to setting a safety target for any future radioactive repository in the United Kingdom. The National Radiological Protection Board of the United Kingdom considers that 1 fatality in 10^5 people is the maximum risk which should be permissible from any radioactive waste repository, but actually recommends a risk target of 1 in 10^6 people per year, because there is a consensus that this is defensible to the public (Kass et al. 1997).

The disposal of radioactive wastes is in fact the most tightly regulated of all types of waste disposal. Therefore, evaluations of the effectiveness of chemical containment in relation to these types of waste have certain important implications for the management of other types of waste. For example, the overall objective of radioactive waste disposal is highlighted by **Frank** as the requirement to 'protect the public health and environment effectively and reliably and to minimize any burden on future generations'. This objective should also be applied to all types of waste disposal in the geosphere.

One important aspect of safety assessment is to predict the degree to which toxins present in the waste will potentially affect living organisms. That is, it must be established to what extent the toxins are bioavailable. This concept of bioavailable waste is closely linked to the concept of chemical containment. Chemical containment will be achieved adequately if chemical processes prevent toxins from being bioavailable. This issue is illustrated by **West et al.**, who consider the mobility of heavy metals at industrially contaminated sites. They point out that the health risks arising from these metals depend largely upon their chemical forms. This is partly because the toxicity of the metals depends upon their forms and partly because their mobility also depends upon these forms. Thus, it is inappropriate to use only the total concentration of a metal present in a piece of contaminated land to evaluate the risk that is posed by the metal. They concluded that simple batch leaching tests offer the best currently available means for incorporating mobility into risk assessment. Unambiguous speciation of contaminant metals in solid requires an integrated approach using both spectroscopic techniques to characterize the solids and leaching tests.

The extent to which chemical containment may contribute to a risk target being met is usually technically demanding and quantitative assessments of risk are difficult. **Alexander & McKinley** consider how a conservative approach can be used to ensure that the presumed impacts on risk are defensible. They describe how chemical containment was taken into account

during a performance assessment in connection with a high-level radioactive waste management programme in Switzerland. The solubilities of key radionuclides were selected to be 'reasonable, but pessimistic in the handling of uncertainties'. A key principle was that '...most phenomena that could be detrimental to safety are excluded or forced to low probability by the repository design or siting concept'. These authors also highlight the value of natural analogue studies in building confidence in performance assessments for potential radioactive waste repositories. These analogue studies have shown that the assumptions underlying performance assessments are indeed conservative.

Numerical evaluations of risk connected with other forms of waste are less sophisticated than those adopted by the radioactive waste management industry. However, in recent years there has been an increasing realization that such evaluations are useful. **Plimmer et al.** describe a method for taking uncertainty into account when estimating risks from landfills. They describe how the computer code LandSim is used by regulatory authorities in the UK to aid the decision making process when considering licence applications for landfills. A feature of LandSim is that it may take chemical containment into account as well as physical containment. The code is a particularly useful tool when comparing different landfill designs and allows optimization of the containment system to permit the maximum use of chemical containment, in both engineered and natural barrier systems.

Inevitably, waste disposal tends to be within low-permeability rock formations, to take advantage of physical containment processes. Assessment of the chemical containment properties of these media presents the challenge of acquiring chemical data from pore waters. By virtue of their low permeability, it will usually be impossible to obtain samples of flowing water directly from the formation. **Reeder & Cave** evaluate alternative methods for acquiring such data, including squeezing, leaching and centrifugation. They conclude that these techniques are most appropriate for determining concentration gradients for major solutes and potentially for evaluating the actual spatial distributions of certain pollutants. This information is most useful for constraining theoretical models for past and future mass-transport. However, the data are rather less valuable for determining the chemical speciation of solutes, owing to the fact that pH and redox constraints are invariably unreliable.

Discussion and conclusions

The term 'containment' is poorly defined in connection with waste management by geological disposal. The term implies the retention of hazardous waste constituents within a given rock volume by some kind of barrier system. It may be true that some contaminants will be locked into immobile phases for geologically significant time periods. However, in practice, over a sufficiently long time, some contaminants will leak from any containment system around any waste form. Therefore, in reality, 'containment' should really be viewed as 'highly restricted leakage'. The efficacy of containment must be judged against criteria laid down by regulatory authorities, which will vary depending upon the social and political circumstances in any particular country and upon the particular waste.

The contributions in this volume demonstrate that many chemical processes contributing significantly to waste containment can be identified and characterized. It has also been recognized increasingly that chemical processes will affect the time-dependency of risks arising from a particular waste. However, to quantify the effects of these processes is rather more difficult and predictions become increasingly unreliable as longer future timescales are considered. In particular, reliable quantitative assessments of the chemical buffering capacity of the geosphere and the rates of relevant geochemical reactions cannot be made in most cases. It follows that accurate predictions of the contribution made by chemical containment to lowering future risks also become increasingly difficult as timescales become greater. In the worse case, chemical containment processes that lead to low risks early in the life of a containment system, may lead to higher risks later on, if chemical conditions change.

Many research programmes, and especially those dealing with the management of radioactive wastes, have made great strides in acquiring the data required to make such predictions. However, reliable, fully quantitative predictions are still impossible for most waste disposal situations. Therefore, evaluations of the reliability of containment tend to take a conservative approach and deliberately aim to underestimate the effectiveness of chemical processes that contribute to containment. Furthermore, in the case of radioactive waste management, where engineered barriers and natural barriers are envisaged, the emphasis is placed upon quantifying processes within the engineered barrier system, rather than the processes within the natural barriers. This approach is taken because the composition and construction of engineered barriers are more readily characterized.

In the cases of other types of waste management, there are usually even less relevant data available with which to predict chemical containment accurately and quantitatively. A particular limitation is the poor understanding of the coupling between relevant chemical partitioning processes. This means that the effects of chemical containment on risk cannot be predicted quantitatively and reliably. The experience of research programmes concerned with the disposal of radioactive wastes implies that a very large research effort will be required before accurate, quantitative, predictions can be made. However, even if fully quantitative assessments are impossible, there is still merit in showing that chemical containment processes will occur and that it will cause the safety of the barrier system to be enhanced. Furthermore, it is essential to ensure that certain critical conditions are not developed, thereby preventing the barrier system failing.

There is also scope for technologies developed in one waste management discipline being applied to another. From the contributions presented in this volume, it is apparent there is a certain synergy between research into the uses of clay barriers in landfills and radioactive waste repositories. In the UK, the view is widely held that clay barriers used as landfill liners function as physical barriers rather than as chemical barriers. Accordingly, the relevant regulator, the Environment Agency of England and Wales, requires that these clay liners have a typical specification of a compacted thickness of at least 1 m and a permeability of no greater than 10^{-9} ms^{-1} (Department of the Environment 1995). However, this view is only strictly applicable where a 'dilute and attenuate' philosophy is followed, as in the UK. The function of the barrier in such cases is to control the flux of leachate. Consequently, an advective transport mechanism is considered more important than diffusion. In contrast, research by the radioactive waste management industry has provided important new insights into transport processes in clay barriers, which could be valuable to the design of liners for landfill sites. In barriers for radioactive waste containment, diffusion, rather than advection, is the principal transport mechanism, if the clay is intact (Horseman et al. 1996). Sorption is the main mechanism by which the clay attenuates (retards) the pollutants and chemical speciation is the major factor controlling contaminant mobility. However, in compact clays, all these factors are related to the physical properties of the clay. Such a clay behaves as a membrane that restricts the movement of charged chemical species through narrow interparticle spaces, by the mechanism of anion exclusion (Fritz 1986). Size exclusion may also lead to the membrane filtration of large organic molecules and colloids and capillarity may be a major factor in the migration of gases and chemical compounds that are immiscible in water. Therefore, the physico-chemical properties of the barrier material are more important (Grauer 1986; Higgo 1987; Horseman *et al.* 1996). When any clay barrier is used other than as a medium to ensure 'dilution and attenuation' its physical properties cannot be separated from its chemical properties. It is important to consider the function of the barrier when evaluating these relative properties.

A further area in which technology from the radioactive waste management industry might be applied elsewhere lies in the field of low-permeability active containments. It has been well-established theoretically that bentonite/cement cutoff walls might be used to retain heavy metals (e.g. Jefferis 1981; Evans *et al.* 1985, 1997). However, the approach has yet to be applied extensively to the management of contaminated sites. Potentially, the extensive research undertaken by the radioactive waste management industry into the long-term behaviour of cementitious barriers could be applicable directly to the future use of these barriers for the containment of heavy metals.

The contributions in this volume also demonstrate that chemical barriers may, in certain cases (notably in the management of contaminated land and mine wastes), prove easier to deploy than physical barriers. An example is the application of reed beds to clean mine waste. Such cases are permeable active containments. At the opposite end of the spectrum, it is possible to rely mainly on physical containment. This strategy is most appropriate where it is difficult to demonstrate convincingly that a chemical barrier will exist over the time span for which the disposal site must remain isolated.

In summary, it is necessary always to take a holistic view of the containment of wastes in the geosphere. That is, the roles of both physical and chemical processes should be evaluated whenever a disposal system is being designed. It is desirable to do this, even when an engineered system is based predominantly on either chemical or physical processes. This approach is necessary to ensure that the containment system is optimized and to prevent unforeseen detrimental coupled interactions between the physical and chemical processes that affect the barrier. It is also necessary to consider the containment system as a combination of any engineered barrier and the geosphere. The reason for this approach is that it is always desirable to make a disposal system as effective as possible, especially when this can be done without affecting the cost of the system adversely. The demonstration of redundancy in the barrier system is bound to build up the confidence of the regulators and public alike. In other words, a disposal system might be designed to meet the regulatory targets using only physical processes for containment, but it should also take advantage of any chemical containment processes that are available.

The work is largely a development of material presented at a conference entitled 'Chemical containment of wastes in the geosphere', held at the British Geological Survey, Keyworth, on 3–4 September 1996 (Williams 1996). We are indebted to the sponsors of that conference, namely United Kingdom Nirex Limited, the Geochemistry Group (a joint group of the Geological Society of London and the Mineralogical Society), Golder Associates (UK) Ltd and the British Geological Survey. We are also grateful to many referees who provided helpful reviews of the papers. This paper is published with the permission of the Director, British Geological Survey (NERC).

References

ANON 1999. Radwaste – geoscience's verdict. *Geoscientist*, **9**, 5, 19.

BENTLEY, S. P. (ed.) 1996. *Engineering Geology of Waste Disposal*. Geological Society, London, Engineering Geology Special Publications, **11**.

BOWDERS, J. J. & DANIELE, D. E. 1987. Hydraulic conductivity of compacted clay to dilute organic compounds. *Journal of Geotechnical Engineering*, **113/12**, 1432–1448.

CHRISTENSON, T. H., KJELDSEN, P., ALBRECHTSEN, H. J. *et al.* 1994. Attenuation of landfill leachate pollutants in aquifers. *Critical Reviews in Environmental Science and Technology*, **24**, 119–202.

DEPARTMENT OF THE ENVIRONMENT 1990. *Landfilling Wastes, Technical Memorandum for the Disposal of Wastes on Landfill Sites*. Waste management paper 26, HMSO, London.

——1995. *Landfill Design, Construction and Operation Practice*. Waste management paper 26B, HMSO, London.

EVANS, J. C., FANG, H. Y. & KUGELMAN, I. J. 1985. Containment of hazardous materials with soil-bentonite slurry walls. In: *Proceedings of the 6th National Conference on the Management of Uncontrolled Hazardous Waste Sites, Washington DC, November 1985*, 249–252.

——, ADAMS, T. L. & PRINCE, M. J. 1997. Metals attenuation in minerally enhanced slurry walls. *In*: *Proceedings of the 1997 International Containment Technology Conference, February 1997.*

FRITZ, S. J. 1986. Ideality of clay minerals in osmotic processes: a review. *Clays and Clay Minerals*, **34**, 214–223.

GRAUER, R. 1986. *Bentonite as a Backfill Material in the High Level Waste Repository*. Nagra Technical Report NTB 86-12E. Nagra, Baden, Switzerland.

HETTIRACHI, J. P. A., HRUDLEY, S. E., SMITH, D. W. *et al.* 1988. A procedure for evaluating municipal solid waste leachate components capable of causing volume shrinkage in compacted clay soils. *Environmental Technology Letters*, **9**, 23–34.

HIGGO, J. J. W. 1987. Clay as a barrier to radionuclide migration. *Progress in Nuclear Energy*, **19**, 173–207.

HORSEMAN, S. T., HIGGO, J. J. W., ALEXANDER, J. *et al.* 1996. *Water, Gas and Solute Movement in Argillaceous Media*. OECD/NE SEDE Working Group on Physical Understanding of Groundwater Flow Through Argillaceous Media. Nuclear Energy Agency, Paris. Report CC-96/1.

JEFFERIS, S. A. 1981. Bentonite-cement slurries for hydraulic cut-offs. *In*: *Proceedings of the Tenth International Conference, International Society for Soil Mechanics and Foundation Engineering, Stockholm 1981*, **1**, 425–440.

KASS, G., BRADSHAW, T. & NORTON, M. 1997. *Radioactive Waste – Where Next?* Report of the Parliamentary Office of Science and Technology, London.

NAGRA 1994. *Kristallin-I Safety Assessment Report*. NAGRA Technischer Bericht NTB, 93-22. Nagra, Wettingen, Switzerland.

NIREX 1997. *An Assessment of the Post-Closure Performance of a Deep Waste Repository at Sellafield*. Nirex Report S/97/012. Nirex, Harwell.

SAVAGE, D. 1995. *The Scientific and Regulatory Basis for the Geological Disposal of Radioactive Waste*. Wiley, Chichester.

STEIF, K. 1989. Strategy on landfilling solid wastes. *In*: BACCINI, P. (ed.) *Landfill – the Reactor and Final Storage*. Lecture notes in earth sciences, Springer, Berlin, 275–291.

WILLIAMS, L. A. (ed.) 1996. *Chemical Containment of Wastes in the Geosphere: Extended Abstracts*. British Geological Survey, Keyworth.

Overview of partitioning and fate of contaminants: retention, retardation and regulatory requirements

RAYMOND N. YONG

Geoenvironmental Engineering Research Centre, Cardiff School of Engineering, University of Wales, Cardiff, CF2 1XH, UK

Abstract: This overview study examines the principal interactions between contaminants and soil fractions responsible for contaminant partitioning, with the aim of developing a further understanding of retardation and retention processes during contaminant transport. The defining mechanisms for sorption and desorption of contaminants are generally known for many types of contaminants in interaction with specific soil fractions. However, contaminant accumulation in soil-engineered barriers or soil substrate material needs to be addressed in terms of contaminant retention or contaminant retardation as a contaminant attenuation process. This is necessary if regulatory requirements regarding *natural attenuation capabilities* are to be fulfilled. The combined processes leading to partitioning of contaminants in complex mixtures of contaminants and soil fractions have yet to be fully defined and understood. However, sufficient information exists concerning chemical and biologically mediated mass transfer to permit one to appreciate the need for distinguishing between irreversible sorption and temporary sorption processes. Failure to properly realize this can make the natural attenuation concept for clay barriers a 'disaster waiting to happen'.

One of the major areas of concern in waste and leachate management is the partitioning of contaminants by chemical and physical mass transfer during contaminant transport in soils. To a very large extent, the permanent or temporary transfer of contaminants to the soil solids or soil matrix, (i.e. removal of solutes from solution or accumulation of contaminant solutes in the soil matrix) is an important factor in the use of containment barriers. This is true whether such barriers are in the form of naturally occurring (i.e. in-place) soil substrate or as 'engineered' barriers constructed with soil material. The determination of the various processes occurring during the course of contaminant transport requires an assessment of the persistence and fate of the contaminants in the leachate plume.

The use of the term *attenuation* in connection with the transport of contaminants in the soil substrate requires processes which result in temporary and/or permanent accumulation of the contaminants by the soil solids (i.e. soil fractions). The result of the contaminant accumulation processes by the soil fractions is generally termed as *partitioning* between 'what is sorbed by the soil fractions' and 'what remains in the pore-water or aqueous phase'. The nature and extent of the interactions and reactions established between contaminants and soil fractions impact directly on:

- the accumulation processes responsible for partitioning
- the establishment of processes which differentiate between attenuation by retention mechanisms and/or attenuation by retardation of the contaminants.

Partitioning by retention mechanisms results from irreversible sorption of contaminants by the soil fractions, i.e. the advancing pollution plume will not be 'recharged' by desorbed contaminants. The term *attenuation* is generally used by soil scientists to indicate reduction of contaminant concentration resulting from retention of the contaminants in the soil. In other words, attenuation has been used to describe chemical mass transfer of contaminants where the contaminants are sorbed by the soil fractions in excess of the cation exchange capacity of the soil. In essence, this means dilution of the contaminants' concentration in the contaminant plume during the transport process. The sorption process or the results of sorption are irreversible in this case, i.e. contaminants are non-extractable with neutral salts and mild acid solutions. In contrast, the term *retardation* which is often used in the context of contaminant transport in the substrate generally refers to 'a retardation (slowing down) of the movement of the contaminants'. Attenuation of contaminants by

From: METCALFE, R. & ROCHELLE, C. A. (eds) 1999. *Chemical Containment of Waste in the Geosphere.* Geological Society, London, Special Publications, **157**, 1–20. 1-86239-040-1/99/$15.00 © The Geological Society of London 1999.

retardation processes or mechanisms generally refers to situations where the contaminants are considered to be non-reactive. Hence no 'permanent' chemical mass transfer is involved in this case. Because sorption processes are reversible, all of the contaminants entering the system will eventually be transported to their final destination, i.e. the total amount of contaminants remains constant. The schematic illustration shown in Fig. 1 portrays the resultant effects caused by the two kinds of processes.

As can be noted from the schematic diagram, if the solute pulse is retarded, the area under each of the retardation pulse curves remains constant as the pulse travels downward through the unsaturated zone toward the aquifer. This indicates that the total contaminant load is transferred to the aquifer. By contrast, the retention pulse shows decreasing areas under the pulse curves. This indicates partitioning by chemical mass transfer and irreversible sorption, and resultant dilution of contaminant concentrations. If proper design procedures are used, no transfer of contaminants to the aquifer should occur, thereby fulfilling the requirements for contaminant containment using the *natural attenuation capability* of the soil substrate liner system.

The consequences arising from a failure to distinguish between attenuation by retention and attenuation by retardation mechanisms can be quite severe. Such a failure can have important implications for transport modelling for prediction of contaminant plume advance. As has been well noted in the discussion on regulatory attitudes given by Yong (1996), the regulatory requirement relating to the use of the 'natural attenuation capacity' of substrate soil material for leachate management relies on the capability of the system to perform according to the requirements of the various permitting criteria and protocols. Various institutions and agencies have developed guidelines as implementing procedures required to meet the intentions of the regulatory requirements (USCFR 1992, CIRIA 1996, NRA 1989, DOE 1995, Ontario MOE 1986, CCME 1991). These guidelines are meant to provide the basis for construction of barriers where, in the case of soil materials, retention mechanisms will be responsible for the 'natural attenuation capacity' of the system. That is, partitioning of contaminants is expected to be by permanent chemical mass transfer and irreversible sorption processes. An implication is that capability presently exists to provide the necessary information to ensure that these processes occur. Failure to provide the proper retention capability by the barrier system will result in disastrous consequences. It is evident that much has yet to be learnt about the fate of contaminants in the substrate, and that because

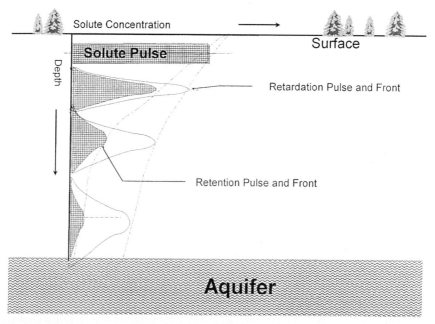

Fig. 1. Schematic diagram showing differences between Retention and Retardation pulses and fronts.

the natures of sites and contaminants differ from region to region, a proper appreciation of the physical and chemical interactions between contaminants and soil fractions is needed if retention and retardation mechanisms are to be distinguished.

In view of the preceding discussion, this overview paper examines the principal interactions between contaminants and soil fractions responsible for contaminant partitioning. The aim is to develop a better understanding of retardation and retention processes during contaminant transport. The role of clays as catalytic agents is seen to be a key factor yet to be fully explored and understood in the determination of the long-term fate of contaminants. To avoid confusion in the use of the term 'attenuation', and to distinguish between processes and resultant effect, the terms 'partitioning' and 'accumulation' are used in the general discussion dealing with contaminant transport processes. 'Partitioning' refers to partitioning of the total contaminant load between soil fractions and the fluid phase, and 'accumulation' refers to the accumulation of contaminants by the soil fractions by various mass transfer mechanisms. The result of partitioning is contaminant attenuation during transport of the leachate. The term 'retention' will be used to denote processes which render irreversible (or near-irreversible) sorption results. The term 'retardation' will be reserved for use in designating temporary accumulation processes, or reversible sorption processes and other processes resulting in the 'slowing-down' of the contaminant transport.

Problem elements

The key elements of the problem of a contaminant plume emanating from a landfill are depicted in Fig. 2 and include:

- soil substrate system, including solid, liquid and gaseous phases
- aquifer
- biological system
- contaminants in the contaminant (pollution) plume
- others, including temperature, climate, regional controls, magnetic properties, etc.

The nature of the questions that need to be addressed, regarding the presence of the pollution plume shown in Fig. 2 are: (a) what is the extent and distribution of the contaminants in the contaminant plume? (b) in what manner are the contaminants 'retained' in the soil system (i.e. what accumulation processes lead to partitioning); retention and/or retardation? (c) what is the nature and propagation rate of the pollution plume? and (d) in what manner should the 'threat' be managed (i.e. by 'containment', by remediation, or by disregarding)?

The basic issues raised concerning the above require determination of the interactions established between the contaminants and soil fractions. In essence, one needs to determine the *persistence and fate of contaminants* in the soil substrate inasmuch as the transport of contaminants in the substrate is intimately linked to the fate of the contaminants. Of particular interest are the partitioning and distribution of the

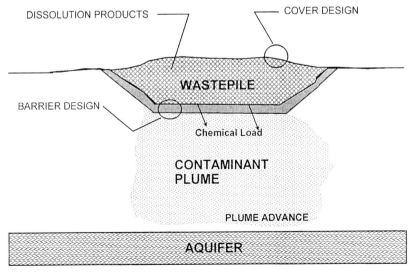

Fig. 2. Illustration of contaminant plume advance in soil substrate.

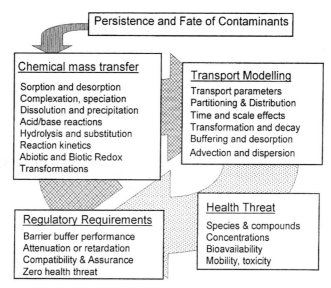

Fig. 3. Interaction processes and required considerations in evaluation of persistence and fate of contaminants.

contaminants, or chemical mass transfer as illustrated in Fig. 3. This figure illustrates the close interplay between chemical mass transfer, transport modelling, health threats and regulatory requirements.

Contaminant interaction and partitioning

As compared to geotechnical engineering considerations, land environmental pollution assessments require a more complete accounting of the various soil fractions and contaminants. Particular emphasis must be placed on their surface properties and characteristics. Interactions between contaminants and soil fractions that lead to partitioning and fate of the contaminants are determined partly by the properties of the surface features (functional groups) of the soil fractions and contaminants. However, the catalytic action of clay soils and the biology of the system are also important factors. As stated previously (Yong 1996, Ontario MOE 1986, CCME 1991) the popular choice of 'natural attenuation' for landfill control of waste disposal favoured by many regulatory agencies places a great demand on a proper understanding and assessment of the interactions between contaminants and soil fractions, and the fate of the contaminants. For the problem shown in Fig. 2 in the absence of dilution in the saturated zone, the natural attenuation option requires contaminant leachates to be chemically buffered in the soil barrier system with no future (or eventual) desorption of the sorbed contaminants. In other words, the 'chemical buffer system' shall ensure that no contaminants will ever be delivered to the aquifer.

Retention and retardation mechanisms

The various mechanisms contributing to chemical mass transfer of contaminants during the transport process are shown in the upper-left box in Fig. 3 under the heading of 'chemical mass transfer'. One of the major concerns in adopting the natural attenuation option as a means for leachate management in the landfill barrier is the significant difference between the control of the fate and partitioning of contaminants through retention as against retardation mechanisms. Considering heavy metal contaminants as an example, Farrah & Pickering (1978) have identified four possible mechanisms that may mobilize heavy metals in soils: (1) changes in acidity; (2) changes in the system ionic strength; (3) changes in the oxidation–reduction potential of the system and (4) formation of complexes. By and large, the principal mechanisms and processes involved in heavy metal retention include precipitation as a solid phase (oxide, hydroxides, carbonates), and complexation reactions (Harter 1979; Farrah & Pickering 1977, 1979; Maguire et al. 1981; Yong et al. 1990). The literature reports on ion-exchange adsorption as a means of 'retention' should be considered as 'retardation' in the present context.

In general, desorption of contaminants occurs where initial sorption is by ion-exchange mechanisms. Taking a kaolinite soil as an example, one notes that the two types of functional groups populating the surfaces of the edges of the particles are both hydroxyl (OH) groups. One type is singly coordinated to the Si in the tetrahedral lattices whereas the other is singly coordinated to the Al in the octahedral lattices that characterize the kaolinite structure. These OH groups are most likely the type 'A' hydroxyl groups which have been protonated to form Lewis acid sites (Sposito 1984). Thus, both types of edges will function as Lewis acid sites, i.e. they are electron acceptor sites. The surfaces of the kaolinite function as nucleation centres for heavy metals. In the case of sorption of heavy metal contaminants by kaolinites, if the metal concentrations are less than the cation-exchange capacity (CEC) of the kaolinite, desorption occurs easily because the initial sorption processes are mainly non-specific, i.e. processes involving outer-sphere surface complexation of ions by the surface functional groups exposed on soil particles. However, if the metal concentrations are greater than the CEC, desorption is more difficult because the initial sorption processes will include both non-specific adsorption and some specific adsorption (inner-sphere complexation). Release (desorption) of the previously sorbed metal ions can occur when saturation sorption occurs and when the ions in the bulk or pore fluid are lower in concentration than the initial sorbed ions (Yong et al. 1992; Bolt 1979).

Desorption of cations can also occur through replacement, and it is useful to take note of the replacement order, given as the familiar lyotropic series – i.e. the cations on the right will displace those towards the left:

$$Na^+ < Li^+ < K^+ < Rb^+ < Cs^+ < Mg^{2+}$$

$$< Ca^{2+} < Ba^{2+} < Cu^{2+}$$

$$< Al^{3+} < Fe^{3+} < Th^{4+}$$

Replacement of the metal ions can also occur as the pH of the system becomes more acidic. Under such circumstances, the hydrogen ions will replace the metal ions. It is however not always easy to fully distinguish between mechanisms responsible for retention and retardation of contaminants in the transport process. In the upper left box (chemical mass transfer box) shown in Fig. 3 the first set of interactions (sorption and desorption) are perhaps the simplest of those which occur, as discussed above. Accumulation processes such as complexation,

speciation and precipitation, for example, can effectively attenuate contaminants. This attenuation occurs through removal of the contaminant solutes from solution via competition with the soil fraction for sorption sites (speciation and complexation), sequestering (complexation), and through formation of new phases (precipitation). However, should the pH environment change, it is possible for these processes to be significantly affected. Dissolution of the precipitates (pH lowering) may occur causing release of the previously precipitated contaminants occurs.

The availability and mobility of heavy metal contaminants are very largely dependent on how they are bonded with the various soil fractions. The presence of inorganic and aqueous organic ligands in the soil-engineered barrier system will compete with the adsorption sites from the soil fractions for sorption of the contaminants. In other words, these ligands will feature prominently in chemical mass transfer. Changes in the immediate soil environment will undoubtedly alter the fate of the contaminants in the soil, and in the case of heavy metal pollutants, redox and pH conditions will control the chemical forms and mobility of the heavy metals. Considerable research information is available on heavy metal retention mechanisms of individual soil components (Yong et al. 1993; Allen et al. 1993; Yong & MacDonald 1998; Jenne 1998). However, accurate predictions of sorption and partitioning in a multi-component contaminant and soil system are hampered by inadequate information concerning the competitive and interactive relationships between soil fractions and contaminants. Retention of heavy metals is only predictable in simple systems such as a single mineral species and simple single oxides. Predictions made on the basis of a representation of the multi-component system by some average collective property or as a collection of discrete pure solid phases, can lead to significant differences in comparison to actual test information, as, for example, in the binary suspension studies shown by Anderson and Benjamin (1990) and Bassett and Melchior (1990). Anderson and Benjamin (1990) show that in some cases the addition of a second adsorbent decreases the overall adsorption capacity of the system, and that analyses of interaction should take into account both adsorption and changes in the physical characteristics of the soil.

For contaminant attenuation by retention mechanisms, the contaminants must be retained by specific adsorption. Other mechanisms for retention include chemisorption, i.e. high affinity specific adsorption in the inner Helmholtz layer through covalent bonding, via hydroxyl groups

Table 1. *Heavy metal retention by some clay minerals*

Clay mineral	Chemisorption*	Chemisorption at edges	Complex adsorption†	Lattice penetration‡
Montmorillonite	Co, Cu, Zn		Co, Cu, Zn	Co, Zn
Kaolinite			Cu, Zn	Zn
Hectorite			Zn	Zn
Brucite			Zn	Zn
Vermiculite			Co, Zn	Zn
Illite	Zn	Zn, Cd, Cu, Pb		
Phlogopite			Co	
Nontronite			Co	

Adapted from Bolt (1979).
* High affinity specific adsorption in the inner Helmholtz layer through covalent bonding, via hydroxyl groups from broken bonds in the clay minerals.
† Adsorption resulting in formation of metal-ion complexes.
‡ Lattice penetration and imbedding in hexagonal cavities.

from broken bonds in the clay minerals, formation of metal-ion complexes, and precipitation as hydroxides or insoluble salts. Using information obtained from Bolt (1979), Table 1 lists some of the above mechanisms responsible for retention of Cu, Co, Zn, Pb, and Cd by some representative clay minerals.

Other soil fractions (i.e. other than clay minerals) also show capabilities for retention of heavy metals and other types of contaminant. The information given in Fig. 4 uses the test results obtained previously by Yong & Phadungchewit (1993) regarding partitioning of Pb by an illitic soil over a pH range from 2 to 7.

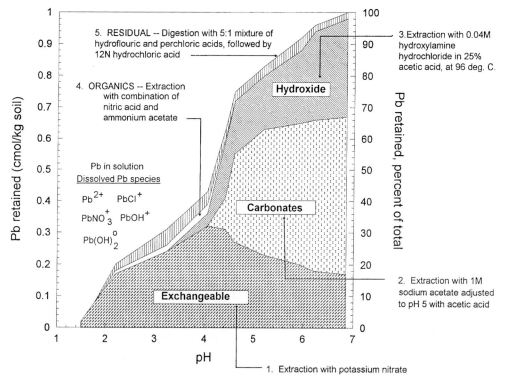

Fig. 4. Protocols for selective sequential extraction demonstrating 'bonding' strength and requirements for extraction of sorbed Pb.

The diagram shows the extraction treatment procedures required to remove the sorbed Pb from the associated soil fractions, using techniques described by Yong et al. (1993) in selective sequential extraction. Because of the possibility of some mineral dissolution occurring at pH values of less than 3 the reliability of information concerning retention below pH 3 should be considered somewhat questionable. However, using the information for values of pH greater than 3 and interpreting the behaviour of the system relating to a Pb leachate stream, it is seen that below the precipitation pH of the Pb, the primary means for partitioning of Pb is by nonspecific adsorption, an exchange mechanism. The extraction procedure given as step 1 shows that the extractable Pb is essentially 'exchangeable'. Under such circumstances, desorption of the Pb can be expected and the process should therefore be classified as one contributing to a perceived retardation of contaminants.

It is interesting to note from all the other steps given in the selective sequential extraction procedures noted in Fig. 4 that as the order of the steps increases, from step 1 to step 5 the extraction procedures become very aggressive. As the level of aggression increases, a corresponding increase in the retention capability of the soil for the contaminants should be expected. This has generally been used as an indicator of the degree of retention capability, and is a qualitative measure of the irreversibility of sorption by the soil.

The difficulties in distinguishing between sorption-retention (physical and chemical adsorption mechanisms) and precipitation-retention (i.e. retention by precipitation mechanisms) as opposed to retention by sorption, is rendered difficult because of experimental constraints. This is a significant problem inasmuch as desorption of physisorbed (i.e. sorption resulting primarily from ion-exchange reactions) contaminant solutes can readily occur. That is, this process falls under the category of a contaminant retardation process. As noted from Fig. 4 and other reported studies, accumulation of heavy metal contaminants increases with increasing pH, and precipitation of the heavy metals at around neutral pH and above would form compounds such as carbonate, hydroxide, sulphate and chlorate species. The distinction between adsorbed and precipitated heavy metal contaminants also depends on concentration of the heavy metals (Yong 1997a)

Yong & MacDonald (1998) studied the relationship between Cu and Pb retention regarding soil pH (using the same soil studied by Yong & Phadungchewit 1993, Fig. 4) and the presence of aqueous OH, HCO_3^-, and CO_3^{2-}. This work demonstrated not only that competition for metallic ions is offered by the sorption sites provided by the soil fractions and the anions, but also that the formation of several precipitation compounds depends on the pH environment. For example, it was shown that the soluble Pb concentration was influenced by the precipitation of $PbCO_3$ (cerrusite) and $Pb(CO_3)_2(OH)_2$ (hydrocerrusite). Since $PbCO_3$ precipitated at lower pH values than both calcite and dolomite, it was possible for the Pb carbonates to precipitate because of the dissolution of Mg and Ca as carbonates. In the case of soluble Cu concentration however, the controls were exercised by the precipitation of CuO (tenorite). It is noteworthy that the variable pH-dependent hydrolysis of the metal cations (Cu^{2+}, $CuOH^+$, $Cu(OH)_2$, Pb^{2+}, $PbOH^+$, $Pb(OH)_2$) changes the Lewis acid strength of the aqueous species of the metals and thus their affinity for the soil particle surfaces. The Lewis definition of acids and bases is used here since it permits one to attribute acidity to a unique electronic arrangement and not necessarily to a particular element (Stumm & Morgan 1981). This is particularly significant in that borderline Lewis acids such as Pb^{2+} and Cu^{2+} can behave as hard or soft acids depending on the environment solution, and since this impacts on affinity relationships between the metals and the reactive soil surfaces, sorption and desorption capabilities are severely conditioned. The Pearson (1963) classification of Lewis acids and bases, and particularly the categorization of these as hard and soft is useful in that the stability of complexes formed between various species can be predicted by the hard-soft-acid-base (HSAB) principle.

Yong & MacDonald (1998) show that upon apparent completion of metal sorption, measurements of the equilibrium pH of the system generally showed a reduction below initial pH. This reduction in pH was attributed to the resultant effect of the many reactions in the system. These reactions included the release of hydrogen ions by metal/proton exchange reactions on surface sites, hydrolyses of metals in the soil solution, and precipitation of metals. It was apparent that more detailed information was needed to distinguish between surface and solution reactions responsible for release of hydrogen ions. However, it was evident that if surface complexation models are to be used, the relationship between metal adsorption and proton release needs to be established. That is, net proton release or consumption is due to all the chemical reactions involving proton transfer.

Differences in the predicted penetration rate and depth of a pollutant plume depend not only on the choice of transport coefficients, but also on whether the contaminants are 'retained' or 'retarded'. The choice and nature of the transport coefficients used in the prediction model will describe the expected distribution and attenuation of the contaminants with time and space. Whether the attenuation is by retention or retardation mechanisms becomes the critical consideration and issue. In the first instance (i.e. retention), the attenuation of contaminants will lead to a successful containment of the pollutants in the substrate. In contrast, in the second instance (retardation), the retardation transport mechanism will result only in a 'postponement' of the time for contamination of the aquifer, i.e. a 'ticking time bomb' problem or a health hazard waiting to happen.

Strictly speaking, it might be argued that the regulatory point of view does not (or need not?) concern itself with the 'retention and/or retardation' problem. It could be considered instead that the assurance of the capability of a soil-engineered barrier or soil substrate to function as a natural attenuation (retention) barrier lies with the practitioner responsible for implementing the project. It is necessary to assess whether the 'accumulation' process which results in partitioning is through retardation or retention mechanisms. Such an assessment requires one to distinguish between contaminant solutes contained in the diffuse ion-layer (and exchangeable), and partitioning through any or all of the following retention mechanisms: chemisorption; specific adsorption; metal-ion complexation; and precipitation. Partitioning of contaminants is caused by the kinetics of reactions. Predictions of contaminant transport cannot be properly undertaken without a knowledge of the fate of contaminants. In addition to the Brønsted acidity (capability for proton transfer), which is significantly high, clay surfaces also posses Lewis acidic sites, as noted previously. The combination of such sets of surface activities and the varied chemical nature of different contaminants is a recipe for a variety of reactions that will control the characteristics of transport and retention of contaminants. Thus, whilst the 'Chemical mass transfer' box shown in Fig. 3 lists many of the processes that are involved in the determination of the fate of contaminants, the problems associated with 'translating' them into appropriate transport modelling inputs and parameters have yet to be fully addressed.

Aside from the inorganic chemical contaminants, organic chemical compounds evoke mechanisms of interactions in respect to retention and retardation processes which deviate somewhat from those described previously. Transport of PHCs (petroleum hydrocarbons) in soils is a case in point. Interactions between oil and soil surfaces are important in predicting the oil retention capacity of the soil and the bioavailability of the oil. The interaction mechanisms are influenced by the mineralogical nature of different soil fractions, the type of oil, and the presence of water. As in the case of inorganic contaminant–soil interaction, the existence of surface active fractions in the soil such as organic, amorphous materials and clays, can significantly enhance oil retention in soils, largely because of large surface areas and high surface charges. The problem of 'first wetting' becomes very important. If the soils are first wetted with water, interference with soil–oil bonding occurs and the amount of oil associated with the soil fractions will decrease in proportion to the amount of first wetting. Because of its low aqueous solubility and large molecular size, penetration into the Helmholtz layers (electrified double-layer adjacent to the clay particles) is not easily achieved. For example, the effective diameter of various hydrocarbon molecules varies from 1 to 3 nm for a complex hydrocarbon type in contrast to a water molecule which has a diameter of approximately 0.3 nm. Thus, the determination of retention of HCs and most non-aqueous phase liquids (NAPLs) must include determination of first and residual wetting of the soil-engineered barriers and soil substrate.

The weakly polar (resin) to non-polar compounds (saturates and aromatic hydrocarbons) that make up 'oil' provide for different reactions and bonding relationships with the soil fractions. Weakly polar compounds are more readily adsorbed onto soil surfaces than non-polar compounds. The adsorption of non-polar compounds onto soil surfaces is dominated by weak bonding (van der Waals attraction) and is restricted to external soil surfaces primarily because of their low dipole moments (less than 1) and also low dielectric constants (less than 3) (Yong & Sethi 1973; Yong & Rao 1991). As discussed in a later section, the aqueous solubility and partition coefficients for these compounds control the interactions. Because most hydrocarbon molecules are hydrophobic and have low aqueous solubilities, partitioning onto soil surfaces occurs to a greater extent than in the aqueous phase, resulting in lower environmental mobility and higher retention.

The desorption of PHCs for example can give some insight into organic chemical retardation phenomena. The studies by Yaron (1989) of desorption of m-xylene, n-decane, t-butylbenzene from soils with different clay and organic matter (OM) contents, showed desorption of these hydrocarbons to be strongly affected by the type of soil and hydrocarbon. The rate of desorption occurred as follows:

m-xylene > n-decane > t-butylbenzene

Desorption of PHCs through leaching processes can be in an aqueous phase or as a separate liquid phase NAPL. In the former, water solubility is an important factor, and in the latter, viscosity and surface wetting properties are critical. Light hydrocarbons are more likely to volatilize and be leached whereas heavier constituents will tend to be retained in the soil fractions.

Transport, fate and regulatory implications

A common procedure presently used is to specify a limiting maximum value of hydraulic conductivity for the engineered clay barrier for control of transport and transfer of contaminants to the substrate (Jessberger 1995; Yong 1996; Manassero *et al.* 1997) However, this approach does not necessarily pay attention to the specific role of chemical mass transfer and/or the natural attenuation phenomenon. Contaminant retardation occurs via control on the limiting hydraulic conductivity. If retention

occurs as a result of the limiting hydraulic conductivity, this is not determined or directly considered in the specification of the Darcy coefficient of permeability k value. This is because no calculations or analyses are provided in the protocols or procedures established for construction of the clay-engineered barrier. Instead, a limiting maximum k value and minimum thickness of the clay barrier are generally specified as construction requirements to provide contaminant attenuation. The differences in thicknesses of required barrier material, and differences in limiting hydraulic conductivities permitted in the clay barrier material between various countries for liner containment of municipal solid waste (MSW) can be seen in Fig. 5. It is interesting to note that the type of soil material is generally not specified, and neither is the governing in-place density or moisture content of the clay-engineered barrier. In addition to the required minimum thickness of the clay barrier, the major item considered is the Darcy coefficient of permeability k, which is generally limited to a maximum value of $10^{-9}\,m\,s^{-1}$.

Hydraulic/fluid conductivity, partitioning and regulatory implications

The information given in Fig. 5 in conjunction with the role of partitioning *vis-à-vis* regulatory

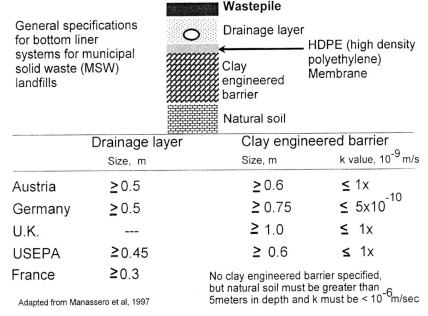

General specifications for bottom liner systems for municipal solid waste (MSW) landfills

Wastepile
Drainage layer
HDPE (high density polyethylene) Membrane
Clay engineered barrier
Natural soil

	Drainage layer	Clay engineered barrier	
	Size, m	Size, m	k value, 10^{-9} m/s
Austria	≥ 0.5	≥ 0.6	≤ 1x
Germany	≥ 0.5	≥ 0.75	≤ 5x10^{-10}
U.K.	---	≥ 1.0	≤ 1x
USEPA	≥ 0.45	≥ 0.6	≤ 1x
France	≥ 0.3	No clay engineered barrier specified, but natural soil must be greater than 5meters in depth and k must be < 10^{-6} m/sec	

Adapted from Manassero et al, 1997

Fig. 5. Clay engineered barrier specifications for some countries.

R. N. YONG

criteria (Jessberger; 1995, Yong 1996; Manassero et al. 1997) which limit maximum hydraulic conductivities for natural soil barriers, suggests some pertinent questions, e.g. (1) How is chemical mass transfer evaluated and factored into the specification for thickness of the barriers? (2) What are the relationships between the k values and chemical mass transfer? (3) How are variations in soil composition considered in respect to partitioning and chemical mass transfer? (4) How is the chemical composition of the leachates considered in the specification of the 'limiting maximum value for hydraulic conductivity'?

Calculations of energies of interaction (Yong et al. 1992) in a soil–water–contaminant system have shown that as the net energies of repulsion increased as a result of transport and partitioning, resultant hydraulic conductivities tended to decrease. Except for significant changes in energies of interaction, the changes in hydraulic conductivity because of a change in the (inorganic) chemistry of the permeant have been noted to be relatively small for clay soils which are not highly surface active. The same however cannot be said for soils with active surfaces, particularly bentonites. Energies of particle interaction which depend on the species and concentration of the ions as much as the type and arrangement of soil particles, contribute directly to the control of soil permeability. Table 2 (from Yong et al. 1992) shows the variation in soil permeability given in terms of the ratio of k_f the final Darcy permeability coefficient with respect to the permeant and k_w the Darcy permeability coefficient with respect to water as the permeant. As can be seen, significant differences in the fluid conductivity values can be obtained as a result of the nature of the permeant. The acid, base and characteristics of the cations in the permeant are important controls on the conductivity.

The information given in Table 2 shows that permeation with the acids leads to an approximate 5 to 12-fold increase in the fluid conductivity of the bentonite specimens as measured by the Darcy permeability coefficients in comparison to that determined using water as the permeant. It is interesting to note that a decrease in acid concentration has little effect on the k_f/k_w ratio (k_f = final k value to k value and $k_w = k$ value when water is the permeant). The differences or modifications in soil permeability in relation to chemistry of the permeating fluid can be attributed to:

Table 2. *Variation in clay permeability permeated with inorganic contaminants*

Permeant			Clay type	Variation in soil permeability
Acids			Bentonite*	k_f/k_w
7%	HNO$_3$	(7%)		5.6–7.7
0.7%	——	(0.7%)		11.6
5%	H$_2$SO$_4$	(5%)		6.9
3.65%	HCl	(3.65%)		10.2
0.36%	——	(0.36%)		10.9
Bases			Bentonite*	k_f/k_w
4%	NaOH	(4%)		0.3
0.4%	——	(0.4%)		0.2
Cations				
Valency effect			Montmorillonite†	$\dfrac{k_{Ca^{2+}}}{k_{Na^+}} = 28$ at $e = 2.5$
			Bentonite‡	$\dfrac{k_{Ca^{2+}}}{k_{Na^+}} = 8$ at $e = 2.0$
				$\dfrac{k_{Fe^{3+}}}{k_{Na^+}} = 33$ at $e = 2.0$
Cation size effect			Bentonite	$\dfrac{k_{K^+}}{k_{Na^+}} = 5$ at $e = 2.0$

k_f = Final Darcy Coefficient of permeability.
k_w = Darcy permeability coefficient when water is the permeant.
* Pavilonsky (1985).
† Mesri & Olson (1971).
‡ Sridharan et al. (1986).

(1) extraction of lattice aluminum ions from the octahedral sheets of the clay minerals,
(2) ion exchange on the surface of the clay minerals due to replacement of naturally adsorbed cations of lower valence (Ca^{2+}, Mg^{2+}, Na^{2+}, K^+) by the extracted aluminum ions which have a valence of 3 and, hence, a reduction in the thickness of the diffuse double layer,
(3) increase in effective pore space and a decrease in the tortuosity factor, resulting thereby in a higher permeability coefficient.

Except for the likelihood of ion exchange as given in (2) above, the information given in Table 2 says very little about the nature of sorption mechanisms. The question of 'retention' and/or 'retardation' is raised. Thus, one is alerted to the fact that changes in the chemistry of the permeant can have significant effects on the fluid conductivities of surface-active clays, but the problem of partitioning via retention and retardation mechanisms is not readily resolved. This raises some important issues concerning the relationship between the limiting k value and desirable retention characteristics of the clay-engineered barrier. There are, in turn, important implications for any regulatory requirement based on a limiting maximum value for hydraulic conductivity, if proper attenuation of contaminants is to be realized.

Pavilonsky (1985) undertook permeability experiments with compacted montmorillonite specimens and nitric, hydrochloric and sulphuric acids, over periods of 200 to 700 days. These experiments confirm the role and effect of ionic species and concentration in the soil pore water on the development of soil permeability. As noted by Yong et al. (1992), inorganic acids have the ability to solubilize some of the constituents that comprise the clay soil structure, e.g. acids have been reported to solubilize aluminum, iron, alkali metals and alkaline earth metals (Grim 1953). The solubility of clays in acids varies not only with the nature of the acid, but also with acid concentration, acid to clay ratio, temperature and degree of interaction between the clay and the acid. Pask and Davis (1954) showed that when several clay minerals were boiled in acid, the percentage of solubilization of alumina was 3% for kaolinite, 11% for illite and greater than 33% for montmorillonite. Interactions between contaminant leachate solutions containing inorganic bases and clay minerals can result in dissolution of the silica in the minerals in addition to increasing the net negative charge on the clay surfaces because of the anion presence.

Results from tests with organic contaminants in leaching and fluid conductivity experiments have often shown significant 'shrinkage' in the soil samples tested, suggesting that the diffuse double layers (DDL) could not fully develop. Interaction of clay minerals with organic chemicals having dielectric constants lower than water will result in the development of thinner interlayer spacing because of the 'contraction' of the soil-particle system. One can consider the transport of organic molecules through the soil substrate as being by diffusion and advection through the macropores. Partitioning will occur between the pore-aqueous phase and soil fractions throughout the flow region. The weakly adsorbed molecules will tend to move more quickly through the connected aqueous channels. Hydrophobic substances such as heptane, xylene and aniline, which are well partitioned onto solid phases at any instant of time, would develop soil-organic chemical permeabilities that will be much lower than the corresponding soil-water permeability. One can conclude that organic fluid transport through a clay barrier is conditioned not only by the hydrophobicity or hydrophilic nature of the fluid, but also by other properties such as the dielectric constant of the substance.

The water solubility of an organic chemical pollutant is of significant importance in controlling the transport and fate of the pollutant, since it determines the partitioning of the pollutant to the soil fractions. For most organic pollutants, the equilibrium partition ratio of the concentration of the pollutant in other media to that in water is considered to be well correlated with the water solubility of the pollutant. Chiou et al. (1982) have reported good correlations between the solubilities of organic compounds and their octanol-water partition coefficient K_{ow} (ratio of solubility in octanol to solubility in water) as shown for example in eqn. (1) where the relationship is given in terms of the solubility S, (S is expressed either as ppm or ppb).

$$\log K_{ow} = 4.5 - 0.75 \log S \text{ (ppm)}$$
$$\log K_{ow} = 7.5 - 0.75 \log S \text{ (ppb)}$$
(1)

Reported experimental measurements show that octanol can be considered to be a good surrogate for organic matter in soils, and that chemicals with low K_{ow} (e.g. less than 10) may be considered relatively hydrophilic; they tend to have high water solubilities and small soil adsorption coefficients. Conversely, chemicals with a high K_{ow} values (e.g. greater than 10^4) are very hydrophobic, and thus demonstrate low water solubilities. As discussed by Yong (1995) solvent systems which are almost completely immiscible (e.g. alkanes–water) conform fairly well to predicted partitioning behaviour. If the departures from ideal behaviour exhibited by the more polar solvent systems are not too large, a thermodynamic treatment of partitioning can be applied to determine the distribution of the organic chemical without serious loss of accuracy.

Figure 6 shows the results reported by Yong (1997a) for tests on two soils concerning relative permeability of the soils. The permeability tests were conducted with various organic chemicals and their respective K_{ow}, and the results expressed in relation to the ratio between final permeability coefficient k_f and k_w (permeability coefficient with respect to water as the permeant). Both illite and kaolinite soils start from the same common point of $k_f/k_w = 1$ and whilst they also terminate at the same (coincidental) point at $\log K_{ow} = 3$ they both show considerable decrease in fluid conductivity through the soils. The higher the value of K_{ow}, the greater is the partitioning. The higher partitioning value shows a corresponding decrease in the fluid conductivity of the soil. This indicates that those substances which show the least compatibility with water will move more slowly through the soil system.

Aqueous concentrations of hydrophobic organics such as polyaromatic hydrocarbons (PAH), compounds such as nitrogen and sulphur heterocyclic PAHs, and some substituted aromatic compounds, tend to partition in proportion to the organic content (natural organic matter NOM) of the soil substrate (soil-engineered barrier). In other words, there is a direct correlation between partitioning of hydrophobic chemical compounds and NOM. The dominant partitioning mechanism involved is the hydrophobic bond established between the organic chemical and the NOM. Information obtained from studies have shown that variations in sorption coefficients between different soils are due to such soil characteristics as surface area, cation exchange capacity, pH, and also the amount and nature of the NOM present. If octanol is considered to be a surrogate for the NOM, the partition coefficient K_{ow} can be related to the organic content coefficient K_{oc} (ratio of the distribution coefficient K_d to the organic carbon fraction of the soil). For most concentrations of organic chemicals encountered in natural systems, a Freundlich type of linear distribution coefficient K_d such as that shown in Fig. 7, (slope of the sorption isotherm) has been found to be applicable for description of the partitioning (Means et al. 1982).

Partitioning of petroleum hydrocarbons (i.e. sorption-bonding) by the active soil fractions' mineral surfaces occurs only when the water solubility of the PHCs is exceeded and the hydrocarbons are accommodated in the micellar form. The accommodation concentration of hydrocarbons in the pore-water has been found to be useful in reflecting the partitioning

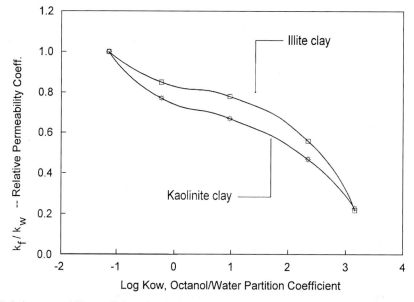

Fig. 6. Relative permeability coefficients in relation to K_{ow}, octanol/water partition coefficient.

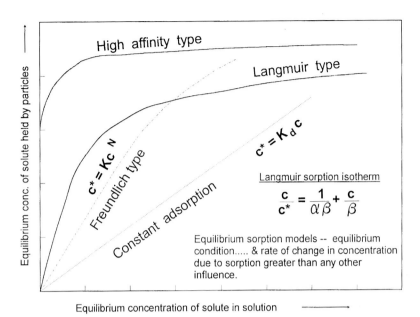

Fig. 7. Batch equilibrium adsorption isotherms showing different types of isotherms. α is the adsorption constant in the Langmuir model and is related to the binding energy, and β refers to the maximum amount of solute that can be sorbed by the solids. N is a constant in the Freundlich model, and K_d and K_c are the slopes of the curves.

tendency of organic substances between the pore-aqueous phase and soil solids. Hydrocarbon molecules with lower accommodation concentrations (i.e. higher K_{oc} values) would be partitioned to a greater extent onto the soil fractions than in the pore-aqueous phase. The results of Meyers and his co-workers (1978, 1973) show an inverse relationship between the accommodation concentration of the hydrocarbons and the proportion (percentage) adsorbed. The lower the accommodation concentration of the hydrocarbon in water, the greater the tendency of the organic compound to be associated with the reactive surfaces of the soil fractions. From a regulatory perspective, the significance of such a relationship is that the aromatic fractions of petroleum products (i.e. the most toxic components) would be least likely to partition to the reactive surfaces associated with the soil fractions. As an example, a study of adsorption data of hydrocarbons showed a high K_{oc} value and a very low solubility of the organic compound anthracene in water. Anthracene is significantly adsorbed, and the MAH/PAH (monoaromatic hydrocarbon/polyaromatic hydrocarbon) association with the reactive surfaces in soils would be:

anthracene > napthalene > ortho-xylene

> toluene > benzene.

Partitioning and saturated transport prediction

Analytical and/or computer prediction models are essential tools in the decision-making process for regulatory agencies. Such models aid assessments of the capability of a soil-engineered barrier (or soil substrate) to function as a natural attenuation barrier system over the period of leachate generation and transport (generally anywhere from one to fifty years and more). Partitioning of contaminants is generally considered through parametric inputs relying on batch equilibrium adsorption isotherm tests for the commonly used sets of saturated transport relationships. However, the process whereby partitioning occurs (i.e. retention, retardation mechanisms or a combination thereof) is not directly addressed. The use of such transport relationships for various scenarios has been reviewed by Shackelford (1997), and will not be repeated herein. The problem under consideration here is the need to distinguish between retention and retardation mechanisms for the observed partitioning of contaminants. These impact directly on the regulatory requirements or expectations. As stated previously, the consequences of rendering a wrong judgement of the accumulation process in modelling predictions of pollution plume advance can be severe.

The prediction of pollution plume transport using analytical models of saturated transport, (i.e. contaminant transport in fully saturated soil media, without considering storage and radioactive decay), is most often cast in terms of the 'advection-diffusion' relationship as follows:

$$\frac{\partial c}{\partial t} = D_L \frac{\partial^2 c}{\partial x^2} - v \frac{\partial c}{\partial x} - \frac{\rho}{n\rho_w} \frac{\partial c^*}{\partial t} \quad (2)$$

where: c = concentration of contaminant of concern, t is the time, D_L is the diffusion–dispersion coefficient, v is the advective velocity, x is the spatial coordinate, ρ is the bulk density of soil media, ρ_w is the density of water, n is the porosity of soil media, and c^* is the concentration of contaminant adsorbed by soil fractions. The c^* term used in Eq. (2) is of particular interest since this is the parameter that deals with the partitioning of contaminants between soil fractions and pore fluid. The examples of some common types of adsorption isotherms shown in Fig. 7 say very little about the process of transfer of contaminants from the fluid phase onto the soil solids. Desorption isotherms can also be obtained from laboratory batch equilibrium tests; these are generally not used unless specifically required. However, regardless of whether adsorption and desorption isotherms are used in the type of relationship depicted in

Eq. (2), it needs to be repeated that these isotherms do not provide information which allows the researcher to directly distinguish between attenuation and retardation mechanisms responsible for the partitioning. Although very seldom undertaken, inferences on retention can be made through simple subtractive relationships between adsorption and desorption phenomena. These however can only be taken as simple approximations.

Defining K_d as the partitioning coefficient, the linear adsorption isotherm shown in Fig. 7 provides one with the relationship $c^* = K_d c$. If this relationship is substituted into eqn. (2), as is the common procedure, the readily recognized popular relationship shown as Eq. (3) is obtained:

$$R \frac{\partial c}{\partial t} = D_L \frac{\partial^2 c}{\partial x^2} - v \frac{\partial c}{\partial x}$$

where
$$R = \left(1 + \frac{\rho}{n\rho_w} K_d\right). \quad (3)$$

It is easily recognized that if a non-linear adsorption isotherm is adopted, the R factor in Eq. (3) will not be obtained, since K_d is not a constant. Under such circumstances, it is necessary to determine the function relating c and c^* and to use eqn. (2). The predictions made from

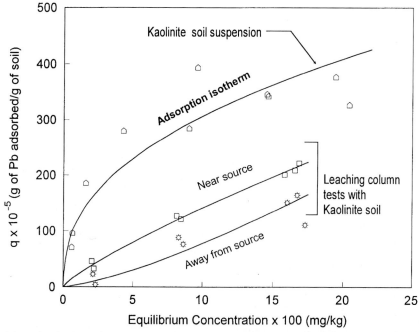

Fig. 8. Adsorption isotherm for Pb contaminant leachate from batch equilibrium tests on soil suspensions, and adsorption characteristics from leaching column (compact soil) tests.

the relationships can only inform one of the concentration of the specific target contaminants used in the model in relation to time and spatial distance. The predicted distribution and attenuation of contaminants can only be considered as accurate as the source and quality of the inputs provided for determination of c^*. The predicted distribution and attenuation of contaminants also depends on the understanding of how the contaminants are partitioned and held in the soil system.

The adsorption isotherm obtained through batch equilibrium tests which is generally conducted with soil suspensions cannot be considered to be representative of the sorption characteristics of intact soils. Figure 8 shows the results of sorption of Pb in a leaching column containing a kaolinite soil. For comparison, the adsorption isotherm of the same soil is also shown. The results show that there are differences not only between the adsorption characteristics of the intact soil sample and the adsorption isotherm, but also with the sorption location. The sorption characteristics between near source (source next to the leachate inlet) and away from source (source furthest away from leachate inlet) are distinctly different, and reflect the attenuation of the Pb.

Transport and partitioning in the unsaturated zone

Transport of pollutants in the unsaturated zone is almost inevitable if proper design and placement of landfills is to be achieved. One attempts to site landfills in regions where the aquifer is some distance below the bottom of the landfill. The partitioning of contaminants (i.e. the fate of contaminants) during transport of leachate through the unsaturated soil-engineered barrier and particularly through the soil substrate that characterizes the unsaturated zone is not well understood or studied. The same can also be said for the development of prediction models which can describe unsaturated flow and distribution of contaminants in the unsaturated zone. In the absence of significant positive (hydraulic) heads, the relationships given previously as Eqs (2) and (3) which refer to transport of contaminants in the saturated zone, cannot be readily applied. This is because moisture transport in the unsaturated soil substrate responds to the soil-water potential y (i.e. a measure of the energy status of the soil-water – see Yong et al. 1992) and generally moves by diffusive means.

If gravitational flow is considered to be insignificant, the governing relationship for one-dimensional transport of moisture is given as:

$$\frac{\partial \theta}{\partial t} = \frac{\partial}{\partial x}\left(k(\theta)\frac{\partial \psi_\theta}{\partial x}\right) \quad (4)$$

where $\theta =$ volumetric water content, $\psi_\theta =$ soil-water potential, and $k =$ Darcy permeability coefficient dependent upon the volumetric water content. A common procedure in casting the contaminant transport relationship for unsaturated flow is to associate the concentration of the contaminant of concern, c, with the volumetric water content θ. Accordingly, one obtains the following:

$$\frac{\partial c\theta}{\partial t} = \frac{\partial}{\partial x}\left(k(c,\theta)\frac{\partial \psi_c}{\partial x}\right) \quad (5)$$

where the chemical potential ψ_c is written with specific reference to the target contaminants, and where the Darcy permeability coefficient k is dependent on both the volumetric water content, θ, and the concentration c of the target contaminants.

One can make several simplifying assumptions, as for example denoting k to be a function only of θ, $\psi_c = \psi_\theta =$ a single-value function of θ, and obtain thereby the following relationship:

$$\frac{\partial c\theta}{\partial} = \frac{\partial}{\partial x}\left(D_c(\theta)\frac{\partial c}{\partial x}\right) - \frac{\partial(\rho'c^*)}{\partial t} \quad (6)$$

where $D_c =$ diffusion coefficient of the target contaminants, and $\rho' =$ bulk density of soil divided by the density of water. There are some very obvious difficulties in proceeding with implementation of Eq. (6). The determination of D_c is difficult when the diffusion rate of the target contaminants is affected by the surface properties of the reactive surfaces (particularly of the properties of the inner and outer Helmholtz layers) and c^*.

The approach taken recently by Yong & el Zahabi (1997c) using principles of irreversible thermodynamics considers the coupling of the moisture and contaminant solutes. Denoting the subscripts θ and c as moisture and concentration of contaminants respectively, the fluxes due to the respective thermodynamic forces are given as:

$$J_\theta = L_{\theta\theta}\frac{\partial \psi_\theta}{\partial x} + L_{\theta c}\frac{\partial \psi_c}{\partial x}$$

$$J_c = L_{c\theta}\frac{\partial \psi_\theta}{\partial x} + L_{cc}\frac{\partial \psi_c}{\partial x} \quad (7)$$

where J_θ and J_c are the fluid and solute fluxes respectively, $\partial\psi_\theta/\partial x$ = thermodynamic force due to the soil-water potential (i.e. soil-water potential gradient), $\partial\psi_c/\partial x$ = thermodynamic force due to the chemical potential (i.e. chemical potential gradient), and $L_{\theta\theta}$, $L_{\theta c}$, $L_{c\theta}$, L_{cc} are the phenomenological coefficients. The various diffusivity coefficients such as $D_{\theta\theta}$ (moisture), D_{cc} (solute), $D_{c\theta}$ (solute-moisture) and $D_{\theta c}$ (moisture-solute) have been obtained (Yong & el Zahabi 1997c) as:

$$D_{\theta\theta} = L_{\theta\theta}\frac{\partial\psi_\theta}{\partial x} \qquad D_{cc} = L_{cc}\frac{RT}{c}$$

$$D_{c\theta} = L_{c\theta}\frac{\partial\psi_\theta}{\partial\theta} \qquad D_{\theta c} = L_{\theta c}\frac{RT}{c} \tag{8}$$

and the final set of coupled relationships given in the following form:

$$\frac{\partial\theta}{\partial t} = \frac{\partial}{\partial x}\left[D_{\theta\theta}\frac{\partial\theta}{\partial x} + D_{\theta c}\frac{\partial c}{\partial x}\right]$$

$$\frac{\partial c}{\partial t} = \frac{\partial}{\partial x}\left[D_{c\theta}\frac{\partial\theta}{\partial x} + D_{cc}\frac{\partial c}{\partial x}\right] - \frac{\rho}{\rho_w\theta}\frac{\partial S_c}{\partial t} \tag{9}$$

where S_c is the sorbed concentration of contaminants. Solution of the coupled flow relationships uses an identification technique for evaluation of the phenomenological coefficients

(Yong & Xu 1988). The choice of functional forms for the phenomenological coefficients has been based on experimental knowledge of the distribution of contaminants along the length of unsaturated leaching column samples, together with moisture contents associated with the distributed contaminants.

Macropores, micropores and partitioning

Evaluation of the accumulation processes (retention and retardation) leading to partitioning in unsaturated zone transport, requires more detailed consideration of the micropores and macropores which characterize the structure of the kinds of clay soil material used in soil-engineered barriers. In saturated transport, the significant energy relationships established in micropores within the soil peds (i.e. soil aggregate groups) provide for high contaminant retention capacities, and most of the contaminant interactions occur in the flow processes through the macropores. However, in contrast, unsaturated transport requires consideration of inter-ped and intra-ped initial moisture contents (prior to wetting) and subsequent degree of wetting. The problem lies in the high potential difference between the micropores (intra-ped pore spaces), which are generally significantly smaller than 25 nm, and a reference pool of

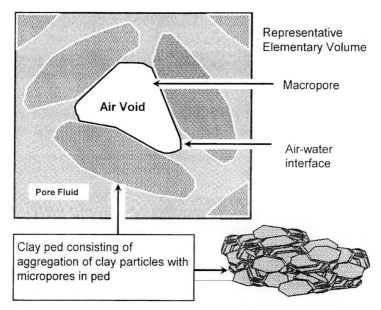

Fig. 9. Schematic diagram showing partly saturated representative elementary volume of clay soil consisting of clay peds (i.e. soil aggregate groups or clusters of clay particles).

water. This contrasts with the potential differences demonstrated by the macropores (inter-ped spaces) and the same reference pool of water. A typical representative elementary volume, REV, (Fig. 9) illustrates the presence of a macropore formed as the inter-ped pore space and the micropores existent in the clay peds. The REV model provides a means for structuring a mechanistic interpretation of contaminant partitioning in unsaturated transport of contaminants.

Figure 10 depicts the mechanistic interpretation of equilibrium in the partly saturated state in the absence of further available fluid input. On first wetting by the contaminant leachate, the higher potential difference in the clay peds will ensure saturation of the peds and contaminants transported to the peds will be more strongly held than in the macropores. At equilibrium, the osmotic potential ψ_π in the clay peds is balanced by the matric (micro-capillary) potential ψ_m in the inter-ped structure. Because further fluid input will only begin to fill the macropores, and because the potential difference in the inter-ped structure is considerably smaller than that exhibited within the clay peds, desorption of sorbed contaminants will occur primarily within the macropores that characterize the inter-ped structure. Retention of contaminants can be visualized as being the dominant accumulation process in the peds. Transport of con-

taminant solutes into and out of the peds is primarily via diffusion mechanisms, within a regime dominated by the surface forces from the clay particles forming the clay peds.

At this stage of development of the mechanics of unsaturated contaminant transport, the 'translation' of the mechanistic picture presented in Figs 9 and 10 into the necessary format for application of the relationships provided in Eq. (9) relies on the information obtained from unsaturated contaminant transport tests in leaching columns. By casting the relationships in terms of soil-water and chemical potentials, the opportunity has been provided for matching of prediction with experimental information.

Concluding remarks

Soil as a catalytic agent

The catalytic action of clay minerals in the promotion of reduction–oxidation reactions, particularly involving organic contaminants, contributes to the fate of contaminants. As an example, the oxidation of phenols in the presence of a montmorillonite soil reported by Yong *et al.* (1997*b*) shows that the environmental mobility of the phenols can be decreased considerably if the phenols can be transformed through oxidative processes to form phenolic

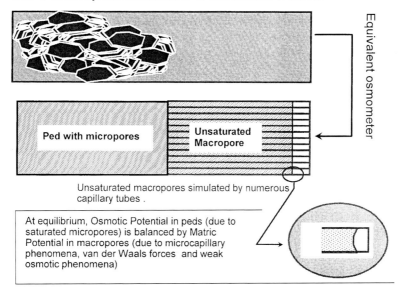

Fig. 10. Representation of micro-macropore interaction in the unsaturated mode in terms of components of soil-water potential.

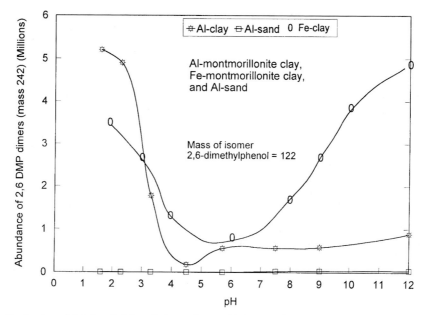

Fig. 11. Production of 2,6-dimethylphenol dimers from oxidative coupling of isomers of 2,6-dimethylphenol.

polymers (oligomers). Figure 11 shows results for the interaction between 26 dimethylphenol (DMP) and 2 initially 'dry' (i.e. 8.2% water content) montmorillonite clays, i.e. Fe^{3+}-saturated and A^{3+}-saturated montmorillonite clays, together with a 'blank' dry Al^{3+} sand. These results indicate that the montmorillonite clays transformed the DMP which had an isomer mass of 122 to dimers of mass 242 and traces of trimers of mass 362 (not shown), and that the abundance of the dimers varied as a function of pH.

At least three different mechanisms contribute to the oxidation of the phenols studied:

• structural elements of clay minerals such as Al, Fe, Zn and Cu transfer electrons to the surface-adsorbed oxygen, which are then released as hydroperoxyl radicals. These latter are capable of abstracting electrons from phenols. The partially oxidized product, a phenoxy radical, couples with unreacted phenols (or other phenoxy radicals) to form phenol dimers, trimers and tetramers (oligomers).
• oxidation–reduction properties of a clay's exchangeable cations such as Fe(III) and Cu(II) contribute to phenol polymerization through coupling of the radical cations with phenols,
• partially coordinated aluminium located on the edge of clay crystals (montmorillonite,

nontronite and kaolinite) are Lewis acids capable of accepting electrons from various aromatic compounds. Aluminium located in the silica sheet as a result of isomorphous substitution can behave in a similar fashion. Phenols with activating groups are oxidized more easily in comparison to phenols with deactivating groups.

Conclusion

Chemical mass transfer responsible for partitioning of contaminants constitutes a significant part of the processes involved in the transport and fate of contaminants. In the longer term, the redox environment (pE) and subsequent reduction–oxidation reactions will ultimately determine the final fate of the contaminants. The assessment of whether retention or retardation processes are responsible for the observed partitioning and hence the attenuation of contaminants within the soil matrix, is vital and critical in the evaluation of the 'natural attenuation capability' of the soil barrier system. The dilemma facing both regulatory agencies and practitioners is obvious: If potential pollution hazards and threats to public health and the environment are to be minimized or avoided, 'How can one ensure that the processes for contaminant attenuation in the substrate are the result of (irreversible sorption) retention mechanisms?' The associated problem of determining the long-term capability of the natural attenuation material to maintain its capability is another critical issue.

The defining mechanisms for sorption and desorption are generally known for many types of contaminants in interaction with specific soil fractions. However, the combined processes leading to partitioning of these contaminants in complex mixtures of contaminants and soil fractions have yet to be fully defined and understood. The preceding notwithstanding, there is sufficient knowledge of chemical mass transfer and biologically mediated mass transfer to distinguish between the following contaminant attenuation phenomena: (a) irreversible sorption responsible for partitioning and hence attenuation of contaminants; (b) temporary sorption processes and (c) physical hindrances which lead to retardation of contaminants.

References

ALLEN, H. E., PERDUE, E. M. & BROWN, D. S. (eds.) 1993. *Metals in Groundwater*. Lewis, Boca Raton.

ANDERSON, P. R. & BENJAMIN, M. M. 1990. Modeling adsorption in aluminum-iron binary oxide suspensions. *Environmental Science and Technology*, **24**, 1586–1592.

BASSETT, R. I. & MELCHOIR, D. C. 1990. Chemical modeling of aqueous systems: An overview. *In: Chemical Modeling of Aqueous Systems II*. American Chemical Society, 1–14.

BOLT, G. H. 1979. *Soil chemistry: B: Physico-Chemical Models*. Elsevier, Amsterdam.

CCME (Canadian Council of Ministers of the Environment) 1991. *National guidelines for the landfilling of hazardous waste*. Report **CCME-WM/TRE-028E**, April 1991.

CIRIA (Construction Industry Research and Information Association) 1996. *Barriers, liners and cover systems for containment and control of land contamination*. CIRIA Special Publication **124**.

CHIOU, G. T., SCHMEDDING, D. W. & MANES, M. 1982. Partition of organic compounds on octanol-water system. *Environmental Science and Technology*, **16**, 4–10.

DOE (Department of the Environment) 1995. *Landfill design, construction and operational practice*. Waste Management Paper **26B**, HMSO, London.

FARRAH, H. & PICKERING, W. F. 1977. The sorption of lead and cadmium species by clay minerals. *Australian Journal of Chemistry*, **30**, 1417–1422.

—— & ——1978. Extraction of heavy metal ions sorbed on clays. *Water, Air and Soil Pollution Journal*, **9**, 491–498.

—— & ——1979. pH effects in the adsorption of heavy metal ions by clays. *Chemical Geology*, **25**, 317–326.

GRIM, R. E. 1953. *Clay Mineralogy*. McGraw Hill, New York.

HARTER, R. D. 1979. Adsorption of copper and lead by Ap and B2 horizons of several northeastern United States soils. *Soil Science Society of America*, **43**, 679–683.

JENNE, E. (ed.) 1998. *Adsorption of Metals by Geomedia: Variables, Mechanisms, and Model Applications*. Academic, New York.

JESSBERGER, H. L. 1995. Waste containment with compacted clay liners. *Proceedings of the International Conference on The Geoenvironment 2000*. ASCE Special Publication, **46**, 463–483.

MAGUIRE, M., SLAVEK, J., VIMPANY, I., HIGGINSON, R. F. & PICKERING, W. F. 1981. Influence of pH on copper and zinc uptake by soil clays. *Australian Journal of Soil Research*, **19**, 217–229.

MANASSERO, M., VAN IMPE, W. F. & BOUAZZA, A. 1997. Waste disposal and containment. *Second International Congress on Environmental Geotechnics*, Japan, **3**, 1425–1474.

MEANS, J. C., WOOD, S. G., HASSETT, J. J. & BANWART, W. L. 1982. Sorption of amino and carboxyl-substituted polynuclear aromatic hydrocarbons by sediments and soils. *Environmental Science and Technology*, **15**, 2–93.

MESRI, C. & OLSON, R. E. 1971. Mechanisms controlling the permeability of clays. *Clays and Clay Minerals*, **19**, 151–158.

MEYERS, P. A. & QUINN, J. G. 1973. Association of hydrocarbons and mineral particles in saline solution. *Nature*, **244**, 23–24.

—— & OAS, T. G. 1978. Comparison of associations of different hydrocarbons with clay particles in simulated sea water. *Environmental Science and Technology*, **12**, 934–937.

NRA 1989. *Earthworks to landfill sites*. National Rivers Authority, North-West Region.

ONTARIO MINISTRY OF ENVIRONMENT MOE 1986. *Incorporation of the reasonable use concept into MOE groundwater management activities*. Ontario MOE Policy **15-08**, April 1986.

PASK, J. A. & DAVIS, B. 1954. *Thermal analysis of clays and acid extraction of alumina from clays. Differential Analysis*, US Bureau of Mines, Denver.

PAVILONSKY, V. M. 1985. Varying permeability of clayey soil linings. *Proceedings of the International Conference on Soil Mechanics and Foundation Engineering*, San Francisco, **2**, 1213–1216.

PEARSON, R. G. 1963. Hard and soft acids and bases. *Journal of the American Chemical Society*, **85**, 3533–3539.

SHACKELFORD, C. D. 1997. Modeling and analysis in environmental geotechnics: An overview of practical applications. *Second International Congress on Environmental Geotechnics*, Japan, **3**, 1375–1404.

SPOSITO, G. 1984. *The Surface Chemistry of Soils*. Oxford University Press.

SRIDHARAN, A., RAO, S. M. & MURTHY, N. S. 1986. Compressibility behaviour of homoionized bentonites. *Geotechnique*, **36**, 551–564.

STUMM, W. & MORGAN, J. J. 1981. *Aquatic chemistry*. 2nd edn. Wiley, New York.

YARON, B. 1989. On the behaviour of petroleum hydrocarbons in the unsaturated zone: Abiotic aspects. *In: Toxic Organic Chemicals in Porous Media*, Springer, Berlin, 211–230.

USCFR (US Code of Federal Regulations) 1992. **40 CFR 258** *Protection of Environment*. Office of Federal Register National Archives and Research Administration, US Printing Office, Washington, DC.

YONG, R. N. 1995. *The fate of toxic pollutants in contaminated sediments*. ASTM Special Technical Publication **1293**, 13–38.

——1996. Waste disposal, regulatory policy and potential health threats. *In:* BENTLEY, S. P. (ed.). Geological Society, London, Engineering Geology Special Publication, **11**, 325–340.

——1997. Multi-disciplinarity of environmental geotechnics. *Second International Congress on Environmental Geotechnics*. Japan, **3**, 1255–1273.

—— & EL ZAHABI, M. 1997. Unsaturated zone transport of heavy metals. *In:* YONG, R. N. & THOMAS, H. R. (eds) *Geoenvironmental Engineering, Contaminated Ground: fate of pollutants and remediation*. Telford, London, 173–180.

—— & MACDONALD, E. M. 1998. Influence of pH, metal concentration, and soil component removal on retention of Pb and Cu by an illitic soil. *In:* JENNE, E. (ed.) *Adsorption of Metals by Geomedia: Variables, Mechanisms, and Model Applications*. Academic, New York, 229–254.

—— & PHADUNGCHEWIT, Y. 1993. pH influence on selectivity and retention of heavy metals in soil clay soils. *Canadian Geotechnical Journal*, **30**, 821–830.

—— & SETHI, A. J. 1973. *Basic concepts in oil-soil interaction and preliminary results*, Research Report **OS-NR-1**, McGill University, Montreal.

—— & RAO, S. M. 1991. Mechanistic evaluation of mitigation of petroleum hydrocarbon contamination by soil medium. *Canadian Geotechnical Journal*, **28**, 84–91.

—— & XU, D. M. 1988. An identification technique for evaluation of phenomenological coefficients in coupled flow in unsaturated soils. *International Journal for Numerical and Analytical Methods in Geomechanics*, **12**, 283–299.

——, DESJARDIN, S., FARANT, J. P. & SIMON, P. 1997. Influence of pH and exchangeable cation on oxidation of methylphenols by a montmorillonite clay. *Applied Clay Science*, **12**, 93–110.

——, GALVEZ, R. & PHADUNGCHEWIT, Y. 1993. Selective sequential extraction analysis of heavy metal retention in soil. *Canadian Geotechnical Journal*, **30**, 834–847.

——, MOHAMED, A. M. O. & WARKENTIN, B. P. 1992. *Principles of contaminant transport in soils*. Elsevier, Amsterdam.

——, WARKENTIN, B. P., PHADUNGCHEWIT, Y. & GALVEZ, R. 1990. Buffer capacity and lead retention in some clay materials. *Water, Air and Soil Pollution Journal*, **53**, 53–68.

Regulatory objectives and safety requirements for the disposal of radioactive wastes in Switzerland

E. FRANK

Swiss Federal Nuclear Safety Inspectorate, CH-5232 Villigen-HSK, Switzerland

Abstract: According to Swiss Law the producers of radioactive waste are responsible for management and safe disposal of all types of radioactive waste. Therefore, in 1972, the electricity utilities involved in nuclear power production formed the National Cooperative for the Disposal of Radioactive Waste (NAGRA) which is responsible for final disposal and related work. The general disposal strategy includes two repository siting programmes, a deep repository for high-level waste in Northern Switzerland (crystalline basement or Opalinus claystone) and a repository with horizontal access for low- and intermediate-level waste in the Alps (marls and shales of the Wellenberg-site).

The safety of waste disposal is assessed on the basis of specific protection objectives formulated by the Swiss regulatory authority. The protection objectives limit the annual dose equivalent or the radiological risk to an individual and they require that post-closure safety must not depend on long-term monitoring or any other interventions to ensure safety. After a final repository has been sealed, it must be possible to dispense with safety and surveillance measures.

In Switzerland, five nuclear power plants are at present in operation with a total net capacity of 3077 MW (e). Most of the radioactive wastes arise from the back-end of the nuclear fuel cycle. These include high-level waste from reprocessing, low- and intermediate-level wastes from reactor operation and reprocessing, and low-level decommissioning wastes. Additionally, there are radioactive wastes resulting from research, industry and medical applications.

Responsibilities for waste management are separated in Switzerland between regulatory and implementing organizations:

- The Federal regulatory authority comprises the Swiss Federal Nuclear Safety Inspectorate (HSK) and the Swiss Federal Nuclear Safety Commission (KSA). Various licences are required for waste management facilities such as a licence for exploratory drilling, a general licence, a construction licence, an operating licence and a closure licence. For each of these licences, a procedure with two public consultations is applied. The licences are granted by the Federal Government on the basis of review reports submitted by the regulatory authorities. In addition, a general licence has to be approved by the Swiss parliament.
- According to Swiss law the producers of radioactive waste are responsible for its man-

agement and safe disposal. To carry out this task the nuclear power utilities and the Swiss Confederation – responsible for the waste from medicine, industry and research – have formed the National Cooperative for the Disposal of Radioactive Waste (NAGRA) which is charged with preparing and implementing waste disposal in Switzerland. The legislation further prescribes that Swiss radioactive waste has to be disposed of in Switzerland.

Regulatory objectives and safety requirements

The main goal of safe and permanent disposal of radioactive waste is to protect the environment and human health adequately against ionizing radiation without imposing undue burdens on future generations. Long-term safety of a disposal facility shall not depend on surveillance and maintenance. Therefore only geological disposal in deep underground repositories is considered in Switzerland. The safety concept of geological disposal of radioactive waste is to be based on a three-pillar strategy 'Isolation, Immobilization and Retention'. This concept assumes that groundwater transport is the mostly likely mechanism by which radionuclides might return from

From: METCALFE, R. & ROCHELLE, C. A. (eds) 1999. *Chemical Containment of Waste in the Geosphere*. Geological Society, London, Special Publications, **157**, 21–25. 1-86239-040-1/99/$15.00 © The Geological Society of London 1999.

a deep geological repository to the biosphere. Since groundwater transport is a three-stage process, the safety of waste disposal must cover each stage as follows:

- *Isolation*: the waste is isolated as efficently as possible against groundwater
- *Immobilization*: the radionuclides are fixed in a leach-resistant solid matrix which restricts dissolution in the groundwater.
- *Retention*: the migration of radionuclides through the technical barrier system and the geosphere is characterized by various retention or retardation processes such as chemical adsorption, ion exchange, precipitation, chemical interactions, molecular filtration, dispersion and diffusion. This causes the nuclide velocity to be lower than the groundwater velocity.

Safety assessment of the geological disposal system is based upon the multiple barrier principle including physical containment (waste container), engineered containment (lining, backfill, high-integrity seals), chemical containment (waste form, buffer materials) and geological containment (low-flow geological formation with efficient retardation properties).

The principal objectives and requirements for long term disposal of radioactive wastes were issued by the Swiss Nuclear Safety Authorities in 1980 and updated in 1993 in the Swiss guideline HSK-R-21 'Protection Objectives for the Disposal of Radioactive Waste' (HSK & KSA 1993). The guideline takes into account the recommendations of other international organizations, in particular the International Commission on Radiological Protection (ICRP 1985, 1991), IAEA's Safety Principles (IAEA 1989) and OECD's Nuclear Energy Agency (NEA) publications (e.g. NEA 1984). The HSK-R-21 guideline outlines the *overall objectives* and *principles* which apply to the disposal of radioactive waste. The safety requirements which must be satisfied by disposal projects are then set out in the form of *protection objectives*. In the licensing procedure, the applicant must demonstrate by means of a safety analysis that his project fulfills these requirements.

In the following the most important features of guideline HSK-R-21 are summarized.

Overall objectives of disposal

The overall objectives of radioactive waste disposal are to protect public health and the environment effectively and reliably, and to minimize any burden on future generations.

Principles of disposal

In order to achieve the overall objectives, the following main principles are to be applied to the disposal of radioactive waste:

- The long-term safety of a repository must be ensured by a system of passive multiple safety barriers. Different types of engineered and natural barriers should be considered in order to guarantee effective and reliable containment of the radionuclides.
- When disposing of radioactive waste, the multiple barrier system must give assurance that any release of radioactive material to the environment is restricted to an acceptably low rate. The resulting additional radiation dose to the population shall not exceed a low limit corresponding to a small fraction of the dose from natural background radiation.
- Environmental protection shall be ensured in a way that living species will not be endangered and the use of mineral resources will not be unnecessarily restricted.
- The risks to humans and the environment arising from radioactive waste disposal in Switzerland shall not, at any time in the future and anywhere abroad, exceed the levels which are permissible today in Switzerland.
- In final disposal in geological formations, there is no intention of retrieval or long-term institutional control. Any measures which would facilitate surveillance and repair of a repository or retrieval of the waste must not impair the functioning of the safety barriers.
- The provisions for radioactive waste disposal are the responsibility of current society which benefits from nuclear technology. This responsibility shall not be passed on to future generations.

Safety requirements

The safety requirements which must be satisfied by a disposal project are set out in the form of three specific protection objectives:

Protection objective 1. The release of radionuclides from a sealed repository subsequent upon processes and events reasonably expected to happen, shall at no time give rise to individual doses which exceed 0.1 mSv/year.

The dose limit of 0.1 mSv/year was arrived at after a comparison with the level of average natural background and man-made radiation in Switzerland and its seasonal and spatial variation (see Table 1). This limit corresponds to

Table 1. *Average annual radiation dose received by an individual living in Switzerland*

Natural Radiation	dose (mSv a^{-1})
Terrestrial radiation (including radon and daughter isotopes)	2.1
Body internal dose	0.4
Cosmic radiation	0.3
Manmade radiation	
Nuclear fallout (Weapon testing, Tschernobyl, NPP, ...)	0.2
Medical diagnosis	1.0
Total individual dose	4.0

Völkle (1996).

about 3% of natural background radiation and was considered as reasonable low.

From the system of dose limitations recommended by the International Commission on Radiation Protection and applying the risk factor of 5% per Sv (ICRP 1991), a radiation dose of 0.1 mSv per year leads to a radiological risk of fatality of five per million per year. Similar quantitative dose limits are set for geological waste disposal facilities by regulators in other countries (Table 2).

Protection objective 2. The individual radiological risk of fatality from a sealed repository, subsequent upon unlikely processes and events not taken into consideration in protection objective 1 above, shall at no time exceed one in a million per year.

Protection objective 3. After a repository has been sealed, no further measures shall be necessary to ensure safety. The repository must be designed in such a way that it can be sealed within a few years.

Monitoring or long-term surveillance of a sealed repository is not ruled out by this requirement.

Table 2. *Comparison of individual dose limits for geological waste disposal as defined by regulatory authorities in different countries*

Individual dose limit (mSv a^{-1})	
Canada	0.05 mSv
Belgium	0.10 mSv
Finland	0.10 mSv
Sweden	0.10 mSv
Switzerland	0.10 mSv
France	0.25 mSv
Germany	0.30 mSv
USA (Yucca Mountain Site)	0.25 mSv

Warnecke (1994).

However, the dose, limited by protection objective 1 above, must on no account depend upon surveillance measures.

According to present planning, before final sealing of a repository, technical safety evaluation by *in situ* experiments will take place during an adequate time to ensure satisfactorily the long-term performance of the disposal system. The application for and the decision on closure of the repository will then be based on this demonstration phase.

Swiss regulations stipulate that the multiple barrier approach to disposal safety must primarily be seen in the context of an overall system approach which considers the safety of the disposal system as a whole. Specific safety criteria on single barrier elements are not formulated because it may lead to a loss of flexibility. Although further criteria may give some advice as to whether the protection objectives can be reached they do not replace a complete safety analysis of the repository system.

Safety analysis

The applicant has to show in a safety analysis addressing all relevant features, processes and future events of the repository system that the repository can achieve the protection objectives. The applicant has to submit a safety analysis for each stage of the licensing procedure (general-, construction-, operating- and closure-licence). Safety relevant data on the repository system obtained from preliminary investigations should be supplemented by ongoing investigations during construction of the repository. The safety analysis for the post-closure phase has to be refined accordingly with improved knowledge of the repository system.

Every model relevant to safety assessment has to be validated. Validation should give reasonable assurance that the model is applicable to the specific disposal system. It will largely depend on the barrier concept and the specific geological situation for which models have to be validated. As a tentative list of models that are prime candidates for validation, the following processes and models can be mentioned:

- groundwater flow (regional–local)
- groundwater chemistry
- container integrity
- waste form degradation/dissolution
- retardation processes (sorption, matrix diffusion, ...)
- radionuclide solubility and speciation
- long-term performance of buffer/seal materials (mostly bentonite)

Any calculation of the potential consequences of a sealed repository involves long-term predictions and inevitably contains a degree of uncertainty which has to be taken into account. The more long-term the predictions are, the greater the uncertainty. The individual components of the repository system (engineered barriers, host rock, far-field geology, biosphere) differ in their degree of predictability. Given this situation, long-term dose calculations are not to be interpreted as effective predictions of radiation exposures for a defined population group. They are much more indicators for evaluating the impact of a potential release of radionuclides into the biosphere. In this latter sense, long-term dose and risk calculations should be carried out at least for the maximum potential consequences from the repository.

Planned repositories in Switzerland

Currently NAGRA is in the site selection and conceptual stage of designing geological repositories for radioactive wastes. Two types of repositories are planned:

A repository for low- and intermediate-level waste. The repository concept foresees a mined system of horizontally accessed disposal caverns in a low-permeable host rock covered by some hundreds of metres of overburden. The safety barrier system comprises the waste matrix (mostly cement), steel drums placed in large containers of cement, backfilling of the remaining space with concrete, concrete lining of the disposal caverns and finally low permeable host rock and its overburden. In 1993 NAGRA decided to focus the site investigation for this type of repository at Wellenberg (Cretaceous shales of the Drusberg nappe in Central Switzerland). A preliminary safety analysis was performed in 1994 based on experimental results as well as on-site specific borehole data (NAGRA 1994a). For safety assessment most emphasis is placed on chemical containment of the radionuclides in the huge mass of cement imposing strongly buffered chemical conditions to the system as well as on the geological containment providing very low groundwater fluxes and high retardation capacities by clay minerals. Subsequently, the licensing procedure was started by applying for a general licence, but this is currently blocked due to lack of public acceptance at cantonal level.

A deep repository for high-level and long-lived intermediate-level waste. This type of reposi-tory is planned to be located at a depth of 600 to 1000 m below the surface in Northern Switzerland, in a region with rather low topographic relief. The concept includes two vertical shafts for access to the repository area which consists of a system of horizontal drifts containing the wastes. The high-level waste will be isolated by a multiple barrier system consisting of a leach-resistant glass matrix in a steel mould, a massive corrosion-resistant cast steel overpack, a large buffer-ring of highly compacted impermeable bentonite and finally a low-permeable host-rock. The safety assessment (NAGRA 1994b) is strongly based on physical containment (steel canister), chemical containment (waste form, bentonite-buffer, reducing groundwater) as well as geological containment (low-flux geological environment, efficient retardation processes). The potential host rocks currently under investigation are crystalline basement rocks (Granites and Gneisses) and the Opalinus claystone (argillaceous rocks).

Outlook

During the last 15 years much progress has been made in radioactive waste management in Switzerland including development of specific disposal concepts and implementation of different site evaluation programmes. For the realization of final repositories for low- and high-level wastes, substantial amounts of further work are still required.

There is general agreement among experts that the problem of low-level waste disposal can be solved adequately by the geological repository concept as proposed by NAGRA at the site of Wellenberg. The safety of this repository system, which must be demonstrated for some hundreds to some thousands of years, relies mainly on the different technical barriers in combination with limited slow groundwater flow and efficient retardation in the geosphere.

Considering the much longer time-scales of high-level disposal system performance (some hundreds of thousands to millions of years) and the considerable uncertainties associated in quantifying the long-term behaviour of the engineered components of the repository system, the natural barriers become much more important. For achieving long-term safety, the geological situation and the host-rock properties play a dominant role. The reconstruction of past flow systems and geochemical fluxes (palaeohydrogeology) become a key element in understanding the future evolution of a repository site. An essential quality of a geological site includes reliable predictability of its lithological, structural, geochemical and hydrogeological properties and traceable evidences of its tectonic long-term stability. This means that there must be sufficient and adequate results from drilling and underground investigations showing that the low permeable repository 'block' – the rock mass that contain the maze of repository tunnels – is fairly homogenous, has no major permeable discontinuities and no preferential direct pathways to the human environment. In order to fulfil this requirement, the groundwater regime in and around the repository must be well understood in terms of where the deep groundwater is coming from, where it is going and which travel times can be estimated.

References

HSK & KSA. 1993. Guideline for Swiss Nuclear Installations **HSK-R-21** *Protection objectives for the disposal of radioactive waste*. Swiss Federal Nuclear Safety Inspectorate (HSK) and Federal Commission for Safety of Nuclear Installations (KSA), Villigen.

IAEA. 1989. *Safety Principles and Technical Criteria for the Underground Disposal of High Level Radioactive Wastes*. Safety Series No. **99**, IAEA, Vienna.

ICRP. 1985. *Radiation Protection Principles for the Disposal of Solid Radioactive Waste*. ICRP Publication **46**, Pergamon, Oxford.

ICRP. 1991. *Recommendations of the International Commission on Radiological Protection*. ICRP Publication **60**, Annals of the ICRP Vol **21**(1–3), Pergamon, Oxford.

NAGRA. 1992. *Nukleare Entsorgung Schweiz–Konzept und Realisierungsplan*. Nagra Technischer Bericht **NTB 92-02**, Wettingen.

NAGRA. 1994a. *Bericht zur Langzeitsicherheit des Endlagers SMA am Standort Wellenberg*. Nagra Technischer Bericht **NTB 94-06**, Wettingen.

NAGRA. 1994b. *Kristallin-I Safety Assessment Report*. Nagra Technischer Bericht **NTB 93-22**, Wettingen.

NEA. 1984. *Long-Term Radiological Protection Objectives for Radioactive Waste Disposal*. Nuclear Energy Agency, OECD, Paris.

VÖLKLE, H. 1996. *Umweltradioaktivität und Strahlendosen in der Schweiz, Jahresbericht 1995*. Bundesamt für Gesundheitswesen, Fribourg.

WARNECKE, E. 1994. Disposal of radioactive waste – a completing overview. *Journal for Nuclear Engineering, Energy Systems, Radiation and Radiological Protection, Zeitschrift für Kerntechnik*, **59**(1–2), Carl Hanser Verlag, München.

Geochemical factors in the selection and assessment of sites for the deep disposal of radioactive wastes

DAVID SAVAGE,[1] RANDOLPH C. ARTHUR[2] & SHIGEYUKI SAITO[3]

[1] QuantiSci Ltd., 47 Burton Street, Melton Mowbray, Leicestershire LE13 1AF, UK
(Present address: Quintessa Ltd, 24 Trevor Road, West Bridgford,
Nottingham NG2 6FS, UK)
[2] Monitor Scientific, 3900 South Wadsworth Boulevard, Denver, Colorado 80235, USA
[3] Mitsubishi Materials Corporation, Nuclear Ecosystem Department, Global Ecoindustry
Center, Koishikawadaikoku Bld., 1-3-25, Koishikawa, Bunkyo-ku, Tokyo 112, Japan

Abstract: Geochemical parameters which are important to waste isolation include (with optimum conditions in parenthesis): groundwater pH ($10 > pH > 8$), redox ($Eh < -200$ mV), P_{CO_2} ($<10^{-4}$ bars), and ionic strength (<0.5 M); the colloid (<0.01 ppm), organic (<5 ppm) and inorganic ligand ($Cl^- > HCO_3^-$, low abundances of phosphate) content of groundwaters; and the mineralogical composition and structure of groundwater flowpaths (zeolites, clays, iron oxyhydroxides with appreciable diffusive porosity). pH, redox, and in some environments, P_{CO_2}, will be controlled by water–rock reaction; other parameters are influenced more by geological history or environment. pH and redox buffer intensities of likely mineral assemblages and groundwater compositions under possible physical conditions of deep disposal have been calculated. For pH, reactions between aluminosilicates, carbonates and water should dominate over those in the fluid phase alone. The reaction of small amounts of ferrous iron-bearing minerals (≥ 5 vol %) is sufficient to maintain chemically reducing conditions under likely hydraulic conditions at depth. Consideration of laboratory data suggest that pH and redox reactions should not be hindered by kinetics under likely physical conditions for deep disposal, but further field and laboratory experimental data are required to verify these conclusions. The most suitable geochemical environments for waste isolation are basic or intermediate igneous rocks in tectonically stable, low-lying terrain away from sedimentary basins. Geochemical criteria should not be used in isolation of other factors for the selection of sites, but suitable geochemical conditions may help relax the requirements for optimum performance dictated by other geological and hydrogeological factors.

Geochemical factors have taken a secondary role in the selection and characterization of sites for the geological disposal of radioactive wastes, with the primary role usually defined by geological (e.g. Gray et al. 1976) or physical hydrogeological (e.g. Chapman et al. 1986) criteria only. This emphasis has developed from the recognition of the importance of groundwater travel time and physical dispersion processes in the migration of radionuclides from a deep geological repository and, hence, the impact upon calculated radiation doses in safety assessment calculations. However, geochemical conditions at a site for the disposal of radioactive wastes are intrinsic to the optimum performance of the so-called 'engineered barrier system' (EBS – wasteform, waste canisters, backfill, etc.) of the repository and in delaying the transport of radionuclides through the geosphere, and may

thus impact upon safety assessment calculations. Consequently, geochemical/mineralogical attributes such as the capacity of the rock–groundwater system to maintain suitable redox conditions and pH, and to sorb, ion exchange, filter, precipitate or trap radionuclides in dead-end pore space, become potentially as important as the physical hydrogeological properties of a site.

An important role of the EBS is the control of chemical conditions to enhance barrier preservation and radionuclide immobilization, and maximize its capacity to withstand external perturbations to these conditions. In the geosphere, chemical characteristics cannot be engineered and must be assessed via site characterization efforts. In this regard, it is important to distinguish between simply characterizing ambient chemical conditions in a particular water–rock system and defining the control or 'buffer' of

From: METCALFE, R. & ROCHELLE, C. A. (eds) 1999. Chemical Containment of Waste in the Geosphere. Geological Society, London, Special Publications, **157**, 27–45. 1-86239-040-1/99/$15.00 © The Geological Society of London 1999.

the measured chemical conditions. We can thus distinguish between 'intensity' (absolute values of pH, redox, salinity etc.) and 'capacity' factors (total amounts of reactive chemical species) (Scott & Morgan 1990). Understanding the mechanism of chemical control enables prediction of the potential for geochemical variables to change if other parameters (e.g. groundwater flow rate, chemical composition of recharge) are perturbed during natural phenomena such as glaciations or earthquakes. However, site hydrogeochemical characterization tends to be directed more towards establishing the groundwater flow regime at a site, rather than deriving data which can be used in safety assessment (e.g. Richards *et al.* 1996).

Another issue which is germane to the consideration of geochemical criteria for site selection is whether the rock–groundwater system should be chemically compatible with the EBS design. In general in radioactive waste investigations, the selection of EBS design has been conducted independently of the choice of site. Clearly, some sites may be better than others for the preservation of the EBS. For example, the preservation of hyperalkaline pore fluid conditions in repository designs employing large quantities of cementitious materials may best be preserved in rock–groundwater systems with minimal amounts of reactive chemical components in groundwater such as Mg and CO_2. However, the reaction of cementitious pore fluids with the rock–groundwater system could conceivably produce more favourable conditions for waste isolation by reducing the porosity and permeability of the geosphere (e.g. Savage & Rochelle 1993) and/or increasing the capacity of the geosphere to retard radionuclide migration (e.g. Berry this volume). The relative advantages and disadvantages of chemical reactivity of EBS design need to be considered in site selection and characterization.

This paper attempts to address the above issues and discusses the following: which geochemical parameters are of importance to radioactive waste disposal and why; what natural controls and chemical buffering mechanisms exist for these parameters; and lastly, which geological environments may provide the most suitable geochemical conditions for waste isolation.

Important geochemical parameters

The preservation of the repository EBS and the migration of radionuclides are sensitive to at least one or more of the following geochemical/ mineralogical parameters: pH; redox; P_{CO_2}; the types and amounts of inorganic and organic ligands and colloids in groundwaters; the ionic strength of the groundwater and the mineralogical composition of groundwater flowpaths. Although not a geochemical parameter as such, microbial material in groundwater may, depending on characteristics and amounts, also impact upon waste isolation either directly or indirectly. The following text discusses the relevance and potential natural controls of these parameters in more detail.

pH

pH, defined as the negative (base ten) logarithm of the hydrogen ion activity of a water, is a key variable in groundwater and water–rock systems and, in terms of safety assessment, has important effects on: radioelement solubility, speciation and sorption; metal (canister) corrosion; and buffer/backfill behaviour. In addition to affecting the overall magnitude of heterogenous reactions, pH can also impact upon their rates through surface sorption and desorption processes (e.g. Blum & Lasaga 1988).

The pH of most groundwaters of interest to waste disposal is in the range 6–10 (Fig. 1). There are a number of factors which are important in governing the pH of groundwater: reactions with minerals in the rock, particularly aluminosilicates such as feldspars, micas and clays; the partial pressure of carbon dioxide (P_{CO_2}); the concentration of so-called 'conservative' or 'mobile' ions such as chloride (Michard 1987) and temperature. Reactions with the rock mineralogy and dissolved carbon dioxide can affect pH through temperature-dependent equilibria such as:

$$CaAl_2Si_2O_8 + H^+ + HCO_{3-} + H_2O$$

$$= Al_2Si_2O_5(OH)_4 + CaCO_3 \qquad (1)$$

pH can be calculated using thermodynamic data for minerals and aqueous species with charge balance governed by varying amounts of a mobile ion such as chloride. Figure 1 compares pH values measured in groundwaters in fractured hard rocks from a variety of localities worldwide and those calculated assuming equilibrium between a suite of low-temperature minerals frequently found lining fractures in such rock types. Figure 1 shows that although pH is variable over 5 units at low salinities, at salinities above 10 000 mg/l Cl^-, pH may be controlled by thermodynamic mass action relationships between aluminosilicates, carbonates and water. Similar conclusions have been reached for cold groundwaters at Stripa, Sweden (Grimaud *et al.* 1990) and warm groundwaters in sedimentary

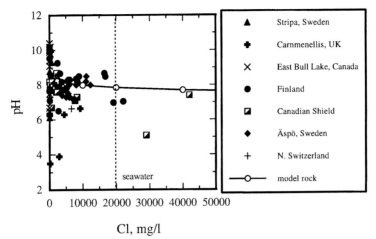

Cl, mg/l

Fig. 1. Variation of pH with Cl⁻ in groundwaters from fractured hard rocks from the Canadian (Bottomley *et al.* 1990, 1994) and Fennoscandian (Nurmi *et al.* 1988; Nordstrom *et al.* 1989*a*; Smellie & Laaksoharju, 1992) Shields, the northern Swiss basement (Pearson *et al.* 1989) and the Carnmenellis granite, UK (Edmunds *et al.* 1984), together with those generated using EQ3/6 (version 7.2a, Wolery 1992) and thermodynamic data (from Johnson *et al.* 1992) for an assemblage of low temperature minerals in equilibrium with water and different concentrations of Cl⁻. The mineral assemblage (with controlled element in parenthesis) employed was: chalcedony (Si); albite (Na); K-feldspar (K); laumontite (Ca); chlorite (Mg); kaolinite (Al); calcite (HCO₃⁻).

basins (Hutcheon *et al.* 1993; Hanor 1994). Reactions buffering pH in groundwaters are discussed in more detail in a later section.

Redox

Redox, like pH, is a key variable in groundwater systems and has important effects upon the solubility, speciation and sorption of redox-sensitive radionuclides, and metal corrosion. Despite the common linkage between pH and redox (in *Eh*–pH diagrams, for example), there are marked physical and conceptual distinctions between the two variables (Hostettler 1984). The redox state of a groundwater is generally expressed as *Eh*, p*E* or f_{O_2}. *Eh* is defined by a relation known as the Nernst equation:

$$Eh \text{ (volts)} = Eh^{\circ} + \frac{2.3RT}{nF} \log\left(\frac{[\text{oxidant}]}{[\text{reductant}]}\right) \quad (2)$$

where Eh° is a standard or reference condition at which activities of oxidants and reductants are unity, n is the number of electrons transferred, F is the Faraday constant (96.42 kJ per volt gram equivalent), R is the gas constant (8.314×10^{-3} kJ/deg mol), and T is the absolute temperature (K). p*E* is related to *Eh* by the following expression:

$$pE = \frac{[nF]}{2.303RT} Eh \quad (3)$$

Geochemists familiar with high temperature systems also refer to oxygen fugacity as a measure of redox. Oxygen fugacity is related to *Eh* by:

$$\log f_{O_2} = \frac{4F}{2.303RT} Eh - \log a_{H^+}$$
$$+ 2\log a_{H_2O} + \log K_{Eh} \quad (4)$$

where a_{H^+} and a_{H_2O} are the activities of hydrogen ions and water respectively and K_{Eh} is the equilibrium constant for a $O_2(g)/H_2O$ half reaction. Redox levels in groundwater are controlled by the relative rates of introduction of oxidizing water in recharge, the rate of groundwater flow and circulation, and the rate of consumption of oxygen by (often microbially mediated) reduction reactions with reduced aqueous species or reduced chemical species in the rock. Potential reductants include: ferrous iron aqueous species, ferrous silicates, sulphides, oxides and carbonates; reduced sulphur aqueous species and sulphides; reduced manganese aqueous species and oxides; and organic materials such as bitumen, graphite or aqueous organic compounds such as dissolved methane, carboxylic and humic acids etc.

There are three important aspects to natural controls of redox conditions. Firstly, depending upon the dominant redox reaction in the system, the absolute value of redox potential may vary considerably. Iron is usually the most important chemical component in this regard, but sulphur,

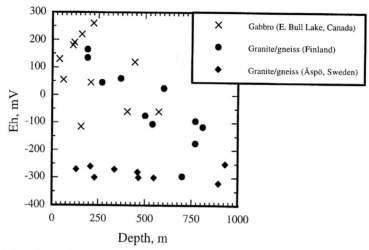

Fig. 2. Variation of groundwater redox state (expressed as *Eh*) versus depth for granitic (*sensu lato*) and basic rock types. It may be seen that groundwaters in basic rock types do not have intrinsically lower redox states as compared with those in granitic rocks, although there is a general decrease in redox state in all groundwaters with increasing depth.

carbon and manganese may also be relevant. Redox levels in groundwater may vary with lithology. This was demonstrated (albeit under extreme physical conditions) by Grassi *et al.* (1992) who measured different redox states in groundwater–rock autoclave experiments at 300°C for tuff, granite and basalt, with log f_{O_2} for each rock type decreasing in the order: rhyolitic tuff (log $f_{O_2} = -27$ bars) > granite (log $f_{O_2} = -30$ bars) > basalt (log $f_{O_2} = -33$ bars). In other words, one may expect rock types rich in ferromagnesian minerals (basalts, gabbros etc.) to develop lower absolute redox values in groundwater than ferromagnesian-poor lithologies. However, the link between *absolute* redox conditions and rock type need not always be apparent, e.g. Fig. 2. Secondly, the *amount* of reduced chemical component present in the rock and groundwater is relevant.

Different rock-groundwater systems will contain different amounts of reduced chemical species according to rock type and concentrations of aqueous ligands. For example, basic igneous rocks such as gabbro or basalt may contain up to 20 times more ferrous oxide than granite (Ahlbom *et al.* 1992). Lastly, the rate of reaction of the reduced component of the rock is important. The rate of removal of dissolved oxygen from groundwater will in general be dependent upon surface complexation reactions and the rate of dissolution of the mineral containing the reduced chemical species and thus the rate of release of reduced chemical species to the groundwater. The rates of dissolution of various reduced iron-bearing solids may vary considerably (Table 1). It may be seen from Table 1 that rocks containing olivine (e.g. basalt, gabbro) would react appreciably quicker than those containing epidote (e.g. metamorphosed basic rocks).

Table 1. *Dissolution rates and mean lifetimes of 1 mm grains of various iron-bearing solids under dissolution at far from equilibrium conditions, 25°C and neutral pH*

Mineral	log dissolution rate (mol m^{-2} s^{-1})	Lifetime (a)
epidote	−12.61	923 000
prehnite	−12.41	579 000
chlorite	−12.46	436 000
clinopyroxene	−10.15	6 800
olivine	−9.5	2 300

Data from Lasaga *et al.* (1994) and Rochelle *et al.* (1995).

P_{CO_2}

P_{CO_2}, or the partial pressure of carbon dioxide, represents the activity or thermodynamic concentration of carbon dioxide in the aqueous phase. It is a parameter which is not measured directly in groundwaters without a coexisting gas phase, but is calculated from the following relationship:

$$P_{CO_2} = \frac{a_{H_2CO_3}}{K_{CO_2} a_{H_2O}} \qquad (5)$$

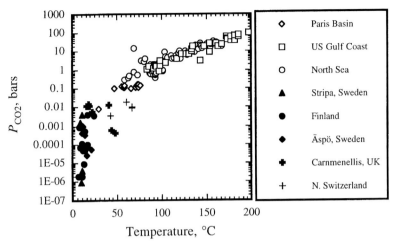

Fig. 3. Variation of P_{CO_2} with temperature for pore fluids from sedimentary basins (open symbols) and groundwaters in fractured hard rocks (closed symbols). Open diamonds – Paris Basin (Michard & Bastide 1988), open squares – US Gulf Coast (Smith & Ehrenberg 1989), open circles – North Sea (Smith & Ehrenberg 1989). Closed symbols and other sources of data are as for Fig. 1.

where $a_{H_2CO_3}$ and a_{H_2O} are the activities of aqueous H_2CO_3 and water, respectively, and K_{CO_2} is the equilibrium constant for the solubility of CO_2. This 'apparent' P_{CO_2} has a direct effect upon the abundance of inorganic carbon in groundwaters and impacts upon the speciation and solubility of some actinides and the stability of cementitious backfill materials.

P_{CO_2} in groundwater varies according to geological environment and temperature (Fig. 3). P_{CO_2} in sedimentary basins increases regularly with temperature and has been suggested to be a product of temperature-dependent reactions between aluminosilicates, carbonates and water

(Smith & Ehrenberg 1989; Coudrain-Ribstein & Gouze 1993; Hutcheon *et al.* 1993). P_{CO_2} in clays chosen as potential repository host rocks may be expected to follow this trend. Groundwaters in regions subject to active magmatism/tectonism or near sedimentary gas reservoirs could conceivably have P_{CO_2} buffered externally from the rock–groundwater system of interest (e.g. Toulhoat *et al.* 1993). P_{CO_2} in groundwaters in fractured hard rocks show considerable variation at constant temperature (Fig. 3).

P_{CO_2} in groundwaters from fractured hard rocks in Shield regions tends to decrease with increasing depth (Fig. 4). Although carbon

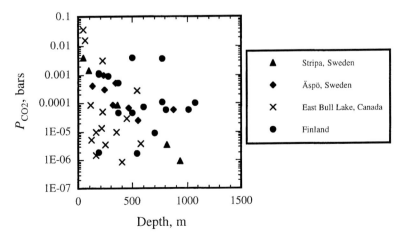

Fig. 4. Variation of P_{CO_2} with depth for groundwaters in fractured hard rocks. Sources of data are as for Fig. 1.

dioxide can be added to deep groundwaters via microbial oxidation of organic matter (Laaksoharju 1995), the principal source of carbon dioxide in groundwaters in hard fractured rocks is from atmospheric recharge ($P_{CO_2} = 10^{-3.5}$ bars). As recharging groundwaters percolate downwards, weathering reactions of calcium aluminosilicates such as Eq. (1) remove CO_2 from groundwater as calcite, so that pH increases and P_{CO_2} decreases with increasing depth. At Äspö in Sweden, this process correlates with an absence of calcite at shallow depths (0–40 m), a zone of calcite precipitation (40–250 m depth), and a zone of calcite absence beneath 250 m (Smellie & Laaksoharju 1992). The lower P_{CO_2} values typical of groundwaters found at depth in fractured hard rocks in comparison with those found in sedimentary basins reflect the variations in pH of the two types of system (pH 5–7 in sedimentary basins and pH 8–10 in fractured hard rocks), which in turn, are a reflection of different mineral buffers of pH (illite, kaolinite and chlorite in sedimentary basins, e.g. Coudrain-Ribstein & Gouze 1993; smectites, zeolites and feldspars in fractured hard rocks, e.g. Savage et al. 1987).

Inorganic ligands

Cations in groundwater will be complexed to a greater or lesser degree by inorganic ligands. The variety and abundances of these ligands in groundwater will therefore affect: radionuclide solubility and sorption, canister corrosion and backfill behaviour. For example, the corrosion of steel is minimized by the alkaline conditions of a cementitious EBS, but studies have shown that chloride ions tend to destroy the protective film on the steel developed at high pH and encourage corrosion (Verbeck 1975).

An important consideration in understanding inorganic ligand behaviour is that of so-called 'hard' and 'soft' acids and bases (Pearson 1969; Langmuir 1997). Soft refers to the deformability of the aqueous species' electron cloud, indicating that the electrons are mobile and easily moved. These species tend to form covalent chemical bonds. Hard species are rigid and non-deformable and tend to form ionic bonds. In geochemistry, this behaviour is evident in the 'Goldschmidt classification' of the elements. The *lithophilic* elements (Li, Be, B Na, Mg, Al, Si, K Ca, Sc, Ti, Rb, Sr, Y Zr, Nb, Cs, Ba, REE, Hf, U Th) are hard acids and tend to bind with hard bases such as O^{2-}, OH^-, CO_3^{2-}, SO_4^{2-}, F^- and PO_4^{3-}. On the other hand, the *chalcophilic* elements (Cu, Ag, Cd, Pd, Hg, Tl) are soft acids and

are found in combination with soft bases such as sulphide (S^{2-}), selenide (Se^{2-}) and telluride (Te^{2-}). Some elements (Ni, Co, Ga, Sn, Zn) show both chalcophilic and lithophilic behaviour ('borderline acids') and will thus complex with either hard or soft bases. This concept is pertinent to understanding and predicting the speciation and solubility behaviour of radioelements of interest to safety assessment.

In sedimentary basins, evaporite deposits are important sources of halides and sulphate. Evaporite minerals are considerably more soluble than silicates or carbonates and dissolve readily on contact with groundwater. This may lead to the development of saline groundwaters or brines. The salinities of groundwaters in Mesozoic sedimentary basins in the UK (Edmunds 1986) and in sedimentary rocks of the North Sea basin (Bjorlykke & Grant 1994) have been linked directly with the dissolution of evaporites. Evaporite deposits in the vicinity of a radioactive waste repository thus become important sources of groundwater solutes. In fractured hard rocks, sources of inorganic ligands are more numerous and more controversial (e.g. Fritz & Frape 1987). The following sources can be considered: via atmospheric recharge (inorganic carbon and halides in marine aerosols); the decrepitation of fluid inclusions (as proposed for groundwaters at the Stripa Mine, Sweden, Nordstrom et al. 1989b); halogen-bearing silicates such as biotite (as proposed for the groundwaters of the Carnmenellis granite, Edmunds et al. 1984); adjacent sedimentary rocks (e.g. groundwaters in the Borrowdale Volcanic Group at Sellafield, Cumbria, UK, Savage et al. 1993; Bath et al. 1996); marine incursions (certain groundwaters at Äspö, Sweden, Smellie & Laaksoharju 1992); the oxidation of organic carbon (Laaksoharju 1995) or invasion by evaporitic brines (e.g. Canadian Shield groundwaters, Bottomley et al. 1994).

Organic ligands

The effects of dissolved organics are four-fold: they can be sources or sinks of H^+ and thus affect absolute pH and the pH buffering capacity of groundwaters; they can act as reducing agents and thus affect redox and the concentrations of redox-sensitive elements; they can be decarboxylated (e.g. $CH_3COOH \rightarrow CH_4 + CO_2$) into carbon dioxide and hydrocarbon gases and they can form soluble complexes with metals.

A range of natural organic compound occurs in groundwaters. Humic and fulvic acids occur in shallow groundwaters and result from the biodegradation of materials such as peat. Humic substances are polymers of 300 to 30 000 Da molecular weight containing phenolic, carboxylic and aliphatic groups (Stumm & Morgan 1981). Humic substances can be split into three fractions: humic acid is soluble in alkaline, but not in acid, solution; fulvic acid is that humic fraction which is soluble over the entire pH range; and humin is the fraction which can be extracted by neither acid nor base. These organic materials have a tendency to trap other organic materials, such as alkanes and fatty acids, sorb on hydrous oxides and clays, form complexes with cations and precipitate in the presence of Ca^{2+} and Mg^{2+}.

In sedimentary rocks, the presence of organic molecules in groundwaters is related to the maturation of hydrocarbons. At temperatures lower than 80°C, the concentrations of organics are generally less than 100 mg/l due to bacterial degradation (Kharaka et al. 1986). Short-chain (C_2–C_5) aliphatic acid anions (acetate, propionate, butyrate and valerate) form as much as 90% of the dissolved organics. Solid hydrocarbons (bitumen) may occur in sandstones and occasionally in crystalline rocks and reflect the migration of hydrocarbons (e.g. Parnell and Eakin 1987). Solid hydrocarbons may be associated with enrichments of uranium. Methane occurs dissolved in groundwaters both in sedimentary basins and in crystalline rocks. Gas contents of up to 600 cm^3 kg^{-1} (gas/water) have been measured in saline groundwaters of the Canadian and Fennoscandian Shields (Sherwood et al. 1988; Sherwood-Lollar et al. 1989; Nurmi et al. 1988). Dissolved gases were mainly CH_4 (up to 86%) and N_2 (up to 78%). Isotopic compositions indicate an abiotic origin and may represent trapped 'metamorphic' or hydrothermal gas. In ultramafic rocks, the association of hydrogen with methane (Sherwood-Lollar et al. 1989) suggests that abiogenic reduction of carbonate may also be important and takes place by reactions such as:

$$15Fe_2SiO_4 + 6H_2O + 2CO_2$$
$$= 10Fe_3O_4 + 2H_2 + 2CH_4 + 15SiO_2 \quad (6)$$

Although organic acids may not occur naturally in groundwaters in a repository host rock, the rock–groundwater system may be required to withstand attack by acids derived from breakdown of organic materials within the repository itself. Any migration of organic acids from the repository has the capacity to destabilize iron oxyhydroxides (e.g. Shock 1988) and aluminosilicates in the rock, potentially decreasing the capability of the rock to retard radionuclide migration.

Ionic strength of groundwaters

Definition of radioelement solubility and sorption behaviour requires information concerning effective concentrations or thermodynamic activities of aqueous species (a_i) and not simply their concentrations (m_i). In very dilute solutions, $a_i = m_i$, but as ionic strength increases, there is increasing departure from this ideal behaviour, so that an activity coefficient (γ_i) must be defined to determine a_i:

$$a_i = \gamma_i m_i. \quad (7)$$

Ionic strength is defined as:

$$I = \frac{1}{2} \sum m_i z_i^2 \quad (8)$$

where m_i is the molality (moles kg^{-1}) of species i, and z_i is the charge of species i. At ionic strengths up to that of seawater, the Debye-Hückel (applicable for $I < 10^{-2.3}$ M), modified Debye-Hückel ($I < 10^{-1}$ M), and Davies ($I < 0.5$ M) theories enable activity coefficients to be calculated. For more concentrated solutions, methods of calculating activity coefficients such as that of Pitzer and co-workers (e.g. Pitzer 1979) are required. Pitzer's model requires interaction coefficients which can be estimated from measurements on solutions of simple salts. Although this technique has enabled the solubilities of evaporite salts to be calculated accurately in solutions up to 20 M there is an absence of relevant coefficients to enable the calculation of the solubilities of aluminosilicates or heavy metal salts. Thus the occurrence of brines at a potential site limits the capacity to model the speciation of aqueous solutions and the solubility of radionuclide-bearing solids.

The salinity of a groundwater can be ascribed to 3 factors: water–rock reactions, geological history and geological environment. Salinity derived from water–rock reaction depends upon the rock (mineral) type. Silicate and aluminosilicate minerals dominate fractured hard rocks and clays, but are sparingly soluble ($< 10^{-6}$ M at 25°C) and thus do not contribute significantly to salinity. Carbonates are more soluble (of the order of 10^{-4} M), and sulphates and halides have solubilities in excess of 10^{-3} M. Geological history impacts the ionic strength of groundwaters through processes such as glaciation, marine incursions and earthquakes. Geological environment is important, e.g. recharge of meteoric waters may dominate at the margins of a sedimentary basin, whereas dissolution of evaporites may be actively occurring at the basin centre.

Colloids

Colloids (particles with diameters less than $10\,\mu m$) have the potential to sorb radionuclides from aqueous solution and be transported with groundwater flow. Colloids may be mineral particles, organic, or biological matter and may be hydrophobic (e.g. metal oxides) or hydrophilic (e.g. bacteria). As ionic strength increases, the tendency for colloids to flocculate (aggregate) increases. This phenomenon results from decreased stability of the so-called 'Gouy' layer around the colloid particles as ionic strength increases, leading to increased van der Waals forces on the particles, and flocculation (Drever, 1988). Currently, there are no data with regard to colloid abundances in different rock types to enable an assessment of the degree to which certain rock types are more susceptible to colloid formation than others.

In deep, warm (15–70°C) groundwaters in granites, the concentrations of trivalent and tetravalent metals (REE, Zr, Sc, Hf, Th) tend not to be solubility-limited but are mostly sorbed on colloidal particles (Alaux-Negrel *et al.* 1992; Toulhoat *et al.* 1993), whereas in groundwaters with higher P_{CO_2}, Alaux-Negrel *et al.* (1992) note that carbonate complexation of these metals may be more important. Vilks *et al.* (1991) studied colloids in spring waters derived from a granite formation at the Whiteshell Research Area, Canada. The average colloid concentration was $0.34\,(\pm0.34)\,\mu g\,ml^{-1}$ for sizes ranging from 10 to 450 nm. Colloid sampling work has been carried out by Nagra (the Swiss radioactive waste disposal organization) in groundwaters from the northern Swiss basement (Degueldre *et al.* 1996). Degueldre *et al.* (1996) concluded that colloids in groundwaters in crystalline rocks were dominantly composed of phyllosilicates and silica originating from the parent rock and suggested that colloid concentrations were unlikely to exceed $0.1\,\mu g\,ml^{-1}$ under 'normal' hydrogeochemical conditions.

Mineralogical and textural characteristics of groundwater flow paths

The retardation of radionuclide migration through the geosphere by sorption and ion exchange processes is dependent to a large extent upon the mineralogical character of the rock. Sorption depends upon mineral type and the radionuclide concerned, e.g. the following sequence of preference was observed for Sr sorption: clays, zeolites, Fe-hydroxides > micas > hornblende, Fe-Ti oxides > feldspars > carbonates > quartz (Landström *et al.* 1982).

In fractured rocks, such as granites or gabbros, there may be large differences between the mineralogical composition of the rock matrix and that lining groundwater flow paths. This is because most water–rock interaction will take place along flowpaths and the primary high temperature mineralogy will be replaced by minerals more stable in low temperature hydrous environments. Consequently, feldspars and micas will be replaced by clays and zeolites, and ferromagnesian minerals by clays and oxyhydroxides. Usually, this alteration will improve the retardation characteristics of the rock. It may be expected that rocks such as gabbro which is full of unstable minerals such as feldspars and ferromagnesians will develop thicker alteration zones around fractures than granites. This is because the mineralogy which typifies gabbros weathers at faster rates than those in granitic rock types (e.g. anorthitic plagioclase which characterizes a gabbro dissolves three orders of magnitude faster than albitic plagioclase which characterizes granite; Lasaga *et al.* 1994). In addition to the mineralogical composition of a rock, the fabric (texture) is also important, as this affects diffusion of aqueous species into the rock matrix. Although much of the rock may be available for diffusion in a mudrock, in fractured crystalline rocks the accessibility of the matrix from groundwater flowpaths is an important consideration. The application of complex mineralogical information to repository safety assessment modelling may be difficult. It is thus important to simplify complex features for the purpose of modelling. Nagra have been successful in transferring complex mineralogical data from investigations of the crystalline basement of northern Switzerland to a form suitable for modelling (e.g. Mazurek *et al.* 1992).

Aspects of water–rock buffering

The two most important geochemical parameters for waste isolation, pH and redox, are controlled by reaction between rock and groundwater. In order to understand how these parameters may respond to perturbations (both internal and external to the repository system) over timescales relevant to waste isolation, it is germane to consider aspects of water–rock buffering.

The chemical concept of buffering is linked with pH control in homogeneous solutions of acids and bases in open systems. Buffer reactions absorb changes in intensive thermodynamic parameters (e.g. masses of H^+, Na^+ etc.) in response to changes in extensive parameters, such as temperature, pressure etc. Buffering in geological systems differs from classical buffering equilibria in aqueous solutions since heterogeneous equilibria between several phases usually dominate the thermodynamic properties of geological systems (Rosing, 1993). *Buffer capacity* is defined as the total mass of protons (or other reactive chemical species of interest) in the chemical system, whereas *buffer intensity* is a measure of the response of the system to changes in concentration of the reactive chemical species.

The control, or chemical buffering, of a chemical component in a water–rock system is dependent upon three factors (Hutcheon *et al.* 1993): the capacity of the system to respond to external addition or removal of that component (mass action constraint); the quantities of buffer materials available in the system (mass balance constraint); and the rate at which the system responds to changes in the abundances of the chemical component of interest (kinetic constraint). For groundwaters in warm (100°C) sedimentary basins, Hutcheon *et al.* (1993) concluded that mass action was the most important constraint governing buffering of pH. For groundwaters in the cooler ($T \leq 50°C$) systems relevant to waste disposal, kinetic constraints will be relatively more important. Calculations to investigate issues concerning pH and redox buffering are presented below.

pH

Mass action relationships for pH can be evaluated by consideration of the pH buffer index, β, which is a quantitative measure of a system's ability to withstand changes in pH and represents the inverse slope of the titration curve resulting from the addition of strong acid or base to the buffer system (Stumm & Morgan 1981; Hutcheon *et al.* 1993):

$$\beta = \frac{-dC_i}{dpH} \qquad (9)$$

where C_i is the amount of acid or base added to the buffer system. Higher values of β indicate greater pH buffer intensities. In terms of analysis by reaction-path computer software such as EQ3/6 (Wolery 1992), then equation (9) can be written as:

$$\beta = \frac{-d\xi_i}{dpH} \qquad (10)$$

where ξ_i is the progress variable for an irreversible reaction, defined as:

$$d\xi_i = -\frac{dx_{i,j}}{n_{i,j}} \qquad (11)$$

where $x_{i,j}$ is the total number of moles of the ith component in the jth phase in the system, and $n_{i,j}$ is the stoichiometric coefficient for the ith component in the jth phase in the reaction of interest.

Titration calculations using EQ3/6 have been carried out for a range of possible homogeneous and heterogeneous reactions in rock–groundwater systems. The homogeneous reactions in the aqueous phase considered were: 100 and 1000 mg/l dissolved HCO_{3^-}, and the heterogeneous reactions were: calcite–100 mg/l HCO_3^-; K–feldspar–illite–water; and calcite–Ca-beidellite–muscovite–water. Pure water was used as a starting fluid in these calculations since Hutcheon *et al.* (1993) demonstrated that variations in ionic strength of the initial fluid had negligible impact upon calculated values of β. Values of β have been calculated from the titration calculations using equation (10) and are presented in Fig. 5. It may be seen from Fig. 5 that buffering minima for all reactions are in the pH regime 7–9.

In the pH range of interest in groundwater (pH 6–10), the highest values of β are calculated for heterogeneous reactions involving calcite and aluminosilicates. Figure 5 demonstrates that these heterogeneous reactions have 2–3 orders of magnitude greater buffer intensity than homogeneous reactions in the aqueous phase in the pH range 6–10. This indicates that from a perspective of mass action only, heterogeneous reactions involving aluminosilicates and calcite are much more important buffers of pH in groundwaters than homogeneous reactions in the aqueous phase. In comparison with reaction with the rock, there is negligible buffer intensity in likely reactions in the aqueous phase alone.

Figure 5 is also relevant to understanding how the elevated pH due to the migration of hyper-alkaline pore fluids from a cementitious repository may be retarded in the geosphere. Figure 5 shows that at pH > 10, in addition to homogeneous reactions in the groundwater, calcite is relatively unimportant as a buffer of pH in comparison to reactions involving aluminosilicates.

Mass action relationships for aluminosilicate–carbonate–water reactions will be the most important constraint in buffering pH if their reaction rates are fast enough to accommodate changing environmental conditions (kinetic constraint), and there are sufficient amounts of these minerals in the rock (mass balance constraint). To attempt to address the kinetic constraint first, we can consider how fast a mineral typical of low temperature alteration in both sediments and fractured hard rocks (albite) behaves. Albite (Na-plagioclase) has a congruent dissolution

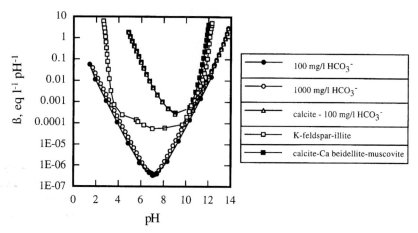

Fig. 5. Variation of the pH buffer index, β with pH at 25°C for various potential pH buffers in groundwaters Values of β were calculated using equation (10).

rate at 25°C greater than that of K-feldspar or muscovite, but less than that of anorthite (Lasaga *et al.* 1994), and thus has a median rate of reaction for minerals typical of low-permeability rocks. Calculation of the distance required for groundwater to reach equilibrium by travelling along a fracture in rock consisting of 100% albite has been carried out using EQ3/6. A transition-state rate equation (e.g. Helgeson *et al.* 1983) was employed, together with laboratory experimental data for the rate of albite dissolution at 25°C at neutral pH under conditions far from equilibrium from Knauss & Wolery (1986). These calculations suggest that the distance for groundwater to travel along this hypothetical fracture before albite equilibrates with the groundwater will be less than 1 m for potential groundwater flow rates (of the order of $<10^{-8}\,\mathrm{m\,s^{-1}}$) and fracture apertures ($<100\,\mu\mathrm{m}$) in deep systems (Fig. 6). However, there is uncertainty associated with the laboratory characterization of mineral reaction rates and some authors have suggested that natural systems weather at rates some 1–3 orders of magnitude slower than in laboratory systems (e.g. Velbel 1986). These lower rates of reaction would then imply that equilibration distances for albite at 25°C and neutral pH would be of the order of 100–1000 m. These slower reaction rates would then bring into question whether water–rock reactions could buffer pH sufficiently fast during major perturbations to the repository system to maintain pH at levels prior to disturbance. More laboratory and *in situ* experiments are necessary to investigate the rates of potential buffer reactions.

The mass balance constraint should not be the most important factor governing pH buffering.

Aluminosilicates abound in most rocks considered suitable for the disposal of radioactive wastes (acidic–basic igneous and metamorphic rock types, and clays), so that the abundance of these minerals as potential buffers of pH should not be a limiting factor. Calcite may be less abundant in fractured hard rock systems than sedimentary rocks such as clays.

pH buffering mechanisms have not been explicitly investigated in most radioactive waste site characterization programmes worldwide. The most relevant *in situ* experiment carried out

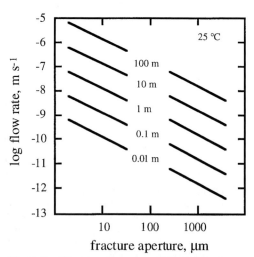

Fig. 6. Equilibration distances calculated for albite at different groundwater flow rates and fracture apertures in a hypothetical parallel plate fracture. Distances were calculated at 25°C and neutral pH in pure water using a transition-state theory rate equation and experimental data from Knauss & Wolery (1986).

Table 2. *Compositions of groundwaters from the 'water-rock interaction experiment' carried out by Nagra at the Grimsel Test Switzerland. Concentrations in* $\mu eq\, l^{-1}$

Parameter	Fracture water	Injected water	Measured final composition	Modelled final composition
pH	9.6	8.9	9.4	–
Na	692	264	524	509
K	3.84	24	10	13.3
Mg	1.08	7.74	2.89	4.1
Ca	254	422	358	352
Sr	3.67	7.76	5.19	5.6
SO_4	115	88	55.2	109
F	330	77.9	231	221
Cl	155	5.9	90	91
Alkalinity	420	600	520	–

From Eikenberg *et al.* (1991).

thus far is a large scale 'water–rock interaction experiment' carried out by Nagra at the Grimsel Test Site, 450 m beneath the Juchlistock, Central Alps (Eikenberg *et al.* 1991). In this experiment, the water composition in an isolated fracture sampled by a dipole borehole arrangement was perturbed by the injection of a 'foreign' water from another borehole at the site. The ground water compositions were slightly different in terms of pH, cation, and anion content (Table 2). The broad chemical characteristics of the mixed waters could be modelled by a mixing model and consideration of ion exchange equilibria. However, data for Mg and K were not modelled very well, and their concentrations were still changing at the time of experiment cessation. The chemical composition of the mixed water showed some response to mineral buffering and was not simply a mixture of the two waters. Unfortunately, this experiment was very short (13 days) and not long enough to investigate rock–water buffering mechanisms. Further experiments of this type are warranted.

Redox

The redox status of an aqueous system is described by the concentrations of the oxidized and reduced chemical species of all components in the chemical system (Scott & Morgan 1990). Because of the slowness of oxidation–reduction reactions, natural rock-water systems are often not in redox equilibrium and thus the concept of a 'system' *Eh* or p*E* becomes meaningless. Intensity factors such as *Eh* or p*E* are not very useful descriptors of the redox state of the system, but *capacity* factors which reflect the total concentration of redox-sensitive species may be better and more conservative measures of redox state.

Scott & Morgan (1990) define these parameters as the 'OXC' and 'RDC' (oxidative and reductive capacities) of a system, respectively. The rock and groundwater may form two separate systems of OXC and RDC, where the latter may be be much more rapidly accessed than the former (Grenthe *et al.* 1992). However, solutes in the aqueous phase may provide only a finite ability to react with oxidizing or reducing species, particularly in low-porosity rocks. At pH > 5 Eary & Schramke (1990) measured half-lives of the reduced chemical species of Fe, S Cu and Mn in water to be less than a few weeks when oxidized by dissolved O_2. This implies that the RDC of a groundwater can be oxidized rapidly upon introduction of dissolved O_2, so that further consumption of O_2 is then dependent upon reactions with the solid phase. In the following discussion of redox buffering it should not be forgotten that the behaviour of the chemical system is intimately linked with the microbial ecology of the system (Scott & Morgan 1990). Redox reactions are strongly catalysed by microbial processes so that quantification of the thermodynamic potential for a reaction to take place may give a non-conservative estimate of buffer capacity.

What is needed in safety assessments of geological disposal systems for nuclear wastes are quantitative parameters by which the redox-buffering properties of potential host rocks can be characterized. Such buffering is necessary to sustain stable reducing conditions for extended periods of time (thousands to millions of years) over a range of plausible variations in physical properties, such as groundwater flow rates, and possible scenarios (e.g. migration of oxidizing surface waters toward the repository). Ideally, the parameters characterizing a rock's buffering behaviour should be readily deducible from site-specific data, such as host-rock mineralogy.

A modelling approach that could fulfil this need is based on the stationary-state approximation to coupled fluid flow and water–rock interaction (Lichtner 1985, 1988). This model represents the chemical evolution of an open, flow-through system as a sequence of relatively long-lived stationary states of the system, which are linked in time by short-lived transients. The basis for the model is the observation that within a representative elemental volume of a rock–water system, the aqueous concentration of any particular species is generally much less than its concentration in minerals. Long periods of time are therefore necessary to dissolve, or precipitate, minerals such that the spatial distribution of mineral abundances, surface area, porosity and permeability is altered significantly. Each time interval represents a stationary state of the system, in which fluid composition, reaction rates and the distribution of primary and alteration minerals vary only as a function of position in the flow path, not of time.

Under stationary-state conditions, reaction front velocities are fixed relative to the Darcy velocity of the fluid. The migration velocity of a redox front, or any other reaction front, is given by (Lichtner 1988):

$$v_f = \frac{v}{\phi(1 + L_j)} \tag{12}$$

where v_f stands for the velocity of the front ($m\,a^{-1}$), v represents the Darcy velocity of the fluid ($m\,a^{-1}$), and ϕ refers to porosity. The parameter, L_j, is a dimensionless quantity for the jth aqueous reactant:

$$L_j = \frac{\sum_{r=1}^{M} \left(\frac{\nu_{jr}}{\bar{V}_r} \right) \{\phi_r\}}{\phi \{\psi_j\}} \tag{13}$$

where M refers to the number of minerals, r, that react with j, ν_{jr} represents the stoichiometric coefficient for j in a balanced dissolution reaction (which may include oxidation/reduction) involving the rth mineral, $\bar{V}_r$ stands for the molar volume ($cm^3\,mol^{-1}$), and $\{\phi_r\}$ and $\{\psi_j\}$ denote differences in the volume fraction of the rth mineral and generalized concentration of the jth reactant, respectively, as measured between upstream and downstream points on either side of the reaction front. The generalized concentration represents the total concentration among all aqueous species containing j.

The migration velocity of a redox front is retarded relative to the fluid velocity when

$L_j > 0$ and the front's velocity is independent of porosity if $L_j \gg 1$. Under such conditions,

$$v_f = \frac{v\{\psi_j\}}{\sum_{r=1}^{M} \left(\frac{\nu_{jr}}{\bar{V}_r} \right) \{\phi_r\}} \tag{14}$$

and the front velocity is thus proportional to $\{\psi_j\}$, and inversely proportional to the volumetric concentrations of the reactant minerals (ϕ_r). The front is stationary until ϕ_r is reduced significantly, at which time a new stationary state is generated and the front advances.

Equation (14) indicates that the migration velocity of a redox front is controlled by the Darcy velocity of the fluid and is limited by mass-action and mass-balance constraints. Mass-action constraints (i.e. reaction stoichiometry, ν_{jr}) determine a given reaction's 'intensity' in resisting changes in redox potential. Mass-balance constraints (i.e. the volume fractions of reactant minerals, ϕ_r) determine a reaction's 'capacity' to resist those changes. Reaction rates are also accounted for explicitly in this model, but kinetic parameters do not appear in Eq. (14) because these rates determine both the position of the front in a given stationary state and the time required to generate the next stationary state. Because the front velocity is given by the ratio of this distance and time, kinetic parameters cancel and the front propagates at a constant velocity that is reaction-rate independent (Ortoleva et al. 1986; Lichtner 1988). This is important for practical reasons because the rates of heterogeneous and homogeneous redox reactions are often exceedingly slow, and are thus difficult to determine experimentally.

Lichtner & Waber (1992) used the stationary-state model to estimate the velocity of redox fronts associated with pyrite oxidation at the Osamu Utsumi uranium mine, Poços de Caldas, Brazil. For the oxidation reaction:

$$FeS_2 + H_2O(l) + \tfrac{7}{2}O_2(aq)$$
$$\rightarrow Fe^{2+} + 2SO_4^{2-} + 2H^+ \tag{15}$$

the migration velocity of the front is given by Eq. (14):

$$v_f = \frac{v\{\psi_{O_2(aq)}\}}{\nu_{O_2(aq),\,pyrite}\{\phi_{pyrite}\}\bar{V}_{pyrite}^{-1}}. \tag{16}$$

Lichtner & Waber (1992) assumed that oxidant [O_2(aq)] concentrations are negligible immediately downstream of the front, and that

$\phi_{pyrite} = 0$ in previously oxidised regions upstream of the front, such that

$$\{\psi_{O_2(aq)}\} \approx \psi^0_{O_2(aq)} = K_H^{-1} P_{O_2(g)}, \text{ and}$$
$$\{\phi_{pyrite}\} = \phi^0_{pyrite} \quad\quad (17)$$

where K_H denotes the Henry's law constant for $O_2(aq) \rightleftharpoons O_2(g)$ ($10^{2.899}$ mol l^{-1} at 25°C), $P_{O_2}(g)$ refers to $O_2(g)$ partial pressure and ϕ^0_{pyrite} represents pyrite's volume fraction in the unoxidized host rocks. It is also assumed that $P_{O_2(g)} = 0.2$ bar, $v = 1$ m a^{-1} and $\phi^0_{pyrite} = 2\%$ (according to field data for this site). With $\nu_{O_2(aq),pyrite} = 3.5$, based on reaction stoichiometry, and $\bar{V}_{pyrite} = 23.99$ cm^3 mol^{-1} (e.g. Johnson et al. 1992), the calculated front velocity equals 8.6×10^{-5} m a^{-1}. The calculated velocity slightly exceeds estimates based on natural series radionuclide profiles, which suggest that maximum front velocities are 1.0×10^{-5} m a^{-1} (MacKenzie et al. 1991). Cross et al. (1991) concluded that the modelling approach yields crude, yet conservative, estimates of redox-front migration rates at Poços de Caldas.

Arthur (1996) also used the stationary-state model to estimate migration velocities of oxidizing fronts in granitic rocks. This study is relevant to the present discussion because it illustrates how the modelling approach can be used to distinguish among differing redox-buffering properties of several rock types. Analytical models are based on Eq. (14), and are constrained by molar concentrations of ferrous silicate (biotite, chlorite), oxide (magnetite), and/or sulphide (pyrite) minerals in unaltered granites, hydrothermally altered granites and fractures comprising the host rocks beneath Äspö Island, southeastern Sweden (Table 3).

Calculated results suggest that small amounts of these minerals in the granites and fractures will significantly retard the downward migration of oxidizing conditions (Table 4), which could be generated by infiltration of glacial meltwaters during periods of recharge associated with future cycles of glacial advance and retreat in Scandinavia (roughly 10 000 to 20 000 years). Front velocities are retarded relative to observed Darcy fluxes by factors ranging from 10^{-3} to 10^{-4} (Table 5). Buffering efficiency is determined by the sum of intensity (ν_{jr}) and capacity (ϕ_r) factors for all the ferrous minerals and decreases in the order: fractures > altered granite ≥ unaltered granite. The most conductive structures in these rocks are thus also the most robust redox buffers.

Neretnieks (1985) came to similar conclusions from a different conceptual standpoint with regard to the redox buffer capacity of granites in the deep Scandinavian Shield and regarded it unlikely that oxidizing conditions in recharge waters could penetrate more than a few metres or tens of metres into granitic rock over a million year time period.

Models such as those discussed above could be extended for purposes of classifying various rock types (e.g. gabbros, granites, etc.) according to their redox-buffering behaviour. Such behaviour can be quantified in terms of buffer intensity and buffer capacity factors, which in turn are determined by the rock's mineralogy and the stoichiometries of corresponding heterogeneous

Table 3. *Mineralogical constraints on the stationary-state model of redox-front migration in granites*

Mineral	Reaction	$\phi_r(\%)^*$			$\nu_{O_2(aq)}$	$\bar{V}_r$ (cm^3 mol^{-1})
		G	AG	F		
biotite (annite) [KFe$_3$AlSi$_3$O$_{10}$(OH)$_2$]	annite + 13H$^+$+3/4O$_2$(aq) $\rightleftharpoons$ Al^{3+} + K$^+$ + 3Fe^{3+} + 3SiO$_2$(aq) + 15/2H$_2$O	2.3–4.2			0.75	154.3†
magnetite [Fe$_3$O$_4$]	magnetite + 1/4O$_2$(aq) + 9H$^+$ $\rightleftharpoons$ 3Fe^{3+} + 9/2H$_2$O	0.4–1.0			0.25	44.5†
chlorite (daphnite) [Fe$_5$Al$_2$Si$_3$O$_{10}$(OH)$_8$]	chlorite + 21H$^+$ + 5/4O$_2$(aq) $\rightleftharpoons$ 2Al^{3+} + 3SiO$_2$(aq) + 5Fe^{3+} + 29/2 H$_2$O	0.09–0.31	2.1–4.4	15.4	1.25	213.4‡
pyrite [FeS$_2$]	pyrite + 15/4O$_2$(aq) + 1/2H$_2$O $\rightleftharpoons$ Fe^{3+} + H$^+$ + 2SO$_4^{2-}$		0.1	0.2	3.75	23.9†

G, unaltered granite; AG, altered granite; F, fractures.
ϕ_r, $\nu_{O_2(aq)}$, $\bar{V}_r$ are the concentration, stoichiometric coefficient for the dissolution reaction and molar volume for each mineral, respectively.
* based on data from Eliasson (1993).
† Johnson et al. (1992).
‡ Helgeson et al. (1978).

Table 4. *Equations for calculation of redox-front velocities*

Mineral	Front velocity
biotite (annite)	$206\nu\{\psi_{O_2(aq)}\}/\{\phi_{ann}\}$
magnetite	$178\nu\{\psi_{O_2(aq)}\}/\{\phi_{mag}\}$
chlorite (daphnite)	$171\nu\{\psi_{O_2(aq)}\}/\{\phi_{dap}\}$
pyrite	$6\nu\{\psi_{O_2(aq)}\}/\{\phi_{pyr}\}$

Source: Arthur 1996.
Numerical coefficients represent $\bar{V}_r/\nu_{O_2(aq)}$ (see Table 3).

Table 5. *Estimated redox-front velocities in unaltered and hydrothermally altered granites, and in fractures. Ranges correspond to variations in mineral abundances (see Table 3)*

Unaltered granite	Altered granite	Fractures
$0.0033\nu{-}0.0016\nu$	$0.0016\nu{-}0.0011\nu$	0.00038ν

oxidation–reduction reactions. Classification of rock types according to redox-buffering properties could help establish criteria for the selection of suitable rock types to host a radioactive waste repository.

Unfortunately, there has generally been little emphasis placed upon the redox-buffering properties of the geosphere in most site characterization programmes for radioactive waste disposal. The few studies which do exist have generally focused on laboratory investigations of redox-buffering processes (e.g. Grandstaff *et al.* 1990; Pirhonen & Pitknen 1991; Grassi *et al.* 1992; Grenthe *et al.* 1992). A noteworthy exception is the SKB (the Swedish radioactive waste disposal agency) study involving *in situ* tests at Äspö to investigate redox buffering processes (Banwart *et al.* 1994; Banwart & Wikberg, this volume).

Preferred geochemical environments

It is possible to synthesize the information presented above to identify the geochemical requirements of a site for waste isolation purposes. With regard to important geochemical parameters, desirable ranges are summarized in Table 6. These ranges have been selected to optimize conditions for EBS preservation and maximize radioelement retardation in the geosphere. The retardation of some radionuclides of concern to

Table 6. *Summary of optimum requirements for geochemical parameters to maximize both EBS preservation and radionuclide retardation and examples of geochemical environments*

Parameter	Optimum range and conditions	Possible geological environments and lithologies
pH	high ($10 > \text{pH} > 8$); fast reaction kinetics and high buffer capacity	low salinity groundwater; low P_{CO_2}; igneous rock types, particularly of basic or alkalic nature
redox	reducing conditions ($Eh < -200\,\text{mV}$); fast reaction kinetics and high buffer capacity	basic igneous rocks, some clays
P_{CO_2}	low ($< 10^{-4}$ bars); closed system behaviour	fractured hard rocks away from tectonic/magmatic activity, particularly in shield regions
inorganic ligands	$Cl^- > HCO_3^-$; PO_4^{3-} low	areas of tectonic/thermal activity to be avoided; rock types with abundant solid sources of phosphate should be avoided
organic ligands	low (< 5 ppm); carboxylic acids preferred to humic and fulvic acids	petroliferous sedimentary basins should be avoided. Depths $> 500\,\text{m}$ should minimize content of surface-derived humics and fulvics
ionic strength	$< 0.5\,\text{M}$	proximity to sedimentary basins with evaporite deposits should be avoided
colloids	low (<0.01 ppm)	low abundances promoted by high ionic strength of groundwater
groundwater flowpaths	clays, zeolites and iron oxyhydroxides preferred, with high porosity; thick alteration zones desirable	basic and alkaline igneous rock, clays

safety assessment, such as [36]Cl, [129]I, [135]Cs will not be affected to a large extent by variations in many of the geochemical parameters defined in Table 6 although the thickness of rock available for matrix diffusion (considered under 'flowpaths') may affect their migration. It may be seen from Table 6 that EBS preservation and radionuclide retardation are maximized by: alkaline pH; a chemically reducing redox state; low P_{CO_2}; low contents of colloids, inorganic and organic ligands in groundwater; a groundwater ionic strength no greater than that of seawater (to optimize thermodynamic modelling capabilities and colloid stability); and groundwater flowpaths with a high sorptive capacity and high porosity (to maximize matrix diffusion). These ranges are not prescriptive, but serve as benchmarks by which the geochemical characteristics of a site may be judged.

There are a number of properties which are important in controlling the hydrochemical evolution of a site which are not geochemical or mineralogical. Geological history, geological environment and the hydrogeological properties of the rock mass are examples. The geological history of a site will have had, and will have in the future, a large impact upon the hydrochemical evolution. The occurrence of physical perturbations such as earthquakes, relative changes of level between land and sea etc. will affect the chemical composition and magnitude of groundwater recharge, the location of groundwater mixing zones and so on. The timing and frequency of such events in the past and in the future will affect the present state and evolution of groundwater composition. The geological environment of a site is tied to a certain extent to its geological history, but this phenomenon is identified separately to distinguish effects of present-day geological location. Relevant factors are whether the site is in a cratonic shield or in a sedimentary basin, whether there are evaporite deposits nearby; whether the site is an upland or coastal region etc. The hydrogeological properties of a site will control factors such as the residence time of groundwater in the system, how much recharge enters the system, the location of groundwater discharge zones, and whether the system acts as a porous or dual porosity (fractured rock) medium. All these factors will have a large effect upon the hydrochemical evolution of the site. It may be concluded therefore that most of the processes governing the hydrochemical evolution of a site are not intrinsically geochemical or mineralogical. However, the selection of a site with well-buffered geochemical properties may 'insulate' the site from future physical perturbations to the system.

From a perspective of lithology and geological environment, the most suitable geochemical conditions could be found in basic or alkalic igneous rocks, in low-lying terrain away from the influences of a sedimentary basin (thus avoiding brines derived from evaporite dissolution, hydrocarbon-derived organics, and high P_{CO_2} conditions). The low permeability and low hydraulic gradients of such lithologies/environments would serve to minimize groundwater throughput and enhance the stability of the hydrochemical system at depth.

Notwithstanding their apparent favourability, basic and alkalic rocks have not featured to a significant extent in most countries' plans for the disposal of radioactive wastes. UK Nirex Ltd were until recently assessing the feasibility of disposal of radioactive wastes in rocks of intermediate composition (metavolcanics of the Borrowdale Volcanic Group), which from the above criteria, possess better geochemical characteristics than granitic (sensu lato) rocks. In a review of the suitability of gabbros in comparison with granitic rocks as repository host rocks in Sweden, Ahlbom et al. (1992) concluded that gabbros were of generally lower hydraulic conductivity, had considerably greater chemical reducing capacity and had similar rock mechanical behaviour but slightly inferior thermal conductivity characteristics. Despite these apparent advantages, gabbros were not selected as possible repository host rocks because of the lack of bodies of appropriate size and the tendency of basic igneous rocks to be more likely to be associated with economic mineralization. However, these disadvantages may not be ubiquitous; thus it is concluded that basic and alkalic rocks provide the best geochemical environment for waste isolation.

Conclusions

Geochemical parameters which are important to waste isolation include (with optimum conditions in parenthesis): groundwater pH ($10 >$ pH > 8), redox ($Eh < -200$ mV), P_{CO_2} ($<10^{-4}$ bars), and ionic strength (<0.5 M); the colloid (<0.01 ppm), organic (<5 ppm) and inorganic ligand ($Cl^- > HCO_3^-$, low abundances of phosphate) content of groundwaters; and the mineralogical composition and structure of ground-water flowpaths (zeolites, clays, iron oxyhydroxides with appreciable diffusive porosity). pH and redox (and P_{CO_2} in most environments) are likely to be controlled by reactions between groundwater and the rock, whereas other parameters are constrained more by geological history or environment.

The long timescales required for the isolation of many types of radioactive wastes necessitate that potential sites must be characterized not only for their ambient geochemical conditions, but also for their capacity to maintain stable geochemical conditions during perturbations both by natural and repository-induced causes.

The buffer capacity of the two most important geochemical parameters, pH and redox have been evaluated using numerical computations. Minimal buffering of pH is available from the groundwater alone, and the principal buffer intensity is derived from heterogeneous reactions involving aluminosilicates and carbonates. These reactions are unlikely to be limited by kinetics if available laboratory experimental data are representative of behaviour in natural systems. Redox is also maintained by reaction of groundwater with the rock. Calculations using the quasi-stationary steady-state approximation and typical rock and groundwater compositions, groundwater flow rates, and fracture apertures suggest that relatively small amounts of minerals bearing reduced iron will be sufficient to reduce oxidizing groundwaters and maintain ambient redox conditions at depth. Uncertainties associated with the application of laboratory and theoretical data to field situations require that further investigations of reactions in the field and at natural analogues sites are warranted.

The most suitable lithologies and geological environments for the disposal of radioactive wastes from a geochemical perspective are igneous rocks of basic or alkalic composition in terrain of low relief, away from sedimentary basins. Cratonic shield regions may be especially favourable environments. No site selection and assessment process should be focused on geochemical criteria alone, but suitable geochemical conditions may help relax the requirements for optimum hydrogeological performance.

Parts of this work were funded by the UK Environment Agency, the Mitsubishi Materials Corporation (Japan), and the Swedish Nuclear Power Inspectorate (SKi) who are thanked sincerely for their support. We would also like to thank J. Smellie (Conterra AB) and M. Laaksoharju (Intera AB) for giving us access to unpublished data. Constructive comments from reviewers G. Darling, M. Gascoyne and R. Metcalfe improved the content and readability of the manuscript.

References

AHLBOM, K., LEIJON, B., LIEDHOLM, M. & SMELLIE, J. 1992. *Gabbro as a host rock for a nuclear waste repository*. SKB Technical Report **92-25**, Swedish Nuclear Fuel and Waste Management Co., Stockholm.

ALAUX-NEGREL, G., BEAUCAIRE, C., MICHARD, G., TOULHOAT, P. & OUZOUNIAN, G. 1992. Trace metal behaviour in natural granitic waters. *Journal of Contaminant Hydrology*, **13**, 309–325.

ARTHUR, R. C. 1996. *Estimated rates of redox-front migration in granitic rocks (SITE-94)*. SKI Report **96:35**, Swedish Nuclear Power Inspectorate, Stockholm.

BANWART, S. & WIKBERG, P. 1999. Protecting the redox stability of a deep repository: Results and experience from the Äspö Hard Rock Laboratory. *This volume.*

——, GUSTAFFSSON, E., LAAKSOHARJU, M., NILSSON, A-C., TULLBORG, E-L. & WALLIN, B. 1994. Large-scale intrusion of shallow water into a vertical fracture in crystalline bedrock. Initial hydrochemical perturbation during tunnel construction at the Äspö Hard Rock Laboratory. *Water Resources Research*, **30**(6), 1747–1763.

BATH, A. H., MCCARTNEY, R. A., RICHARDS, H. G., METCALFE, R. & CRAWFORD, M. B. 1996. Groundwater chemistry in the Sellafield area: a preliminary interpretation. *Quarterly Journal of Engineering Geology*, **29**, S39–S57.

BERRY, J. A. 1999. The role of sorption onto the Borrowdale Volcanic Group in providing chemical containment for a potential repository at Sellafield. *This volume.*

BJORLYKKE, K. & GRANT, K. 1994. Salinity variations in North Sea formation waters: implications for large-scale fluid movements. *Marine and Petroleum Geology*, **11**, 5–9.

BLUM, A. & LASAGA, A. C. 1988. Role of surface speciation in the low-temperature dissolution of minerals. *Nature*, **331**, 431–433.

BOTTOMLEY, D. J., GASCOYNE, M. & KAMINENI, D. C. 1990. The geochemistry, age, and origin of groundwater in a mafic pluton, East Bull Lake, Ontario, Canada. *Geochimica et Cosmochimica Acta*, **54**, 993–1008.

——, GREGOIRE, D. C. & RAVEN, K. G. 1994. Saline groundwaters and brines in the Canadian Shield: geochemical and isotopic evidence for a residual evaporite brine component. *Geochimica et Cosmochimica Acta*, **58**, 1483–1498.

CHAPMAN, N. A., MCEWEN, T. J. & BEALE, H. 1986. Geological environments for deep disposal of intermediate level wastes in the United Kingdom. Proceedings of a Conference on *Siting, Design and Construction of Underground Repositories for Radioactive Wastes*. IAEA-**SM-289**, International Atomic Energy Agency,Vienna, 311–328.

COUDRAIN-RIBSTEIN, A. & GOUZE, P. 1993. Quantitative study of geochemical processes in the Dogger aquifer, Paris Basin, France. *Applied Geochemistry*, **8**, 495–506.

CROSS, J. E., HAWORTH, A., LICHTNER, P. C. *et al.* 1991. *Testing models of redox front migration and geochemistry at the Osamu Utsumi mine and Morro do Ferro analogue study sites, Poços de Caldas, Brazil*. Nagra Technical Report **90-30**, Nagra, Wettingen.

DEGUELDRE, C., PFEIFFER, H-R., ALEXANDER, W., WERNLI, B. & BRUETSCH, R. 1996. Colloid properties in granitic groundwater systems. I: Sampling and characterisation. *Applied Geochemistry*, **11**, 677–695.

DREVER, J. I. 1988. *The Geochemistry of Natural Waters* (2nd edn). Prentice Hall, Englewood Cliffs, NJ.

EARY, L. E. & SCHRAMKE, J. A. 1990. Rates of inorganic oxidation reactions involving dissolved oxygen. *In*: MELCHIOR, D. C. & BASSETT, R. L. (eds) *Chemical Modeling of Aqueous Systems II*, American Chemical Society Symposium Series 416, 379–396.

EDMUNDS, W. M. 1986. Geochemistry of geothermal waters in the UK. *In*: DOWNING, R. A. & GRAY, D. A. (eds) *Geothermal Energy – the Potential in the United Kingdom*, HMSO, 111–123.

——, ANDREWS, J. N., BURGESS, W. G., KAY, R. L. F. & LEE, D. J. 1984. The evolution of saline and thermal groundwaters in the Carnmenellis Granite. *Mineralogical Magazine*, **48**, 407–424.

EIKENBERG, J., BAEYENS, B. & BRADBURY, M. H. 1991. *The Grimsel Migration Experiment: a hydrogeochemical equilibrium test.* Nagra Technical Report **NTB 90-39**, Nagra, Wettingen.

ELIASSON, T. 1993. *Mineralogy, geochemistry and petrophysics of red coloured granite adjacent to fractures.* SKB Technical Report **93-06**, Swedish Nuclear Fuel and Waste Management Co., Stockholm.

FRITZ, P. & FRAPE, S. K. (eds) 1987. Saline water and gases in crystalline rocks. *Geological Association of Canada Special Publications* **33**.

GRANDSTAFF, D. E., GRASSI, V. J., LEE, A. C. & ULMER, G. C. 1990. Comparison of granite, tuff, and basalt as geologic media for long-term storage of high-level nuclear waste. *Scientific Basis for Nuclear Waste Management*, **XIV**, Materials Research Society, Pittsburgh, PA, 849–854.

GRASSI, V. J., ULMER, G. C. & GRANDSTAFF, D. E. 1992. Water-rock hydrothermal experiments: influence of rock type on solution pH and oxygen fugacity. *In*: KHARAKA, Y. K. & MAEST, A. S. (eds) *Water–Rock Interaction*, Balkema, Rotterdam, 1415–1418.

GRAY, D. A., GREENWOOD, P. B., BISSON, G. *et al.* 1976. *Disposal of highly-active, solid radioactive wastes into geological formations – relevant geological criteria for the United Kingdom.* Report of the Institute of Geological Sciences No. 76/12.

GRENTHE, I., STUMM, W., LAAKSOHARJU, M., NILSSON, A-C. & WIKBERG, P. 1992. Redox potentials and redox reactions in deep groundwater systems. *Chemical Geology*, **98**, 131–150.

GRIMAUD, D., BEAUCAIRE, C. & MICHARD, G. 1990. Modelling of the groundwaters in a granite system at low temperature: the Stripa groundwaters (Sweden). *Applied Geochemistry*, **5**, 515–525.

HANOR, J. S. 1994. Physical and chemical controls on the composition of waters in sedimentary basins. *Marine and Petroleum Geology*, **11**, 31–45.

HELGESON, H. C., DELANY, J. M., NESBITT, H. W. & BIRD, D. K. 1978. Summary and critique of the thermodynamic properties of rock-forming minerals. *American Journal of Science*, **278A**.

——, MURPHY, W. M. & AAGAARD, P. 1983. Thermodynamic and kinetic constraints on reaction rates among minerals and aqueous solutions. II. Rate constants, effective surface area, and the hydrolysis of feldspar. *Geochimica et Cosmochimica Acta*, **48**, 2405–2432.

HOSTETTLER, J. D. 1984. Electrode electrons, aqueous electrons, and redox potentials in natural waters. *American Journal of Science*, **284**, 734–759.

HUTCHEON, I., SHEVALIER, M. & ABERCROMBIE, H. J. 1993. pH buffering by metastable mineral-fluid equilibria and evolution of carbon dioxide fugacity during burial diagenesis. *Geochimica et Cosmochimica Acta*, **57**, 1017–1027.

JOHNSON, J. W., OELKERS, E. H. & HELGESON, H. C. 1992. SUPCRT92: a software package for calculating the standard molal thermodynamic properties of minerals, gases, aqueous species, and reactions from 1 to 5000 bar and 0 to 1000°C. *Computers and Geosciences*, **18**, 899–947.

KHARAKA, Y. K., LAW, L. M., CAROTHERS, W. W. & GOERLITZ, D. F. 1986. Role of organic species dissolved in formation waters from sedimentary basins in mineral diagenesis. *In*: GAUTIER, D. (ed.) *Roles of Organic Matter in Sediment Diagenesis.* SEPM Special Publications, 111–122.

KNAUSS, K. G. & WOLERY, T. J. 1986. Dependence of albite dissolution kinetics on pH and time at 25°C and 70°C. *Geochimica et Cosmochimica Acta*, **50**, 2481–2497.

LAAKSOHARJU, M. (ed.) 1995. *Sulphate reduction in the Äspö tunnel.* SKB Technical Report **95-25**. Swedish Nuclear Fuel and Waste Management Co., Stockholm.

LANDSTRÖM, O., KLOCKARS, C. E., PERSSON, O. *et al.* 1982. A comparison of *in situ* radionuclide migration studies in the Studsvik area and laboratory measurements. *In*: LUTZE, W. (ed.) *Scientific Basis for Nuclear Waste Management*, **V**. North-Holland, Amsterdam, 697–706.

LANGMUIR, D. 1997. *Aqueous Environmental Chemistry.* Prentice-Hall, Englewood Cliffs, NJ.

LASAGA, A. C., SOLER, J. M., GANOR, J., BURCH, T. E. & NAGY, K. L. 1994. Chemical weathering rate laws and global geochemical cycles. *Geochimica et Cosmochimica Acta*, **58**, 2361–2386.

LICHTNER, P. C. 1985. Continuum model for simultaneous chemical reactions and mass transport in hydrothermal systems. *Geochimica et Cosmochimica Acta*, **49**, 779–800.

——1988. The quasi-stationary state approximation to coupled mass transport and fluid–rock interaction in a porous medium. *Geochimica et Cosmochimica Acta*, **52**, 143–165.

—— & WABER, N. 1992. Redox front geochemistry and weathering: Theory with application to the Osamu Utsumi uranium mine, Poços de Caldas, Brazil. *Journal of Geochemical Exploration*, **45**, 521–564.

MACKENZIE, A. B., SCOTT, R. C., LINSALATA, P., MIEKELEY, N., OSMOND, J. K. & CURTIS, D. B. 1991. *Natural radionuclide and stable element studies of rock samples from the Osamu Utsumi mine and Morro do Ferro analogue study sites, Poços de Caldas, Brazil.* Nagra Technical Report **90-25**, Nagra, Wettingen.

MAZUREK, M., SMITH, P. A. & GAUTSCHI, A. 1992. Application of a realistic geological database to safety assessment calculations: an exercise in interdisciplinary communication. *In*: KHARAKA, Y. K. & MAEST, A. S. (eds) *Water–Rock Interaction*. Balkema, Rotterdam, 407–411.

MICHARD, G. 1987. Controls of the chemical composition of geothermal waters. *In*: HELGESON, H. C. (ed.) *Chemical Transport in Metasomatic Processes*. NATO Advanced Study Institute Series C Mathematical and Physical Sciences, **218**, 323–353.

——, & BASTIDE, J-P. 1988. Étude géochimique de la nappe du Dogger du Bassin Parisien. *Journal of Volcanology and Geothermal Research*, **35**, 151–163.

NERETNIEKS, I. 1985. Some uses for natural analogues in assessing the function of a HLW repository. *Chemical Geology*, **55**, 175–188.

NORDSTROM, D. K., BALL, J. W., DONAHOE, R. J. & WHITTEMORE, D. 1989a. Groundwater chemistry and water-rock interactions at Stripa. *Geochimica et Cosmochimica Acta*, **53**, 1727–1740.

——, LINDBLOM, S., DONAHOE, R. J. & BARTON, C. 1989b. Fluid inclusions in the Stripa granite and their possible influence on the groundwater chemistry. *Geochimica et Cosmochimica Acta*, **53**, 1741–1755.

NURMI, P. A., KUKKONEN, I. T. & LAHERMO, P. W. 1988. Geochemistry and origin of saline groundwaters in the Fennoscandian Shield. *Applied Geochemistry*, **3**, 185–203.

ORTOLEVA, P., AUCHMUTY, G., CHADAM, J., HETTMER, J., MERINO, E., MOORE, C. H. & RIPLEY, E. 1986. Redox front propagation and banding modalities. *Physica*, **19D**, 334–354.

PARNELL, J. & EAKIN, P. 1987. The replacement of sandstones by uraniferous hydrocarbons: significance for petroleum migration. *Mineralogical Magazine*, **51**, 505–515.

PEARSON, F. J., JR, LOLCAMA, J. L. & SCHOLTIS, A. 1989. *Chemistry of waters in the Böttstein, Weiach, Riniken, Schafisheim, Kaisten and Leuggern boreholes: a hydrochemically consistent data set*. Nagra Technical Report **86-19**, Nagra, Wettingen.

PEARSON, R. G. 1969. *In*: SCOTT, A. (ed.) *Survey of Progress in Chemistry*. Academic, New York, Chapter 1.

PIRHONEN, V. & PITKÄNEN, P. 1991. *Redox capacity of crystalline rocks. Laboratory studies under 100 bar oxygen gas pressure*. SKB Technical Report **TR 91-55**. Swedish Nuclear Fuel and Waste Management Co., Stockholm.

PITZER, K. S. 1979. Theory: ion interaction approach. *In*: PYTKOWICZ, R. M. (ed.) *Activity Coefficients in Electrolyte Solutions, Vol. I*. CRC, Boca Raton, 157–208.

RICHARDS, H. G., BATH, A. H., NOY, D. J., GILLESPIE, M. R. & MILODOWSKI, A. E. 1996. Effective integration of geochemistry into site characterisation. *Proceedings of 1996 International Conference on Deep Geological Disposal of Radioactive Waste*, Winnipeg.

ROCHELLE, C. A., BATEMAN, K., MACGREGOR, R., PEARCE, J. M., SAVAGE, D. & WETTON, P. D. 1995. Experimental determination of chlorite dissolution rates. *In*: MURAKAMI, T. & EWING, R. C. (eds) *Scientific Basis for Nuclear Waste Management*, **XVIII**. Materials Research Society, Pittsburgh, PA, 149–156.

ROSING, M. 1993. The buffering capacity of open heterogeneous systems. *Geochimica et Cosmochimica Acta*, **57**, 2223–2226.

SAVAGE, D., & ROCHELLE, C. A. 1993. Modelling reactions between cement pore fluids and rock: implications for porosity change. *Journal of Contaminant Hydrology*, **13**, 365–378.

——, CAVE, M. R., MILODOWSKI, A. E. & GEORGE, I. 1987. Hydrothermal alteration of granite by meteoric fluid: an example from the Carnmenellis Granite, United Kingdom. *Contributions to Mineralogy and Petrology*, **96**, 391–405.

——, BATH, A. H., CRAWFORD, M. B. *et al.* 1993. Groundwater relationships in cover and basement at the margins of the East Irish Sea Basin, West Cumbria, UK. *TERRA Abstracts*, **5**, 643–644.

SCOTT, M. J. & MORGAN, J. J. 1990. Energetics and conservative properties of redox systems. *In*: MELCHIOR, D. C. & BASSETT, R. L. (eds) *Chemical Modeling of Aqueous Systems II*, American Chemical Society Symposium Series **416**, 368–378.

SHERWOOD, B., FRITZ, P., FRAPE, S. K., MACKO, S. A., WEISE, S. M. & WELHAN, J. A. 1988. Methane occurrences in the Canadian Shield. *Chemical Geology*, **71**, 223–236.

SHERWOOD-LOLLAR, B., FRAPE, S. K., DRIMMIE, R. *et al.* 1989. Deep gases and brines of the Canadian and Fennoscandian Shields – a testing ground for the theory of abiotic methane generation. *In*: MILES, D. L. *Water–Rock Interaction*. Balkema, Rotterdam, 617–620.

SHOCK, E. L. 1988. Organic acid metastability in sedimentary basins. *Geology*, **16**, 886–890.

SMELLIE, J. & LAAKSOHARJU, M. 1992. *The Äspö Hard Rock Laboratory: final evaluation of the hydrochemical pre-investigations in relation to existing geologic and hydraulic conditions*. SKB Technical Report **92-31**. Swedish Nuclear Fuel and Waste Management Co., Stockholm.

SMITH, J. T. & EHRENBERG, S. N. 1989. Correlation of carbon dioxide abundance with temperature in clastic hydrocarbon reservoirs: relationship to inorganic chemical equilibrium. *Marine and Petroleum Geology*, **6**, 129–135.

STUMM, W. & MORGAN, J. J. 1981. *Aquatic Chemistry*. Wiley, New York.

TOULHOAT, P., BEAUCAIRE, C., MICHARD, G. & OUZOUNIAN, G. 1993. Chemical evolution of deep groundwaters in granites, information acquired from natural systems. *In*: *Paleohydrogeological Methods and their Applications*. Nuclear Energy Agency/Organisation for Economic Cooperation and Development, Paris, 105–116.

VELBEL, M. A. 1986. Influence of surface area, surface characteristics, and solution composition on feldspar weathering rates. *In*: DAVIS, J. A. & HAYES, K. F. (eds) *Geochemical Processes at Mineral Surfaces*. American Chemical Society Symposium Series **323**, 615–634.

VERBECK, G. J. 1975. Mechanisms of corrosion of steel in concrete. *American Concrete Institute Publication*, **SP 49-3**, 21–38.

VILKS, P., MILLER, H. & DOERN, D. 1991. Natural colloids and suspended particles in the Whiteshell Research Area, Manitoba, Canada, and their potential effect on radiocolloid formation. *Applied Geochemistry*, **6**, 564–574.

WOLERY, T. J. 1992. *EQ3/6, a software package for geochemical modeling of aqueous systems: package overview and installation guide (Version 7.0)*. Lawrence Livermore National Laboratory Report **UCRL-MA-110662 PT I**, California.

The chemical basis of near-field containment in the Swiss high-level radioactive waste disposal concept

W. R. ALEXANDER[1] & I. G. McKINLEY[2]

[1] GGWW (Rock-Water Interaction Group), Institutes of Mineralogy-Petrology and Geology, University of Berne, Baltzerstrasse 1 3012 Berne, Switzerland
(Present address: Nagra)
[2] Nagra (Swiss National Co-operative for the Disposal of Radioactive Waste), Hardstrasse 73, 5430 Wettingen, Switzerland

Abstract: Concepts for the disposal of high-level radioactive waste (HLW) vary from country to country but, in Switzerland, a key contribution to safety is the chemistry of the near-field. In this paper, the development of the Swiss repository design is discussed as a basis for explaining the chemical confinement of radionuclides within the engineered barrier system (EBS). The expected performance of the EBS is described and the chemistry of the system explained. Potential perturbations to the EBS are examined and the methodology of assessing the long-term behaviour of the EBS is briefly described. The relevance of HLW near-field design to chemo-toxic wastes is discussed and an extensive reference list on the chemical aspects of the EBS is provided.

In this paper, the chemical basis of containment in one repository design for high-level radioactive waste (HLW), that of the Swiss programme, is examined in detail, focusing in particular on the role of the engineered barriers which constitute the repository near-field (the rock body in which the repository is built constituting the far-field). In addition, the potential relevance of HLW repository design to disposal of long-lived toxic wastes will be discussed.

Before going any further, it is worthwhile briefly considering what constitutes HLW as the precise boundaries between radioactive waste (radwaste) types vary somewhat from national programme to national programme, despite the published IAEA (International Atomic Energy Agency) classifications (IAEA 1981). HLW consists of spent fuel from nuclear reactors, if a direct disposal option is chosen, or the most active solidified (e.g. vitrified) residues from reprocessing of such fuel. Particular characteristics of this waste are its high radiation field which necessitates shielding and/or remote handling, a significant radiogenic heat output, high concentrations of very long-lived radionuclides (half-lives in excess of 1000 a) and relatively low volumes. For example, a typical package of vitrified reprocessing waste with a volume of $0.15\,m^3$ would have a total radioactivity content of approximately $7 \times 10^{15}\,Bq$ of β/γ- and $5 \times 10^{13}\,Bq$ of α-emitters and a heat output of

about 600 W at the time of emplacement in a repository (i.e. after 40 years surface storage for cooling). The planned 40 a lifetime of the current five Swiss nuclear reactors (3GW(e) total) would, however, result in less than 2700 of such packages (around $400\,m^3$ of waste) if all spent fuel was reprocessed. In the UK, in comparison, it is currently estimated that a total of about $2280\,m^3$ (or almost six times the Swiss total) of conditioned HLW will have been produced by 2030 (Electrowatt 1996).

The low volume of HLW relative to the high value of the electricity generated means that rather over-designed disposal concepts have been developed (see below). Such very expensive options are justified not only by the very strict performance measures imposed in most countries, but also by the general fear of all things radioactive by the general public.

As described in detail by Alder (1997), the reference inventory for a HLW repository in Switzerland constitutes vitrified waste from approximately 1000 MTHM (Metric Tonnes Heavy Metal) of spent fuel reprocessed at Sellafield (NW England) and Cap de la Hague (N France) and approximately 2000 MTHM of reactor spent fuel (about 5–10% will be MOX, or mixed U and Pu oxide, fuel) which may be disposed of directly. In addition, a small volume of high-activity, long-lived waste (such as fuel assemblage hulls and ends and as yet unspecified

From: METCALFE, R. & ROCHELLE, C. A. (eds) 1999. *Chemical Containment of Waste in the Geosphere.* Geological Society, London, Special Publications, **157**, 47–69. 1-86239-040-1/99/$15.00 © The Geological Society of London 1999.

waste from medicine, industry and research) may be included in the Swiss repository.

Derivation of the Swiss HLW disposal concept

Deep geological disposal of HLW is the selected option in all countries with significant nuclear power programmes. A wide range of other options has been studied, but most have been excluded, either for political reasons (e.g. subseabed disposal, Antarctic ice shelf disposal) or due to limitations/costs of current technology (e.g. disposal in space, transmutation). Even for geological disposal, a wide range of concepts have been studied, but effort is focused on emplacement in specially constructed underground caverns. A prerequisite is geological stability, but many potential host rocks have been identified, including crystalline basement, salt, basalt, tuff and a range of argillaceous sediments.

In all cases, a multiple barrier concept has been adopted, but the relative weighting of the engineered barriers and the geological barriers varies (Table 1). As stated above, this paper will be restricted to considering the chemical basis of the near-field of the Swiss HLW repository (although a comparison of designs worldwide is available in Witherspoon 1991, 1996).

Table 1. *International HLW disposal concepts: examples of typical key contributions to safety*

Key contribution	Country
Far-field:	
dry (salt)	Germany, Holland
unsaturated	USA (Yucca Mt)
low water flow	Canada, Belgium
Near-field:	
canister longevity	Sweden, Finland
chemistry	Switzerland, Japan

To date, the Swiss National Co-operative for the Disposal of Radioactive Waste (Nagra) has carried out three major safety assessments for a HLW repository in crystalline and argillaceous rocks (Fig. 1). This paper will concentrate on the near-field chemical containment aspects identified in the two crystalline host rock assessments (Projekt Gewähr and Kristallin-1; Nagra 1985 and 1994a respectively). Note that the near-field of a Nagra HLW repository in sediments would differ only in that the backfill thickness may be reduced and that it also would contain steel tunnel liners (Nagra 1989).

It is, however, interesting to note that Nagra's original HLW repository concept was strongly influenced by that of the Swedish Nuclear Fuel and Waste Management Company (SKB) which

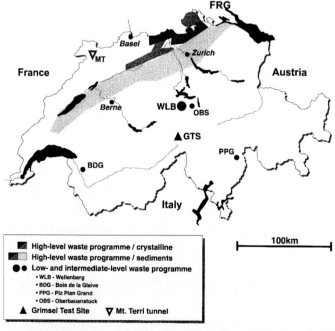

Fig. 1. Areas under consideration as potential high-level waste repository site in Switzerland. Sites examined for L/ILW disposal and the underground test sites at Grimsel and Mt. Terri are also shown.

was designed without any specific regard to the near-field chemistry. Here, it was planned that the repository tunnels would be situated about 500 m deep in the crystalline bedrock of the Scandinavian Shield. A copper canister (encapsulating spent fuel) would be placed axially in a 1.5 m diameter hole drilled vertically into the floor of the tunnel and the canister would then be surrounded by compacted bentonite and the tunnel backfilled with bentonite and quartz sand (SKB's 'in-hole emplacement' concept; KBS 1983).

This design was adapted by Nagra to suit local conditions, resulting in the concept shown in Fig. 2. The repository would be situated 800–1200 m deep in the crystalline basement of northern Switzerland and, because the original design specifications called for temperatures of less than 100°C in significant parts of the backfill, an option with a very thick compacted bentonite backfill (or buffer) was selected.

This ruled out the SKB plan for in-hole emplacement and resulted in a design with horizontal emplacement in the tunnel (subsequently, such in-tunnel emplacement has also been studied in Sweden; SKB 1993).

The original SKB design calls for a 10 cm thick copper canister, filled with lead, to encapsulate the spent fuel and it was assumed that the copper is so inert under the *in situ* conditions that it will last effectively indefinitely (corrosive penetration in 10–100 Ma). Nagra changed to a massive (25 cm thick walls) steel canister for vitrified waste because the deep groundwater chemistry (in particular, the higher sulphate content) in northern Switzerland will be more aggressive to copper than in the Scandinavian Shield (NWGCT 1984) and a greater mechanical strength was required to withstand the greater external swelling pressures (maximum of 30 MPa, the sum of the maximum

bentonite swelling pressure and the hydrostatic pressure) as the thin stainless steel fabrication flask containing the vitrified waste contains a void space. In addition, the initial design called for a canister robust enough to last for at least 1000 years to extend beyond early thermal perturbations in the near-field.

Therefore, although the original design reflects a purely engineering approach to the problems of containment, there have been several very clear chemical spin-offs of the final design, namely:

- massive redox (steel canister) and pH (bentonite) buffers ensuring low solubility of radionuclides released from the waste
- colloid filtration due to the microporous nature of the compacted bentonite after resaturation
- diffusive solute transport dominates over advection because of the extremely low hydraulic conductivity of the resaturated bentonite
- high retardation of radionuclides in the buffer due to the good sorptive properties of the bentonite.

Finally, it is of note that the most recent Swiss HLW safety assessment (Kristallin-1) assumes that the near-field will provide the main constraint on radionuclide release and transport and the main role of the geosphere is to provide a suitable environment for near-field longevity and performance. The three most important safety features provided by the geosphere are '... mechanical protection, adequate geochemical conditions and sufficiently low groundwater flow rates' (Nagra 1994a). Precisely how these features impact on the chemical containment of the waste will be discussed below.

Expected near-field performance

The engineered barrier system (EBS) for the Swiss concept is characterized by the use of large quantities of rather simple, well-understood materials. Design optimization is certainly possible (e.g. McKinley & Toyota 1998) but has not yet been extensively studied, although the consequences of reducing the diameter of the emplacement tunnel has been assessed for the crystalline host rock option. Examples of the inventories of materials for vitrified HLW (for two different tunnel diameters) and spent fuel are presented in Table 2. The inventories for the sedimentary host rock option are similar to those for the 2.4 m diameter tunnel crystalline case with the addition of a tunnel liner.

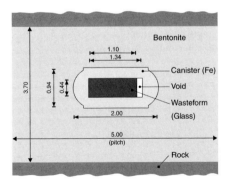

Fig. 2. The multiple component engineered barrier safety system for vitrified HLW which constitutes the repository near-field (dimensions in metres).

Table 2. *EBS material inventories (crystalline host rock) per waste package*

	Vitrified HLW		Spent fuel
Tunnel diameter			
(m)	2.4	3.7	2.4
Waste			
(m^3)		0.15	0.21
(Mg)		0.4	1.9
Steel canister			
(m^3)		0.9	3.6
(Mg)		6.5	28
Compacted bentonite			
(m^3)	52.8	21.3	29.1
(Mg)	110	44	61

Demonstration of the practicality of this concept and development of quality assurance procedures to ensure that emplacement reaches desired standards are amongst the goals of the Spanish implementor ENRESA's (Empresa Nacional de Residuos Radiactivos, S.A., Spain) full-scale engineered barrier system experiment (FEBEX; see McKinley *et al.* 1996 and Huertas & Santiago 1998 for details) currently running at the Grimsel Test Site, Nagra's underground laboratory in the Central Swiss Alps (Fig. 1).

Repository performance assessment

Any repository performance assessment is a large scale undertaking and a wide range of features, events and processes (FEPs) which may affect the overall performance of the repository are identified and linked together into a set of scenarios, each consisting of a sequence of processes and events which describe a possible future evolution of the repository system (see Fig. 3). Nagra opted to employ a hierarchy of deterministic calculations to investigate the different types of uncertainty identified in the FEPs examined, but it is beyond the scope of this paper to detail the three Swiss HLW assessments carried out to date. As an example, the Reference Case (see Fig. 3) for the crystalline host rock is considered in more depth and, here, the near-field is expected to evolve as follows (Nagra 1994*a, b*).

Temperature and water-saturation changes. After EBS emplacement, water will begin to resaturate any drained zones of the surrounding rock and then invade the bentonite. As bentonite is contacted by water (by advection through any gaps and diffusion through the rest), it will swell, seal any gaps present and gradually build up

an isostatic pressure which is in the order of 4–18 MPa (i.e. less than lithostatic pressure at the emplacement depth). At the same time, temperatures will increase in the EBS predominantly due to radiogenic heat from the canister (exothermic corrosion and chemical reactions will contribute less than 1% of the heat production) with maximum values being reached within a few tens of years. Evaluation of the development of temperature and water saturation within the bentonite is a complex, coupled problem but models indicate that, while temperatures in excess of 100°C may be expected at the canister surface, these will last no more than a few decades and the majority of the bentonite will remain <100°C (Sasaki *et al.* 1997). Complete saturation of the EBS will take several (or many) centuries (Andrews *et al.* 1986) and, because the saturated bentonite has an extremely low hydraulic conductivity (10^{-16} m s^{-1} to 10^{-13} m s^{-1}), solute transport will occur predominantly by diffusion (Conca *et al.* 1993). Testing the model assumptions used to develop this description is also an aim of the FEBEX experiment mentioned above.

Near-field water chemistry. The water chemistry in the near field will be established by interaction of groundwater with the engineered barriers. The two most important materials by quantity (see Table 2) are the bentonite (which will tend to buffer pH in the mildly alkaline range) and the steel canister (which will ensure reducing conditions persist). Due to very low rates of solute movement in the bentonite and limited solute exchange with low ambient groundwater fluxes (expected to average less than one litre/waste package/year; Thury *et al.* 1994), the chemistry of porewater is expected to be rather homogeneous and to vary little with time (Nagra 1994*a*). Very slow alteration of bentonite minerals (sodium montmorillonite, minor calcite, pyrite and siderite) will occur but significant effects are not to be expected within the first 10^6 years (see below).

Canister corrosion. The steel canister will corrode slowly, initially consuming any trapped oxygen in a moist atmosphere and then anaerobically in the presence of liquid water after the bentonite saturates. Based on very pessimistic assumptions, it is calculated that canisters fail mechanically after 10^3 years (NWGCT 1984; Steag & Motor Columbus 1985). More recent analyses indicate that both the corrosion rates considered (McKinley 1991) and the safety margins for the mechanical calculations (Attinger &

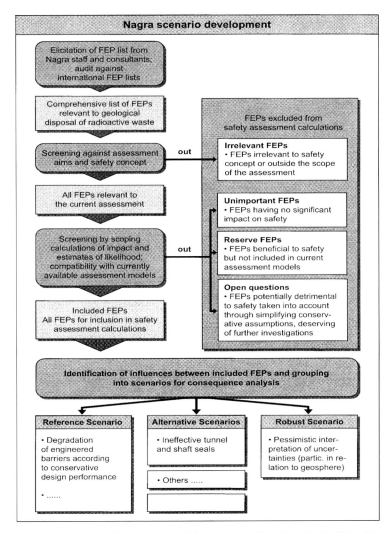

Fig. 3. Overview of the scenario development methodology employed by Nagra in the Kristallin-1 HLW performance assessment (after Smith *et al.* 1995).

Duijvestijn 1994) are extremely conservative and a more realistic estimate of the canister lifetime (which is well supported by extensive natural analogue studies, discussed below) would be about 10^4 years.

Anaerobic corrosion of steel may lead to the production of hydrogen which may form a distinct gas phase. Build up of hydrogen may feed back to further lower the corrosion rate, which can be further encouraged by placing a thin sand layer around the canister (Neretnieks 1985). It is considered unlikely that gas pressures exceeding the bentonite capillary pressure will be produced. Even if this does occur, it has been shown that gas can flow through compacted bentonite without disturbing its solute barrier properties (discussed below).

Waste corrosion. After canister failure, the waste form will begin to corrode in an environment of stagnant porewater. Vitrified waste dissolves at a very low rate and the corrosion rate is assumed to be constant with time (Grauer 1985). Arguments have been advanced for possible acceleration mechanisms (formation of crystalline secondary products which lower silica concentration and thus accelerate glass corrosion) but long-term (*c.* 5 year) experiments carried out at the Paul Scherrer Institute, Switzerland (unpublished data) indicate that rates drop to extremely low levels (probably due to the build-up of secondary products on surfaces). Due to the lower density of alteration products and the limited available space within the canister, it is likely that the assumed glass lifetime of $\sim 10^5$ years is very conservative (McKinley 1991). Radionuclides within the waste are assumed to be released congruently with glass matrix dissolution.

Although not considered in detail in Nagra (1994a), it is worth mentioning here the case where spent fuel is disposed of directly (Schneider et al. 1997). This is somewhat more complex as radionuclides are not homogeneously distributed within the UO_2 (or uranium/plutonium mixed oxide in the case of MOX fuels). During reactor operation, some volatile elements (most importantly caesium and iodine) migrate to crystal grain boundaries and gaps between fuel pellets. On contact with water, this inventory is relatively rapidly leached to give a pulse release of activity. Thereafter, radionuclides are released as the fuel matrix dissolves. Additionally, the fuel rod cladding and the structural parts of the fuel assembly contain some activation products which are released in a complex manner as these materials corrode.

The fuel dissolution process is very dependent on redox conditions which are, in turn, influenced by the extent of radiolysis of water. Radiolysis is dominated by the α-particle flux and is very sensitive to the available surface of fuel, the geometric distribution of water present, the burn-up of the fuel and the time since emplacement. The chemistry of the water, which critically affects the extent and rate of formation of alteration products, is also of importance. Although there is an extensive database of laboratory studies, there is still some controversy on the magnitude of net dissolution rates under expected conditions. Natural analogue studies of uranium ore bodies indicate that dissolution rates would be at the lower end of those derived from laboratory studies (see below).

Release of radionuclides. The release of radionuclides from the waste is constrained, in many cases, by their low solubility. A number of simplifications are made because of the limitations of current models/databases for quantifying solubility limits (McKinley et al. 1999).

- Solubility limits are defined for a water chemistry which would correspond to bentonite porewater with redox conditions buffered by canister corrosion products. Although based on the reference bentonite porewater compositions, uncertainties in host rock groundwater composition are also taken into account.

- A chemical thermodynamic model is used to assess solubility limits for a range of pure mineral phases which could potentially form under the expected conditions. Note that the calculations were carried out at 25°C as, first, relevant data are rare for higher temperatures and, second, other uncertainties in the calculations are much larger than any temperature effect on the solubilities.

- The output of the thermodynamic model is reviewed in terms of literature data on the formation of identified minerals and measured concentrations of the element of interest in relevant natural and laboratory systems.

- The likelihood of formation of pure phases is assessed in terms of the relative inventories of elements with similar chemical properties in the waste.

- Bearing in mind the intention to err on the side of conservatism in the derivation of parameter values (i.e. over-prediction of solubility limits), expert judgement is used to select elemental solubility limits which are defensible as 'realistic-conservative' (reasonable, but pessimistic in the handling of uncertainties) and 'conservative' (worst case based on all available data) values.

The procedure and results of this selection procedure are documented in Berner (1994); a simplified overview of the main considerations involved in the selection of solubility limits for specific elements is given in Table 3 and the resultant limits for the safety assessment are listed in Table 4.

Due to chemical buffering by the canister corrosion products and bentonite minerals (particularly pyrite and siderite), it is not expected that the oxidizing conditions, which may form at the surface of spent fuel, will have a significant influence on the rest of the EBS (Smith & Curti 1995). In this case, the solubility limits derived for the vitrified waste case can be directly applied – a redox front may exist within the canister corrosion products or bentonite but this will not influence releases from the near-field to the geosphere. Extensive precipitation may occur at this redox front which may initially be in the form of colloids. The small pore size (average c. 2–10 nm, Pusch 1980; McKinley 1988) of the compacted bentonite should ensure that any colloids formed are immobile in this

Table 3. *Summary of the main considerations involved in the selection of solubility limits for specific elements*

Element	Rationale for setting solubility limits
Cs, Ni	Very high solubility expected in the absence of co-precipitation processes (arbitrary 'high' value selected which could be set as 10 M)
Pa, Sn, Tc, Th, U Zr, Np, Pu	Main constraint on solubility taken to be formation of the metal oxides (or hydroxides); selected solubilities based on evaluation of uncertainties in thermodynamic data, solid phase crystallinity, laboratory measurements and the likely formation of mixed oxides
Pd	Expected to be effectively insoluble but, conservatively, solubility assessed assuming control by $Pd(OH)_2$
Se	Assumed to be set by $FeSe_2$ although co-precipitation with S might be expected in reality
Am	Assumed to be set by $AmOH (CO_3)$
Cm, Ra	Assumed to co-precipitate with chemically similar elements present in much higher quantities (Am and Ba/Sr/Ca, respectively)

Selection procedure documented in Berner (1994).

Table 4. *Solubility limits for safety-relevant elements*

Element	Solubility (molar)		K_d (m^3 kg^{-1})*	
	Realistic	Conservative	Realistic	Conservative
Am	10^{-5}	10^{-5}	5	0.5
Cm	6×10^{-8}	10^{-5}	5	0.5
Cs	high†	high†	0.01	0.001
Ni	high†	high†	1	0.1
Np	10^{-10}	10^{-8}	5	0.5
Pa	10^{-10}	10^{-7}	1	0.1
Pd	$\sim 10^{-11}$	10^{-6}	1	0.1
Pu	10^{-8}	10^{-6}	5	0.5
Ra	10^{-10}	10^{-10}	0.01	0.001
Se	10^{-8}	6×10^{-7}	0.005	0.001
Sn	10^{-5}	10^{-5}	1	0.1
Tc	10^{-7}	high†	0.1	0.05
Th	5×10^{-9}	10^{-7}	5	0.5
U	10^{-7}	7×10^{-5}	5	0.5
Zr	5×10^{-9}	5×10^{-7}	1	0.1

Derivation described in detail by Berner (1994).
Sorption data for bentonite from Stenhouse (1995).
* In models of solute transport in natural waters, the interactions of the solute with the solid are usually bulked into a simple distribution coefficient K_d (m^3 kg^{-1}) which, in the simplest form, may be defined as $C_r = K_d C_w$ where C_w = the solute concentration in the aqueous phase (mol m^{-3}) and C_r = the concentration sorbed on the solid phase (mol kg^{-1}). For further information on the measurement of K_d values, see Sibley and Myttenaere (1986) and, for a discussion of the limitations of the concept, see McKinley and Alexander (1992, 1993a, b, 1996).
† 'high' indicates that the solubility is sufficiently large that saturation could significantly alter the porewater chemistry. For modelling purposes, this 'unlimited' value could be set to an arbitrary value which is high enough not to be reached during performance assessment calculations (e.g. 10 M).

medium, something which has recently been verified experimentally (Kurosawa *et al.* 1997). It should be emphasised that this situation contrasts to designs with a chemically inert canister and small bentonite annulus (e.g. in Sweden) where the redox front is calculated to pass through the EBS and penetrate the geological barrier (Romero *et al.* 1995).

Diffusion of radionuclides. The diffusion of dissolved radionuclides through corrosion products of the waste form and canister and the surrounding bentonite will be retarded by sorption on the solid phases present (which is distinguished mechanistically from precipitation/co-precipitation processes considered above – cf. McKinley & Alexander 1996). Even though it is likely that canister corrosion products may act as powerful sorbing agents, performance assessment concentrates on the role of bentonite due to its larger inventory and greater confidence in its general availability. Note that canister failure would be localized which would limit the surfaces available to migrating nuclides – although in such a case the transport resistance of the crack involved could be significant (Smith & Curti 1995).

Based on numerous batch sorption and diffusion experiments, several extensive databases exist on solute transport through compacted bentonite (e.g. Stenhouse 1995; Yu & Neretnieks 1997). Although this clearly indicates significant retardation of many key elements, the lack of a simple correlation between sorption experiments carried out on dispersed bentonite and diffusion studies on highly compacted material has caused some confusion (Stenhouse 1995, Appendix A; Conca *et al.* 1993; Yu & Neretnieks 1997). A consensus is presently emerging that compacted bentonite is probably better represented as a microporous charged membrane than a dispersed mineral assemblage. Nevertheless, regardless of the mechanisms involved, a simple sorption model can be applied to the large empirical database and shown to be conservative in that migration is over-predicted. Transport calculations using the performance assessment sorption database (Table 4) indicate that a large proportion of the radionuclide inventory of the waste decays to insignificance within the engineered barriers. Very long-lived radionuclides, which do eventually break through the bentonite barrier, generally do so only in the far future and at levels which are radiologically insignificant (Nagra 1994*a*; Smith & Curti 1995).

Safety considerations

The safety of this disposal concept is illustrated for vitrified HLW in Fig. 4a which shows that, with rather pessimistic assumptions about performance of the geological barrier, releases occur only in the far future and are orders of magnitude below regulatory guidelines (Nagra 1994*a*). Similar performance is indicated for the case of direct disposal of spent fuel (see Forsyth 1995; Schneider *et al.* 1997), although analyses

carried out to date have been less extensive and, as such, will not be described in detail here.

More dramatically, perhaps, Fig. 4b also shows the direct effect of the EBS by 'switching out' the geosphere and allowing releases from the near-field to pass directly to the accessible biosphere. Although releases of less well retarded nuclides occur at earlier times and integrated releases are higher, dose limits remain well below regulatory guidelines at all times. It is much more difficult to carry out the same exercise in reverse: first the geosphere is much harder to define as precisely as the EBS and, second, it is almost impossible to define a sensible source term. Nevertheless, scoping calculations have been carried out assuming unit releases for specific radionuclides and several points are worth noting:

- The choice of geometrical representation of water-conducting features is critical to the calculated performance of the geosphere. Determining which features dominate radionuclide transport and accounting for their complex nature with appropriate models are '. . . key areas for future work'. (Nagra 1994*a*).
- The variability of flow between water-conducting features can also significantly affect the overall performance of the geosphere by adding additional dispersion.
- Currently, whether modelling retardation of radionuclides in water-conducting fractures or assessing the influence of colloids on dose to the biosphere, a lack of detailed data means that conservative representations are always employed. Only with the production of additional data will it be possible to justify a less conservative representation of the geosphere, so increasing supportable safety margins and allowing optimization of the repository concept.

The key parameters which determine near-field performance are:

- canister lifetime
- glass corrosion rate
- elemental solubilities
- elemental sorption on bentonite
- groundwater flux around the engineered barriers

These are now considered in turn, emphasising the effects of selecting more conservative values. The canister lifetime (in terms of loss of containment) of 10^3 years is considered to be very conservative with respect to the assumed corrosion rate which is, in itself very conservative in the light of natural analogue data

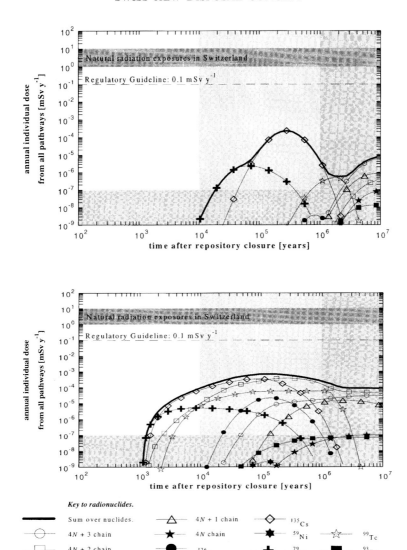

Fig. 4. Time development of the annual individual doses in the safety assessment Reference Case. (**a**) Results of the Reference Case calculations. (**b**) As an illustration of the effect of the EBS alone, the hypothetical case of direct release of radionuclides from the near-field to the biosphere is also presented.

(discussed below). The canister design and subsequent evaluation of the canister performance is discussed in Nagra (1984) and Steag & Motor Columbus (1985) and failure in less than several hundred years is considered extremely unlikely (even then this would have little significant effect on releases due to the constraints set by the slow dissolution rate of the glass waste-form and the powerful barrier role played by the bentonite). In addition, the original analysis has been verified and tested against experiments, including the prediction of failure

of scale-model canisters in the European Community COMPAS exercise (Ove Arup 1990; Attinger & Duijvestijn 1994). Even less likely is the case of failure soon after emplacement (within 100 years, say) due to manufacturing flaws as fabrication quality should be ensured by the use of a simple production methodology (sand-mould casting) and ultrasonic inspection of the completed canister, lid and weld. In any case, early failure would not be expected to increase doses significantly, although the applicability of the simple release model could be questioned due, for example, to the higher temperatures and significant thermal gradients which are present.

The release model would, however, probably be very conservative at such times as the inner zone of the bentonite would still be in the process of resaturation. The waste-form would thus not be exposed to liquid water and, in addition, a net flux of water towards the canister would be expected to prevent outwards radio-nuclide transport.

The glass corrosion rate chosen was based on an extrapolation of laboratory experiments and is believed to be reasonably conservative but calculations were repeated assuming a value two orders of magnitude higher. The release rates of most radionuclides are, in any case, constrained by their low solubilities and hence are unaffected by this change. A notable exception is ^{135}Cs which is not solubility-limited. The maximum ^{135}Cs release concentration from the near-field does not scale linearly with the glass corrosion rate, however, due to dispersion (i.e. spreading of the peak) during transport through the bentonite, which means that the increase in resultant dose is only by less than one order of magnitude. Calculated doses are, therefore, still well below the regulatory guidelines.

Elemental solubility limits are a very impor-tant constraint on the releases of many radio-nuclides and selecting a set of more conservative values of this parameter significantly alters the profile of calculated doses as a function of time. Decay during near-field and geosphere transport minimizes the consequences of such changes for some nuclides as does the fact that, as solubi-lities increase, the glass corrosion rate takes over as a constraint on releases.

There is a fairly large body of experimental data to support selection of 'realistic' elemen-tal diffusivities (derived from both sorption and diffusion measurements) for the bento-nite backfill. Repeating the calculations with more conservative sorption distribution coeffi-cients (generally a factor of ten smaller) has little effect on calculated releases from the geosphere, even though the near-field release profiles change somewhat. A major effect of such increased diffusivity is to allow release from the bentonite of some isotopes which other-wise decay within the near-field (^{59}Ni, ^{126}Sn, ^{239}Pu, ^{242}Pu). However, these radionuclides are not dominant contributors to dose and, in any case, will decay within the far-field.

The rate of diffusion through the backfill is influenced by the rate at which dissolved species can be removed from its outer surface which is related, in turn, to the water flux around the engineered barriers. The sensitivity of releases to this parameter was evaluated by repeating calculations assuming one or two orders of magnitude higher fluxes. Considering the near-field only, calculations show that increasing the flux by a factor of 100 causes an increase of the maximum release rate of ^{135}Cs by a factor of only about five although the maximum is reached at an earlier time ($c.\,10^4$ years rather than $c.\,10^5$ years). Increasing the flux has a more dramatic effect on releases of other radio-nuclides. For example, the release rate of ^{79}Se initially increases almost linearly with water flux, although this drops off somewhat at higher fluxes. Considering the full waste inventory shows that, even for two orders of magnitude higher flux, the regulatory guideline would not be reached.

The general conclusion is, therefore, that the EBS for HLW is 'robust' (McCombie et al. 1991) in that expected performance based only on well-understood processes meets performance guide-lines even if very pessimistic assumptions are made about poorly-understood processes or potential perturbations.

Potential perturbations to the near-field

The previous section makes the case that the system of engineered barriers considered is 'robust'; in order to justify this it is necessary to demonstrate that all conceivable processes which could cause degradation of performance have been assessed. As noted in the sensitivity studies, if the bentonite buffer, canister and waste-form behave as expected, then safety goals can be met even if very pessimistic parameter values are assumed. The critical question is, therefore, can the occurrence of processes which could short-circuit one or more of the multiple safety barrriers be rigorously precluded? It was noted in the Kristallin-1 safety assessment that '...most phenomena that could be detri-mental to safety are excluded or forced to low probability or consequence by the repository design and siting concept' (Nagra 1994a). This is aided by the use of relatively well-understood materials within the EBS which are then further tested to assess their behaviour under repository specific conditions.

In most engineered systems, such as bridge construction, this would include a range of laboratory experiments backed up by expert judgement based on experience with the same or similar systems. Here repository design deviates from standard engineering practice in that no HLW (and only a few L/ILW) repositories yet exist and, when they do, testing their compliance to design limits will be somewhat difficult due to the time scales involved. This being the case, an additional level of testing has been developed within the radwaste industry, that of the natural analogue approach (see Chapman *et al.* 1984; Chapman & McKinley 1987, 1990; Miller *et al.* 1994).

Briefly, a natural analogue may be defined as '...(*a natural or archaeological*) occurrence of materials or processes which resemble (i.e. *are analogous to, but not copies of*) those expected in a proposed geological waste repository' (Côme & Chapman 1986). The advantage of natural analogues over short-term laboratory experiments or medium-term *in situ* field tests is that they enable study of repository-like systems which have evolved over the geological timescales of relevance to a radwaste repository safety assessment (rather than the days to months usual in laboratory tests). However, by their very nature, natural analogues often have ill-defined boundary conditions which means that some specific processes may be better assessed under the well constrained (if less relevant) conditions of a laboratory. Well-designed, realistic *in situ* field tests can bridge the gap between the laboratory and natural analogues by offering repository relevant natural conditions with some of the constraints of the laboratory (and intermediate timescales). In short, combining information from the three sources (long-term and realistic, if poorly defined, natural analogues, medium-term, better constrained, *in situ* field tests and short-term, less realistic but well defined laboratory studies) can provide greater confidence in the extrapolation of laboratory derived data to repository relevant timescales and conditions (Alexander *et al.* 1998). In the following sections, this approach is used to assess the significance of several potential perturbations to the chemical containment of waste in the repository near-field.

Bentonite

The thick layer of compacted bentonite plays a key role in the near field, ensuring that solute transport occurs by diffusion and that colloid migration is precluded. Extensive review efforts (Nagra 1994*a*) have identified several potential perturbations which could, potentially, have a major influence on this barrier function:

- mineralogical alteration of the bulk bentonite so that its plasticity, swelling pressure and microporous structure are lost

- cementation and subsequent fracturing of the bentonite without significant alteration of the bulk bentonite
- sinking of the canister through the bentonite
- physical perturbation of the bentonite due to gas (H_2) produced by anaerobic corrosion of the steel canister.

These processes are now addressed in turn.

Bentonite alteration. A review of the existing literature on laboratory data suggests that the long-term stability of the bentonite (at repository ambient temperatures of 40–55°C) is most likely to be compromised by the possible alteration of montmorillonite to illite (note that alteration to chlorite is extremely unlikely due to the low magnesium levels of the deep crystalline groundwaters). This would have the two-fold effect of reducing both the swelling potential (possibly allowing advective flow through gaps in the bentonite and reducing colloid filtration) and the cation-exchange capacity (CEC) of the bentonite (so decreasing radionuclide retardation in the near-field).

The montmorillonite/illite transformation will only occur when the tetrahedral layer charge of the clay increases and the interlayer ions are replaced by potassium ions. However, increasing the layer charge alone does not change the swelling capacity; the presence of interlayer potassium ions is also necessary (Grauer, 1990), i.e. the supply of potassium to the bentonite can limit illitization. This being the case, the degree of bentonite alteration can be estimated based on the site characterization data on deep groundwater chemistry (potassium concentration), regional hydrology (groundwater flux) and the likely long-term evolution (increased or decreased potassium concentration or groundwater flux) of the deep groundwater system.

In the case of a repository in one of the two potential siting areas in the crystalline basement of northern Switzerland (Fig. 1), total conversion of montmorillonite to illite was calculated to take between 10 Ma (eastern site; McKinley 1985) and 100 Ma (western site; Nagra 1994*a*). For the latter site, even assuming that the groundwater flux is two orders of magnitude higher than the (already conservative) reference flux and assuming that all potassium in the groundwater is taken up by the montmorillonite, the bentonite buffer can still be expected to carry out its barrier role for more than 1 Ma.

Even if it was the case that the potassium supply was unlimited, then the rate of increase in layer charge may be assumed to control illite formation. Illitization in this case is strongly kinetically hindered and is difficult to study in the laboratory under relevant conditions. It has been stated that illitization does not occur at all below 60°C and is negligibly slow below 100°C (Anderson 1983). Although it was noted in Nagra (1994a) that such an extrapolation is not easy to defend due to the large degree of uncertainty regarding the temperature dependence of the activation energy (E_a) of illitization, data from natural analogue studies (Yusa et al. 1991) have increased confidence in the laboratory data, producing E_a values of 27 kcal mol^{-1}, close to the 30 kcal mol^{-1} of Roberson & Lahann (1981).

Nevertheless, several open questions remain on the long-term behaviour of the bentonite buffer and so Nagra has considered natural analogue studies of montmorillonite-rich systems of relevant ages to lend support to the predictions. On the question of the likely degree of illitization of montmorillonite, studies of diagenetic illitization in the clays of the Gulf of Mexico (e.g. Eberl & Hower 1976; Roberson & Lahann 1981; Müller-Vonmoos & Kahr 1985; Freed & Peacor 1989) and elsewhere (Pusch & Karnland 1988) indicate that the rate of illitization in nature is much slower than predicted by kinetic models (e.g. Anderson 1983) and that even if the supply of potassium was much higher than expected, complete illitization of the buffer material can be ruled out within the timescales of relevance to the performance assessment (about 1 Ma) for the given repository conditions.

Thermal effects. In certain repository designs, the canister/bentonite contact may reach several hundred degrees Celsius but, as noted above, in the Nagra concept the maximum temperature should peak at about 150°C and no more than a third of the total thickness of the bentonite should be exposed to temperatures in excess of 100°C for more than a few decades. During this heat pulse, some alteration of the bentonite may be expected: for example, the experimental work of Pusch & Karnland (1990) shows that between 130°C and 150°C all interlayer water is expelled and shrinkage cracks may form. However, Oscarson et al. (1990) report that such shrinkage cracks are reversible and heal following resaturation, as would be the case in the Nagra repository following the short, initial heat pulse.

The effect could be significant only if the CEC of the montmorillonite is saturated with potassium during the maximum temperature pulse (temperatures >110°C are necessary) but this is excluded due to the low potassium flux at the proposed site (see comments above). In addition, the heat pulse will, in any case, coincide with the period of resaturation of the compacted bentonite following repository closure (i.e. there is unlikely to be significant groundwater in the high temperature zone of the bentonite at that time).

It is, however, possible that water vapour may exist in the bentonite if a significant thermal gradient still exists during resaturation. Interaction of the vapour with the compacted bentonite has been observed to cause cementation with an associated loss of some of the swelling capacity (Couture 1985). This has been studied further in the laboratory (eg Meike 1989; Pusch & Karnland 1990; Pusch et al. 1992) and, while significant cementation can occur within the first few centimetres of the heat source, compacted bentonite outwith this zone shows no appreciable change in properties.

The effect can be minimized by careful choice of appropriate densities of the dry bentonite, so ensuring the remnant swelling capacity is sufficient to ensure low permeability of the bentonite (see Grauer 1990). Interestingly, the cementation appears to be caused by secondary sulphates in the hottest (near canister) zone, followed by amorphous silicates further away from the canister, with no evidence of secondary iron phases resulting from bentonite/canister interaction (Pusch et al. 1992). This work backs up that of Müller-Vonmoos et al. (1991) who carried out short-term (i.e. six to seven months' duration) experiments with iron– and magnetite–bentonite mixtures at 80°C and found no significant changes to the bentonite. Nevertheless, canister/bentonite reactions could occur in theory (although insufficient thermodynamic data exist to fully assess the problem), leading to the production of the iron silicates chamosite, greenalite and nontronite (Grauer 1990). However, even assuming all of a canister reacted with the bentonite to form such phases, only a maximum of 20% of the montmorillonite in the buffer in the immediate vicinity would be affected and the changes to the total swelling capacity and sorption capacity would be insignificant (Nagra 1994a).

Heating effects have also been examined via natural analogue studies, the most detailed being that of Pusch et al. (1987) on smectite-rich clays (as an analogue for montmorillonite). The authors concluded that, despite high temperature cementation, a remnant smectite content of 15–25% will ensure a reasonable self-healing ability. Pusch & Karnland (1988) also noted that, at around 150°C, cementation can occur, work which has since been verified in the laboratory (Pusch & Karnland 1990). Further natural analogue studies of high temperature cementation in clays (in NW Scotland and

N Italy) are currently ongoing within the rad-waste programme of the European Community (the results of which will be published within 1999) and ENRESA has begun work on natural bentonites in southern Spain.

Finally, it is of note that ENRESA are also currently (the experiment officially began in November, 1996) carrying out a five year, full-scale field test of the near-field multi-barrier system in Nagra's underground test site at Grimsel in the Swiss Alps (McKinley *et al.* 1996; Huertas & Santiago 1998). It is hoped that this experiment and a complementary 2/3 scale laboratory study, both of which use heaters to mimic the heat flux from a real canister, will be able to confirm some of the small-scale, shorter-term laboratory experiments (e.g. on the effects of water vapour on the compacted bentonite) and act as a bridge between the short-term laboratory data and the information obtained from natural analogues.

Physical perturbations. Although this paper addresses mainly chemical processes, there are a couple of physical perturbations of the bentonite buffer which are worth briefly mentioning: canister sinking and gas disruption.

If the bentonite is to be an effective part of the EBS, it must completely surround the canister over a very long period of time. If the canister were to sink through the bentonite, under gravity, the full effectiveness of the bentonite could be compromised. In Kristallin-1, it was calculated, on the basis of the short-term laboratory data of Börgesson *et al.* (1988), that the canister would sink no more than 1–5 mm in 10 000 years, an amount that is negligible for the safety of the repository. However, potential shortcomings in the calculations, including use of a model developed to study creep in soils and use of parameters from non-bentonitic clays (S. Horseman, pers. comm., 1994), led Smith & Curti (1995) to re-examine the question. Rather than attempt to re-calculate the extent of sinking using another model, their approach was simply to assume significant sinking (69 cm, or halfway to the tunnel floor) and calculate the likely impact on radionuclide release. For this scenario, they found that, for certain critical radionuclides, a maximum 20% increase in release from the near-field to the repository host rock was possible, resulting in negligible radiological consequences.

Nevertheless, due to the importance of the bentonite buffer to the near-field performance and lack of evidence that the short-term laboratory results can be safely extrapolated to repository timescales (it has so far not proved

possible to find an appropriate natural analogue to provide long-term data in support of the laboratory results, although ENRESA are currently examining a possible site in S Spain), design modifications (e.g. incorporating a sand-rich layer around the canister or placing stone supports under the canister) have been considered (Nagra 1994*a*; McKinley & Toyota 1998).

The production of hydrogen by anaerobic steel corrosion is the most significant source of gas in the near-field. Recent experimental work indicates a gas production rate of 0.02 to $0.2\,mol\,m^{-2}\,a^{-1}$ while the maximum loss of hydrogen from the canister surface by aqueous diffusion is approximately $0.02\,mol\,m^{-2}\,a^{-1}$ (Nagra 1994*a*). Thus it is possible that a free gas phase could be produced at the canister/bentonite interface and that this phase could be vented when the gas pressure reaches 30–70% of the bentonite swelling pressure (Pusch *et al.* 1985). The gas is presumably released by the creation of channels through the bentonite and this behaviour has been recreated by the model of Grindrod *et al.* (1994) which shows that the gas probably passes through connected capillary pores of 10 nm in diameter, consistent with the estimate of minimum effective diameter of connected porosity in the bentonite buffer (McKinley 1988).

If this is the case, no significant physical disruption of the bentonite seems likely (and thus no loss of the buffer's radionuclide retention properties). Certainly, laboratory experiments on the process in consolidated clays (e.g. DeCanniere *et al.* 1993; NEA, 1992) have indeed shown that the channels are difficult to detect afterwards other than by X-ray tomography. Nevertheless, it was mentioned in Kristellin-1 that the study of hydrogen evolution and potential disruption of the bentonite should be kept under review and it is worth noting that similar tests on compacted bentonite are currently underway in Japan (PNC 1993) and in the UK (funded by the so-called GAMBIT Club) to further assess the extent of the process.

Canister

The main perturbations associated with the massive steel canister are canister failure through corrosion, hydrogen gas production due to anaerobic corrosion and redox changes around the canister following canister failure (Nagra 1994*a*). Gas has already been treated above and the discussion here will concentrate on corrosion processes during the first 1000 a after repository closure (when complete containment is assured) and the role of corrosion products as redox buffers.

In Nagra's Projekt Gewähr safety assessment (Nagra 1985), three particular corrosion processes were considered to be of potential significance:

(1) Reaction of the canister with oxygen trapped in the EBS (mainly in the bentonite pore spaces) at the time of emplacement in the repository.

$$Fe + 1/2O_2 + H_2O \rightarrow Fe(OH)_2 \quad (1)$$

Per canister, approximately 9.3 kg of oxygen are available at repository closure (NWGCT 1984) and, assuming that all of the oxygen trapped reacts only with the canister (and not with, for example, pyrite in the bentonite) within the 1000 a period (discussed above), a maximum of 32 kg of iron (as cast steel) can be corroded. This corresponds to an average depth of corrosion over the entire canister of 0.7 mm.

(2) Reaction with dissolved sulphides produced from reduction of sulphates in the groundwater.

$$Fe + HS^- + H_2O \rightarrow FeS + H_2 + OH^- \quad (2)$$

When it is conservatively assumed that all sulphate available in the groundwater and the bentonite will be reduced to sulphide by bacteriological action and is thus available for corrosion (Nagra 1984), and that the reaction rate is extremely fast (rate limiting step assumed to be transport of corrodant to the container), then a maximum amount of 173 kg of steel per container will be corroded in the first 1000 years (NWGCT 1984). This corresponds to an average depth of corrosion over the entire canister of 3.8 mm. Re-examined today, these calculations are seen to be even more conservative than was realized at the time as improved deep groundwater sampling techniques indicate groundwater sulphate contents 3–6 times lower than stated in the original work (cf. NWGCT 1984; Pearson & Scholtis 1993).

(3) Anaerobic reaction with water.

$$3Fe + 4H_2O \rightarrow Fe_3O_4 + 4H_2$$

The long-term corrosion rate of iron in the anaerobic, sulphide-free, saline waters representative of the host rock at repository depth was calculated on the basis of a variety of laboratory tests (e.g. ONWI 1983; Marsh et al. 1983; Simpson 1983, 1984).

The experiments generally showed an initial high corrosion rate ($30–70 \mu m a^{-1}$) which rapidly decreased to $5–10 \mu m a^{-1}$ and is independent of the radiation doses expected at the glass surface in the Swiss design. Simpson (1984) concluded that a conservative corrosion rate, including an allowance for pitting, was $20 \mu m a^{-1}$, giving a maximum depth of 20 mm in 1000 years. In addition to this, NWGCT (1984) included a pitting allowance of a factor of two in the corrosion depths calculated for the reactions with oxygen and sulphide (i.e. 9 mm rather than 4.5 mm in 1000 years) giving a maximum conservative corrosion depth of 29 mm in the critical 1000-year period.

Canister stress and stability analyses have shown that a canister suffering pitting corrosion to a depth of 50 mm will still survive the maximum external isostatic pressure of 30 MPa (Steag & Motor Columbus 1985), indicating that the conservative assessment of 30 mm canister corrosion in 1000 years still leaves a high margin of safety in the design.

As an additional check on the long-term performance of the canister, Nagra has carried out natural and archaeological analogue studies of iron artefacts from a range of environments (see NWGCT 1984; Miller et al. 1994). Despite the fact that most material studied came from oxic to sub-oxic environments and would therefore be expected to corrode to a much greater extent than in the oxygen-free repository setting, a maximum corrosion depth of 10 mm in 1000 years has been calculated (Fig. 5), so increasing confidence in the results from the short-term experimental data and the Base Case assumption of 29 mm.

The Kristallin-1 assessment assumes that, following failure, the canister offers no retardation to radionuclides leaching out of the vitrified waste. In reality, of course, failure of the canister will be localized and the resultant crack or hole may be the only route from the glass to the bentonite. This will be a beneficial effect (compared to the assumptions made in Kristallin-1) as the crack will offer transport resistance to the leached radionuclides (Smith & Curti 1995), but a possible detrimental effect would be the release of oxidants produced by the radiolysis of water in contact with the vitrified waste. If the radiolytic oxidants were to pass through a crack in the canister which was coated by non-porous ferric oxides, so minimizing reaction with the bulk of the canister steel, the oxidants could penetrate to the bentonite so leading to a loss of reducing conditions in the bentonite pore waters

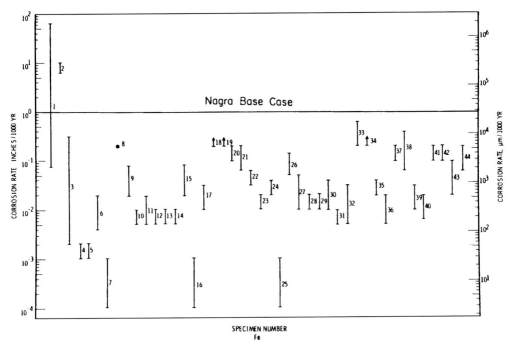

Fig. 5. Corrosion rate data from natural and archaeological analogue studies (after Miller *et al* 1994). The corrosion rates for the archaeological artefacts range from 0.1 to 10 μm a^{-1} (note that samples 1 and 2 are from oxidizing marine conditions; details of all other samples included in Johnson & Francis 1980). The Nagra base case corrosion rate for steel canisters from Projekt Gewähr is also shown for comparison.

and consequently to higher solubilities (see Wanner 1985; Berner 1994) and lower sorption (see Stenhouse 1995) for certain redox sensitive radionuclides.

This scenario has been examined by Smith & Curti (1995) who found that in the worst case (i.e. using high estimates of radiolytic oxidant production, low estimates of iron in the bentonite to soak up the oxidants etc.) the oxidation front would be unlikely to penetrate more than 20 mm into the bentonite buffer and so would not lead to any greater loss of radionuclides from the near-field. Even this assumption is arguably over-conservative on two grounds: first, this does not take into account the likely uptake of at least some of the radionuclide flux at the redox front in the bentonite. Such redox retardation at fronts is well documented in natural conditions (see the review of Hofmann 1999), a good example being the concentration of uranium (up to ore grade) in the so-called roll-front deposits which are so common in southern USA. Second, the canister itself will likely act as a redox buffer and can readily consume all possible oxidants produced by radiolysis of water by vitrified waste (McKinley 1985).

Waste-form

Vitrified waste. The Kristallin-1 safety assessment notes that the most likely perturbation of the glass would be the production of colloids, containing radionuclides, as the glass degrades/dissolves. However, as long as the bentonite buffer remains functional, the colloids should be filtered out in the very fine pore structure of the bentonite. Both laboratory and natural analogue data (reviewed in Lutze 1988 and Miller *et al.* 1994, respectively) suggest that the rate of devitrification of the glass is too slow to be a significant problem (although, arguably, the effect of radiation on long-term devitrification rates should be checked by appropriate natural analogue studies). Existing laboratory data suggest that the glass lifetime of 10^5 years assumed in Kristallin-1 is conservative and natural analogue studies indicate up to 10^7 years. However, differences in the chemistry, thermal history and radiation dose rates of natural glasses and vitrified waste means that, according to Miller *et al.* (1994), '...natural glasses offer no unambiguous evidence for the time at which devitrification begins, or the rate at which it proceeds, although the existence of natural glasses millions of years old may suggest that devitrification is not a problem'.

Spent fuel. Although several major assessments of spent fuel disposal concepts have been carried out (e.g. Anttila *et al.* 1982; KBS 1983; SKB 1992; Vieno *et al.* 1992), Nagra have carried out only preliminary assessments to date (Nagra 1985; Schneider *et al.* 1997). Other than the potential scenarios already discussed above for other components of the EBS which may have an impact on the spent fuel (e.g. redox front propagation through the bentonite buffer), the most significant potential perturbation identified so far relates to radiolytically induced dissolution of the fuel matrix (Forsyth 1995). Despite an extensive laboratory based data-set (e.g. Forsyth & Werme 1987, 1992; Bruno *et al.* 1995; Loida *et al.* 1995), the fact that much of the data have been produced using only model solutions means that uncertainties remain in the likely rate of radiolysis under near-field conditions (Johnson *et al.* 1994).

Consequently, the safety assessments carried out to date have been based on very conservative assumptions. For example, in the Finnish programme, the rate of radionuclide release from the spent fuel matrix is based on data measured under oxidizing conditions, even though more realistic experiments (with reducing conditions in the test vessel, if not at the fuel surface) show release rates which are a factor of ten lower (Vieno *et al.* 1992). The release of radionuclides from the gaps and the grain boundaries is generally modelled as being instantaneous following canister failure (e.g. Schneider *et al.* 1997), but there is some uncertainty in the radionuclide inventories at these locations, especially for high-burnup UO_2 and MOX fuels.

Although in all studies to date the calculated doses to the public remain below the respective national exposure limits, use of more realistic data would undoubtedly boost public confidence in the validity of the assessments. Once again, the problem is obtaining data which realistically can be extrapolated to repository conditions. To this end, many uranium ore bodies have been examined as natural analogues of spent fuel disposal (see review in Miller *et al.* 1994). Arguably, the most relevant of these studies has been the work carried out at the 1.3 Ga Cigar Lake ore body in Canada (see Cramer & Smellie 1994) and at the *c.* 2 Ga Oklo natural reactors in the Gabon (Blanc 1996). The conclusions of a recent reappraisal of the Cigar Lake study (Smellie & Karlsson 1996) were that 'Studies of water radiolysis at Cigar Lake showed that the net changes in the ore and near-field were

considerably overestimated by models currently used by performance assessment to calculate the yield of radiolysis products.' and 'These results appear to support the (*apparent*) inadequacy of current performance assessment radiolysis models as applied to Cigar Lake where the calculated oxidant production rates are significantly higher (*than was observed at the site*).'

Simple studies of the magnitude of the effects caused by radiolysis have been carried out also at Oklo (Curtis & Gancarz 1983). Much information is also available on the fate of radionuclides associated with the uranium ore 'fuel' (e.g. Gancarz *et al.* 1980; Curtis 1985; Brookins 1990; Blanc 1996). Some dissolution and elemental remobilisation occurred, but the fact that more than 90% of the uranium has remained in the same place since criticality and that little uranium ore/groundwater interaction has been observed under the present day conditions (as opposed to the much higher temperatures during criticality) lends qualitative support to the conclusions of the Cigar Lake study that the current performance assessment models of spent fuel dissolution are extremely conservative.

Other potential perturbations

Microbiology. The early ideas that a radwaste repository would provide too harsh an environment for microbial life were quickly proven wrong for both HLW (e.g. Mayfield & Barker 1982; West *et al.* 1982; West & McKinley 1985) and L/ILW (e.g. Francis *et al.* 1980; Francis 1982; McGahan 1987; West *et al.* 1992; Coombs *et al.* 1998). Nagra, among others (e.g. Stroess-Gascoyne 1989, in Canada; Colasanti *et al.* 1991, in the UK), has attempted to assess quantitatively the likely impact of microbes on the chemistry of the repository near-field (McKinley *et al.* 1985). Microbiological activity, with the associated production of organic complexants, could reduce the effectiveness of the EBS by increasing the solubility of radionuclides in the near-field (by organic complexation, for example) and decrease sorption on the buffer and canister.

In the case of a HLW repository, the relative simplicity of the EBS allows reasonable constraints to be placed on the maximum likely microbial activity, which will be tightly constrained by the low supply rates of nutrients and the low availability of usable energy. According to McKinley & Hagenlocher (1993), the maximum steady-state biomass production for a Swiss HLW repository in northern Switzerland would be 10^{-5} kg (dry) a^{-1} per waste package and, in the worst case, this biomass production would be balanced by a similar rate of production of organic complexants. This would be some three orders of magnitude more organic carbon than would be available from the groundwater. However, even when building worst case upon worst case by assuming that the organic carbon is present only as extremely strong specific complexants such as siderophores, that these are not filtered by the bentonite due to their very small size and that all the organics complex only with actinides in the waste, this would only approximately double the maximum release rate of actinides. Considering the other uncertainties involved in estimating near-field solubility limits, this is an insignificant increase.

One potential weakness in the above approach is that the effects of interfaces (pH/Eh) and heterogeneities in the EBS are ignored. This has been tackled by McKinley et al. (1997) who noted that their more thorough analysis '...gives a more realistic evaluation of possible microbial perturbations (of the HLW near-field) which indicate that the net effect of such populations may be positive, leading to decreased radionuclide release (due to uptake in and around the fronts). The critical assumption, however, is that microbes cannot utilise the very low energy density of the reaction between steel and water.' The conclusion that microbial activity could contribute to radionuclide retention is well supported by natural analogue studies which have shown that trace element traps, such as redox fronts or haloes, are established by microbially catalysed reactions (e.g. West et al. 1992; Hofmann 1998).

The extremely stable Cigar Lake uranium ore body also exists in an environment where relatively high microbial populations have been measured (Stroess-Gascoyne et al. 1994) and Smellie & Karlsson (1996) noted that the bacteria present are probably playing a role in stabilizing the ore via consumption of oxidants produced by radiolysis.

Groundwater disturbances. It was noted in the introduction that the geosphere must offer mechanical protection and hydrogeochemical stability to the EBS to ensure performance of the system. The potential HLW sites in Switzerland (Fig. 1) are felt capable of offering such protection (Nagra 1994a, b) but perturbations to the groundwater (reflecting changes in the regional hydrogeology and hydrogeochemistry) have to be examined (Nagra 1994a). For example, the potential effects of the next ice age could include a very thick permafrost layer which could displace shallow groundwater flow to deeper levels. This could bring younger, less evolved, groundwaters into contact with the EBS, so altering the near-field chemistry. Although it has been shown that such an effect could have a significant radiological impact on a site in the UK (Sumerling 1992), the geothermal gradient in the Swiss crystalline basement is such that it would prevent a permafrost layer getting deep enough to significantly alter flow paths (Nagra 1994b; Thury et al. 1994).

Other potential perturbations considered include the effects of long-term climatic change, increased surface erosion, hydrothermal activity and intrusion of saline groundwaters from the north Switzerland Permo-Carboniferous Trough into the repository zone. The low hydraulic conductivity of the crystalline basement and the presence of overlying aquifers means that the first two perturbations are extremely unlikely to make significant differences to the EBS, because the deep groundwater flux will remain low and the residence time long enough to ensure significant rock/water interaction to buffer any extremes in groundwater chemistry. The third, hydrothermal effects, can be avoided by careful siting (the probability of any significant activity at an unexpected site is, in any case, very low). In the fourth case, the intrusion of saline waters (maximum total dissolved solids of $13 \mathrm{g \, l^{-1}}$; Pearson et al. 1989) into the repository zone, the analysis in the Kristallin-1 report is unclear. If the concern is only about saline water intrusion, then the example of the Swedish and Canadian programmes, where the groundwater at the repository horizon will be saline, indicates no significant problems. However, brines (maximum total dissolved solids of $117 \mathrm{g \, l^{-1}}$; Pearson et al. 1989) are also present in the Permo-Carboniferous Trough and if these waters were to intrude into the repository horizon, then the effects could be more significant (the interaction of saline groundwaters and brines with compacted bentonite is currently being studied by SKB in Sweden and PNC in Japan).

Brines could alter the chemistry of the near-field pore fluids but, as noted above, would have little effect on Eh/pH which would remain in the reducing/alkaline range due to buffering by the canister and bentonite. Higher salinity may affect canister or waste-form corrosion rates, but there is a sufficiently large empirical database of experimental and natural analogue data to indicate that any increases are unlikely to be large enough to have a significant impact on releases. An increase in salinity could also influence elemental solubility; effects are expected to be relatively small, but the modelling approach used to support the selection of solubility data breaks down at ionic strengths of about 0.1 M. Alternative formalisms are available (e.g. the Pitzer approach) but appropriate databases are currently lacking. Nevetheless, empirical databases developed by the waste management programmes considering disposal in salt (e.g. Germany, USA), can be used to support the assumption of no major influences.

The final possible influence of increased salinity is on the transport barrier role of the bentonite. Empirical data indicate that the stability and pore structure (colloid filtration) will not be significantly altered by saline fluids. The retardation of some nuclides will be decreased (e.g. Muurinen et al. 1985) but would not be expected to fall below the conservative values currently used to assess repository performance.

As noted in the sensitivity analysis above, higher fluxes of water will not cause unacceptable degradation of the EBS performance as long as the barriers are not directly disturbed (releases of radionuclides are low even for a zero concentration boundary at the outer edge of the bentonite, corresponding to an infinitely high water flow). High flow rates of water could, however, potentially erode the bentonite which, as noted above, is a key component of the EBS. Major water fluxes through the crystalline basement are, however, localized in major fracture zones (Thury et al. 1994) which would be avoided during repository location and construction. The water flux in the smaller fractures and microfractures of the blocks of rock between such fracture zones is much less and has been assessed to be below critical values for erosion to be significant (Nagra 1994a). This conclusion is supported by extensive rheological studies of compacted bentonite and is currently undergoing laboratory tests in Japan (PNC 1993). Were erosion to be shown to be a problem, it could be alleviated by adding sand to the outer layers of compacted bentonite (Coons et al. 1987).

Relevance for other wastes

When compared to the disposal systems considered for other types of toxic waste, the HLW repository design presented here appears to be completely over-specified. The rationale for this over-design is primarily socio-political – reflecting the concern of the general public about radioactivity and the extremely high value of the product (electricity) relative to the low volume of waste produced for the case of nuclear power.

Nevertheless, the principles of defence in depth via a multi-barrier system and robust long-term performance of an engineered barrier system might be taken over to other waste management systems.

It is likely that custom-built, deep geological repositories could not be justified for most chemical/toxic wastes on financial grounds, but many potentially suitable cavities exist as a result of mining activities. For toxic wastes which do not decay, or decay only slowly on a geological timescale, the advantages of being able to isolate the containment system from surface/near-surface perturbations is considerable.

In a dry or low flow environment, very long lifetimes of engineered structures can be assumed if they are chosen to be compatible with the geochemistry of their surroundings. Multiple barriers can provide a low solute transport environment even in the event of localized perturbations of the host rock (e.g. using plastic clays) while chemistry is buffered in an Eh/pH range conducive to low release of waste components; for many metals, alkaline (pH 8–10), reducing conditions are particularly favourable. The use of micro-porous materials to ensure that a colloid source term does not exist is clearly advantageous.

Barriers will eventually degrade and waste will be released to the accessible environment. The EBS should delay such releases as long as possible for wastes which decay and, for all wastes, spread releases over as long a time period as possible (temporal dilution – e.g. due to low solubilities, sorption, transport resistance).

The geological barrier can provide further delays of release and temporal dilution due to retardation/dispersion processes, of which sorption and matrix diffusion are particularly important. Experience in the radwaste industry has shown, however, that characterization of the geological barrier in sufficient detail to confidently quantify such processes is particularly challenging. The hydraulic contrast between the host rock and surface aquifers, on the other hand, might be easier to predict with confidence and such dilution might be a valuable component of a safety case. This leads to the rather counter-intuitive conclusion that, for many waste types, although a dry environment (e.g. unsaturated zone in a desert) might have advantages from the pure safety viewpoint, it may be easier to demonstrate the safety of a wet, low permeability host rock which is overlain by a massive aquifer, as is planned in Switzerland among other countries.

Finally, it is worth noting that the natural analogue approach to building confidence in the long-term behaviour of a repository is of considerable importance in making a clear safety case for radwaste isolation systems. This is slowly being taken up by the chemo-toxic waste industry (see Brasser 1997; Côme 1997) and will surely gain pace in the coming years as the regulations for toxic waste disposal almost inevitably become more and more strict.

Conclusions

Although initially designed purely from the engineering viewpoint, the current design of the EBS of the Swiss HLW repository also utilizes chemical containment processes to retard releases of radionuclides to the far-field. In most cases, due to the decision to use only well understood materials in a robust design, the chemistry of the system is reasonably well understood. Arguably, the geochemical conditions in the far-field are less well understood, but this is less of a problem so long as the host-rock can be shown to provide the near-field with:

- mechanical protection
- adequately stable geochemical conditions
- sufficiently low groundwater fluxes past the near-field.

When this can be shown to be the case, then the chemical containment in the repository near-field can be seen to depend mainly on the bentonite buffer via:

- the bentonite pore water chemistry ensuring low solubility of radionuclides
- the pore structure of the bentonite filtering out colloids
- low hydraulic conductivity/strong sorption in the bentonite ensuring slow diffusion.

In terms of potential perturbations to the EBS, the greatest effects are those which have an impact on the bentonite. Careful siting of the repository will remove most, such as illitization due to a high potassium flux, and most others identified to date are unlikely to have a significant impact on repository performance. Some open questions remain on other potential perturbations, the answers to which would increase confidence in the performance of the repository. These include:

- a better evaluation of chemical profiles in the near-field with the associated development of a comprehensive model of solute transport in the near-field (along with laboratory/natural analogue verification). This will need, among other information, better data on the bento-

nite/water and canister/bentonite reactions to further assess the role of chemical fronts in the near-field
- further experimental verification of the assumed absence of colloid transport through compressed bentonite
- unambiguous experimental solubility measurements for selected radionuclides under near-field conditions
- fuller assessment of the likelihood of brines accessing the repository horizon. If this appears significant, then further study of the impact of brines on the bentonite buffer would be useful.
- the likely effect of hydrogen production due to canister corrosion: as noted in the text, the effects could be positive (gas build up stopping corrosion) or negative (escaping gas producing channels which disrupt the bentonite structure).

Lastly, it should be emphasised that the current design for the Swiss HLW disposal system is a 'first generation' design, aimed predominantly at concept demonstration. Considerable opportunities exist for optimization/improvement before repository construction in 2050 and ongoing experiments such as ENRESA's FEBEX will allow improvement in the design of the EBS materials and emplacement methodologies.

Although details are not transferable, the concepts and approaches described here may be relevant to other types of highly toxic waste. Such waste constitutes a much longer potential hazard in absolute terms and its management is certainly a much more urgent problem.

The authors would like to thank Nagra for permission to publish this work and A. Bath, M. Crawford and R. Metcalfe for constructive reviews which have improved the paper.

References

ALDER, J-C. 1997. Definition of waste types in Switzerland. *In*: Voss, J. W. (ed.) *International Radioactive Waste Management: a compendium of programs and standards*. WM Symposia Inc, Tucson.

ALEXANDER, W. R., GAUTSCHI, A. & ZUIDEMA, P. 1998. Thorough testing of performance assessment models: the necessary integration of *in situ* experiments, natural analogue studies and laboratory work. *Scientific Basis for Nuclear Waste Management*, **XXI**, 1013–1014.

ANDERSON, D. M. (ed.) 1983. *Smectite alteration*. Swedish Nuclear Fuel and Waste Management Co. (SKB) Technical Report **TR 83-03**, SKB, Stockholm, Sweden.

ANDREWS, R. W., LAFLEUR, D. W., PAHWA, S. B. 1986. *Resaturation of backfilled tunnels in granite*. Nagra Technical Report Series **NTB 86-27**, Nagra, Wettingen, Switzerland.

ANTTILA, M. *et al.* 1982. *Safety analysis of disposal of spent nuclear fuel*. Nuclear Fuel Waste Commission of Finnish Power Companies, Report **YJT 82-41**, Helsinki, Finland (in Finnish).

ATTINGER, R. & DUIJVESTIJN, G. 1994. *Mechanical behaviour of high-level nuclear waste overpacks under repository loading and during welding*. Nagra Technical Report Series **NTB 92-14**, Nagra, Wettingen, Switzerland.

BERNER, U. 1994. *Kristallin-1: Estimates of solubility limits for safety relevant radionuclides*. Nagra Technical Report Series **NTB 94-08**, Nagra, Wettingen, Switzerland.

BLANC, P. L. 1996. *Oklo – Natural analogue for a radioactive waste repository (Phase 1), Vol.1 – Acquirements of the project – Final Report*. European Commission Nuclear Science and Technology Report series, **EUR 16857/1EN**, Luxembourg.

BÖRGESSON, L., HOEKMARK, H. & KARNLAND, O. 1988. *Rheological properties of sodium smectite clay*. Swedish Nuclear Fuel and Waste Management Co. (SKB) Technical Report **TR 88-30**, SKB, Stockholm, Sweden.

BRASSER, T. 1997. Comparison of disposal and safety assessment methods for toxic and radioactive wastes with the view to natural analogue application to toxic wastes. *In*: VON MARAVIC, H. & SMELLIE, J. A. T. (eds) *CEC Nuclear Science and Technology Report*, **EN18734**. CEC, Luxembourg.

BROOKINS, D. G. 1990. Radionuclide behaviour at the Oklo nuclear reactor, Gabon. *Waste Management*, **10**, 285–296.

BRUNO, J., CASAS, I., CERA, E., DE PABLO, J., GIMÉNEZ, J. & TORRERO, M. E. 1995. Uranium (IV) dioxide and simfuel as chemical analogues of nuclear spent fuel matrix dissolution. A comparison of dissolution results in a standard NaCl/NaHCO$_3$ solution. *In*: MURAKAMI, T. & EWING, R. C. (eds) *Scientific Basis of Nuclear Waste Management*, **XVIII**, 601–608.

CHAPMAN, N. A. & MCKINLEY, I. G. 1987. The geological disposal of nuclear waste. Wiley, London.

—— & ——1990. Radioactive wastes: back to the future? *New Scientist*, **1715**, 54–58.

——, MCKINLEY, I. G. & SMELLIE, J. A. T. 1984. *The potential of natural analogues in assessing systems for deep disposal of high-level radioactive waste*. Nagra Technical Report Series **NTB 84-41**, Nagra, Wettingen, Switzerland.

COLASANTI, R., COUTTS, D., PUGH, S. Y. R. & ROSEVEAR, A. 1991. *Microbiology and radioactive waste disposal – Review of the Nirex research programme – January 1989*. Nirex Safety Studies **NSS/R131**, UK Nirex Ltd., Didcot, UK.

CÔME, B. 1997 *Natural analogues in toxic waste disposal: an example in the French context. In*: VON MARAVIC, H. & SMELLIE, J. A. T. (eds) *CEC Nuclear Science and Technology Report*, **EN18734**. CEC, Luxembourg.

—— & CHAPMAN, N. A. (eds) 1986. *Natural analogue working group; second meeting, Interlaken, June 1986*. CEC Nuclear Science and Technology Report, **EUR10671EN**, CEC, Luxembourg.

CONCA, J. L., APTED, M. & ARTHUR, R. 1993. Aqueous diffusion in repository and backfill environments. *Scientific Basis of Nuclear Waste Management*, **XVI**, 395–402.

COOMBS, P., GARDNER, S. J., ROCHELLE, C. A. & WEST, J. M. 1998. Natural analogue for geochemistry and microbiology of cement porewaters and cement porewater host rock/near-field interactions. *In*: LINKLATER, C. (ed.) *A natural analogue of cement-buffered groundwaters and their interaction with a repository host rock II*. Nirex Science Report, **S-98-003**. Nirex, Didcot, UK.

COONS, W., BERGSTRÖM, A., GNIRK, P. *et al.* 1987. *State of the art report on potential materials for sealing nuclear waste repositories*. Nagra Technical Report Series **NTB 87-33**, Nagra, Wettingen, Switzerland.

COUTURE, R. A. 1985. Steam rapidly reduces the swelling pressure of bentonite. *Nature*, **318**, 50–52.

CRAMER, J. J. & SMELLIE, J. A. T. (eds) 1994. *Final report of the AECL/SKB Cigar Lake Analogue Study*. AECL Tech. Rep. **AECL-10851**, Pinawa, Canada; Swedish Nuclear Fuel and Waste Management Co. (SKB) Tech. Rep. **TR 94-04**, Stockholm, Sweden.

CURTIS, D. B. 1985. *The chemical coherence of natural spent fuel at the Oklo nuclear reactors*. Swedish Nuclear Fuel and Waste Management Co. (SKB) Technical Report series, **TR 85-04**, Stockholm, Sweden.

CURTIS, D. B. & GANCARZ, A. J. 1983. *Radiolysis in nature: evidence from the Oklo natural reactors*. Swedish Nuclear Fuel and Waste Management Co. (SKB) Technical Report Series, **TR 83-10**, Stockholm, Sweden.

DECANNIERE, V., FIORAVANTE, P., GRINDROD, P., HOOKER, P., IMPEY, M. & VOLKAERT, G. 1993. *MEGAS: modelling and experiments on gas migration in repository host rocks*. CEC Nuclear Science and Technology Report, **EUR14816EN**, 153–164, CEC, Luxembourg.

EBERL, D. D. & HOWER, J. 1976. Kinetics of illite formation. *Bulletin of the Geological Society of America*, **87**, 1326–1330.

ELECTROWATT, 1996. *The 1994 UK radioactive waste inventory*. UK Nirex Ltd Report No. **695**, UK Nirex, Harwell, UK.

FORSYTH, R. S. 1995. *Spent nuclear fuel. A review of properties of possible relevance to corrosion processes*. Swedish Nuclear Fuel and Waste Management Co. (SKB) Technical Report series, **TR 95-23**, Stockholm, Sweden.

—— & WERME, L. O. 1987. *Corrosion tests on spent PWR fuel in synthetic groundwater*. Swedish Nuclear Fuel and Waste Management Co. (SKB) Technical Report series, **TR 87-16**, Stockholm, Sweden.

—— & ——1992. Spent fuel corrosion and dissolution. *Journal of Nuclear Materials*, **190**, 3–19.

FRANCIS, A. J. 1982. Microbial transformation of low-level radioactive waste. *Proceedings of the IAEA Symposium on environmental migration of long-lived radionuclides.* IAEA, Vienna, Austria. 415–429.

——, DOBBS, S. & NINE, F. J. 1980. Microbial leachates from shallow land low-level radioactive waste disposal sites. *Applied Environmental Microbiology,* **40**, 108–113

FREED, R. L & PEACOR, D. R. 1989. Variability in temperature of the smectite/illite reaction in Gulf sediments. *Clay Minerals,* **24**, 171–176.

GANCARZ, A. J., COWAN, G., CURTIS, D. & MAECK, W. 1980. Tc-99, Pb and Ru migration around the Oklo natural fission reactors. *Scientific Basis for Nuclear Waste Management,* **II**, 601–610.

GRAUER, R. 1985. *Synthesis of recent investigations on corrosion behaviour of radioactive waste glasses.* Nagra Technical Report Series **NTB 85-27**, Nagra, Wettingen, Switzerland.

——1990. *The chemical behaviour of montmorillonite in a repository backfill – selected aspects.* Nagra Technical Report Series **NTB 88-24E**, Nagra, Wettingen, Switzerland.

GRINDROD, P., IMPEY, M., SADDIQUE, S. & TAKASE, H. 1994. Saturation and gas migration within clay buffers. *Proc. 5th Int. Conf. High-Level Rad. Waste. Manag.* Las Vegas.

HOFMANN, B. A. 1999. *Geochemistry of natural redox fronts – a review.* Nagra Technical Report Series **NTB 99-05**, Nagra, Wettingen, Switzerland.

HUERTAS, F. & SANTIAGO, J. L. 1998. The FEBEX project: general overview. *Scientific Basis for Nuclear Waste Management,* **XXI**, 343–349.

IAEA 1981. *Underground disposal of radioactive waste: Basic guidance, recommendations.* International Atomic Energy Agency Safety Series **No. 54**, IAEA, Vienna, Austria.

JOHNSON, A. B. & FRANCIS, B. 1980. *Durability of Metals from Archaeological Objects, Metal Meteorites and Native Metals.* Batelle Pacific Northwest Laboratory, PNL-3198, USA.

JOHNSON, L. W., SHOESMITH, D. W. & TAIT, J. C. 1994. *Release of radionuclides from spent LWR UO2 and mixed-oxide (MOX) fuel.* Nagra Technical Report series, **NTB 94-40**, Nagra, Wettingen, Switzerland.

KBS 1983. [*Final storage of spent nuclear fuel –KBS-3].* Swedish Nuclear Fuel and Waste Management Co. (SKB), Stockholm, Sweden [in Swedish].

KUROSAWA, S., YUI, M. & YOSHIKAWA, H. 1997. Experimental study of colloid filtration by compacted bentonite. *Scientific Basis for Nuclear Waste Management,* **XX**, 963–970.

LOIDA, A., GRAMBOW, H., GECKEIS, H. & DRESSLER, P. 1995. Processes controlling radionuclide release from spent fuel. *Scientific Basis for Nuclear Waste Management,* **XVIII**, 577–584.

LUTZE, W. 1988. Silicate glasses. *In:* LUTZE, W. & EWING, R. C. (eds) *Radioactive Waste Forms for the Future.* North-Holland, Amsterdam.

—— & EWING, R. C. (eds) 1988. *Radioactive Waste Forms for the Future.* North-Holland, Amsterdam.

MARSH, G. P., BLAND, I. W., DESPORT, J. A., NAISH, C., WESTCOTT, C. & TAYLOR, K. J. 1983. Corrosion assessment of metal overpacks for radioactive waste disposal. *Nuclear Science and Technology,* **5**, 223–252.

MAYFIELD, C. I. & BARKER, J. F. 1982. *An evaluation of microbiological activities and possible consequences in a fuel waste disposal vault – A literature review.* Atomic Energy of Canada Ltd. Report **TR-139**, Chalk River, Canada

McCOMBIE, C., ZUIDEMA, P. & McKINLEY, I. G. 1991. Sufficient validation – the value of robustness in performance assessment and system design. *Proc. OECD GEOVAL Conference 1991,* NEA/OECD, Paris.

McGAHAN, D. J. 1987. *Survey of microbiological effects in low-level waste disposal to land.* Atomic Energy Research Establishment Report **AERE R 12477**, Harwell, UK.

McKINLEY, I. G. 1985. *The geochemistry of the near-field.* Nagra Technical Report Series **NTB 84-48**, Nagra, Wettingen, Switzerland.

——1988. *The near-field geochemistry of HLW disposal in an argillaceous host rock.* Nagra Technical Report Series **NTB 88-26**, Nagra, Wettingen, Switzerland.

——1991. Integrated near-field assessment. *Proc. from the technical workshop on near-field performance assessment for high-level waste, Madrid 1990,* Swedish Nuclear Fuel and Waste Management Co. (SKB) Technical Report series **91-59**, Stockholm, Sweden.

—— & ALEXANDER, W. R. 1992. Constraints on the applicability of *in situ* distribution coefficient values. *Journal of Environmental Radioactivity,* **15**, 19–34

——1993a. Assessment of radionuclide retardation: uses and abuses of natural analogue studies. *Journal of Contaminant Hydrology,* **13**, 249–259.

——1993b. Comments on responses to our paper 'Assessment of radionuclide retardation: uses and abuses of natural analogue studies'. *Journal of Contaminant Hydrology,* **13**, 271–275.

——1996. On the incorrect derivation and use of *in situ* retardation factors from natural isotope profiles. *Radiochimica Acta,* **74**, 263–267.

—— & HAGENLOCHER, I. 1993. Quantification of microbial activity in a nuclear waste repository. *Proceedings of the International Symposium on Sub-Surface Microbiology (ISSM-93) 19–24 September 1993,* Bath, UK.

—— & TOYOTA, M. 1998. Increasing the transparency of the safety case during optimisation of repository design. *Proceedings of the Waste Management '97 Conf. in Tucson, USA.* WM Symposia Inc., Tucson.

——, WEST, J. M. & GROGAN, H. A. 1985. *An analytical overview of the consequences of microbial activity in a Swiss HLW repository.* Nagra Technical Report series, **NTB 85-43**, Nagra, Baden/Wettingen, Switzerland.

——, KICKMAIER, W., DEL OLMO, C. & HUERTAS, F. 1996. The FEBEX project: full-scale simulation of engineered barriers for a HLW repository. Nagra Bulletin No. 27, 56–67, June 1996, Nagra, Wettingen, Switzerland.

——, HAGENLOCHER, I., ALEXANDER, W. R. & SCHWYN, B. 1997. Microbiology in nuclear waste disposal: interfaces and reaction fronts. *FEMS Microbiology Reviews*, 20, 545–556.

——, ALEXANDER, W. R., GAUTSCHI, A. & WABER, N. H. 1999. An approach to validation of solubility databases for performance assessment. *Radiochim. Acta*, 87, 412–418.

MEIKE, A. 1989. *Transmission electron microscope study of illite/smectite mixed layers.* Nagra Technical Report Series NTB 89-03, Nagra, Wettingen, Switzerland.

MILLER, W., ALEXANDER, R., CHAPMAN, N., McKINLEY, I. & SMELLIE, J. 1994. *Natural analogue studies in the geological disposal of radioactive wastes.* Nagra Technical Report Series NTB 93-03, Nagra, Wettingen, Switzerland/ Studies in Environmental Science 57, Elsevier Science, Amsterdam, The Netherlands.

MÜLLER-VONMOOS, M. & KAHR, G. 1985. *Langzeitstabilität von Bentonit unter Endlagerbedingungen.* Nagra Technical Report Series NTB 85-25, Nagra, Wettingen, Switzerland.

——, BUCHER, F., KAHR, G., MADSON, F. & MAYOR, P. A. 1991. *Untersuchungen zum Verhalten von Bentonit in Kontakt mit Magnetit und Eisen unter Endlagerbedingungen.* Nagra Technical Report Series NTB 91-14, Nagra, Wettingen, Switzerland.

MUURINEN, A., RANTANEN, J. & PENTILLÄ-HILTUNEN, P. 1985. Diffusion mechanism of strontium, cesium and cobalt in compacted bentonite. *Scientific Basis for Nuclear Waste Management*, IX, 617–624.

NAGRA 1984. *An assessment of the corrosion resistance of the high-level waste containers proposed by Nagra.* Nagra Technical Report Series NTB 84-32, Nagra, Wettingen, Switzerland.

——1985. *Projekt Gewähr 1985. vols. 1–8 (vol. 9 English summary).* Nagra Gewähr Report Series NGB 85-01/09. Nagra, Wettingen, Switzerland.

——1989. *Sediment study – intermediate report 1988. Executive summary.* Nagra Technical Report Series NTB 88-25E, Nagra, Wettingen, Switzerland.

——1994a. *Kristallin-1. Safety assessment report.* Nagra Technical Report Series NTB 93-22, Nagra, Wettingen, Switzerland.

——1994b. *Conclusions from the regional investigation programme for siting a HLW repository in the crystalline basement in northern Switzerland.* Nagra Technical Report Series NTB 93-09E, Nagra, Wettingen, Switzerland.

NEA 1992. Gas generation and release from radioactive waste repositories. *Proceedings of a workshop, Aix-en-Provence, 23–26 September 1991.* NEA/OECD, Paris.

NERETNIEKS, I. 1985. *Some aspects of the use of iron canisters in deep lying repositories for nuclear wastes.* Nagra Technical Report Series NTB 85-35, Nagra, Wettingen, Switzerland.

NWGCT (Nagra Working Group on Container Technology) 1984. *An assessment of the corrosion resistance of the high-level waste containers proposed by Nagra.* Nagra Technical Report Series NTB 84-32, Nagra, Wettingen, Switzerland.

ONWI 1983. *Engineered waste package conceptual design. Appendix B: corrosion analysis* Office of Nuclear Waste Isolation Report, ONWI-438, Columbus.

OSCARSON, D. W., DIXON, D. A. & GRAY, M. N. 1990. Swelling capacity and permeability of an unprocessed and a processed bentonitic clay. *Engineering Geology*, 28, 281–295.

OVE ARUP 1990. *COMPAS project: stress analysis of HLW containers, intermediate testwork.* CEC Nuclear Science and Technology Report, EUR12593EN, CEC, Luxembourg.

PEARSON, F. J. & SCHOLTIS, A. 1993. *Chemistry of reference waters of the crystalline basement of northern Switzerland for safety assessment studies.* Nagra Technical Report Series NTB 93-07, Nagra, Wettingen, Switzerland.

——, LOLCAMA, J. L. & SCHOLTIS, A. 1989. *Chemistry of waters in the Böttstein, Weiach, Riniken, Schafisheim, Kaisten and Leuggern boreholes.* Nagra Technical Report Series NTB 86-19, Nagra, Wettingen, Switzerland.

PNC 1993. *Proceedings of an international workshop on research and development of geological disposal.* 15–18 Nov., 1993, Tokai, PNC, Tokyo, Japan.

PUSCH, R. 1980. *Water uptake, migration and swelling characteristics of saturated and unsaturated, highly compacted bentonite.* Swedish Nuclear Fuel and Waste Management Co. (SKB), KBS TR 80-11, Stockholm, Sweden.

—— & KARNLAND, O. 1988. *Preliminary report on longevity of montmorillonite clay under repository-related conditions.* Swedish Nuclear Fuel and Waste Management Co. (SKB) Technical Report TR 88-26, Stockholm, Sweden.

——1990. *Geological evidence of smectite longevity. The Sardinian and Gotland cases.* Swedish Nuclear Fuel and Waste Management Co. (SKB), Technical Report TR 90-44, Stockholm, Sweden.

——, RANHAGEN, L. & NILSSON, K. 1985. *Gas migration through MX-80 bentonite.* Nagra Technical Report Series NTB 85-36, Nagra, Wettingen, Switzerland.

——, BÖRGESSON, L. & ERLSTRöM, M. 1987. *Alteration of isolating properties of dense smectite clay in repository environment as exemplified by seven Pre-Quaternary clays.* Swedish Nuclear Fuel and Waste Management Co. (SKB), Technical Report TR 87-29, Stockholm, Sweden.

——, KARNLAND, O., LAJUDIE, A. & ATABEK, R. 1992. Hydrothermal field experiment simulating steel canister embedded in expansive clay – physical behaviour of the clay. *Scientific Basis for Nuclear Waste Management*, XV, 547–556.

ROBERSON, H. E. & LAHANN, R. W. 1981. Smectite to illite conversion rates: effects of solution chemistry. *Clays and Clay Minerals*, 29, 129–135.

ROMERO, L., MORENO, L. & NERETNIEKS, I. 1995. Movement of a redox front around a repository for HLW. *Nuclear Technology*, **110**, 238–249.

SASAKI, T., ANDO, K., KAWAMURA, H., SCHNEIDER, J. W. & McKINLEY, I. G. 1997. Thermal analysis of options for spent fuel disposal in Switzerland. *Scientific Basis for Nuclear Waste Management*, **XX**, 1133–1140.

SCHNEIDER, J., ZUIDEMA, P., SMITH, P., GRIBI, P. & NIEMEYER, M. 1997. Preliminary calculations of radionuclide release and transport from a repository for spent nuclear fuel in Switzerland. *Proceedings of the Waste Management '97 Conf. in Tucson, USA*. WM Symposia Inc., Tucson.

SIBLEY, T. H. & MYTTENAERE, C. 1986. *Application of distribution coefficients to radiological assessment models*. Elsevier, Amsterdam.

SIMPSON, J. P. 1983. *Experiments on container materials for Swiss HLW disposal projects. Part I*. Nagra Technical Report Series **NTB 83-05**, Nagra, Wettingen, Switzerland.

——1984. *Experiments on container materials for Swiss HLW disposal projects. Part II*. Nagra Technical Report Series **NTB 84-01**, Nagra, Wettingen, Switzerland.

SKB 1992. *Final disposal of spent nuclear fuel – Importance of the bedrock for safety*. Swedish Nuclear Fuel and Waste Management Co. (SKB), Technical Report series **TR 92-20**, Stockholm, Sweden.

——1993. *Project on Alternative System Study (PASS)*. Final report. Swedish Nuclear Fuel and Waste Management Co. (SKB), Technical Report series **TR 93-04**, Stockholm, Sweden.

SMELLIE, J. & KARLSSON, F. 1996. *A re-appraisal of some Cigar Lake issues of importance to performance assessment*. Swedish Nuclear Fuel and Waste Management Co. (SKB), Technical Report series **TR 96-08**, Stockholm, Sweden.

SMITH, P. A. & CURTI, E. 1995. *Some variations of the Kristallin-1 near-field model*. Nagra Technical Report Series **NTB 95-09**, Nagra, Wettingen, Switzerland.

——, ZUIDEMA, P. & McKINLEY, I. G. 1995. *Assessing the performance of the Nagra HLW disposal concept*. Nagra Bulletin **No. 25**, March 1995. Nagra, Wettingen, Switzerland.

STEAG & MOTOR COLUMBUS 1985. *Behälter aus Stahlguss für die Endlagerung verglaster hoch-radioaktiver Abfälle*. Nagra Technical Report Series **NTB 84-31**, Nagra, Wettingen, Switzerland.

STENHOUSE, M. J. 1995. *Sorption databases for crystalline, marl and bentonite for performance assessment*. Nagra Technical Report Series **NTB 93-06**, Nagra, Wettingen, Switzerland.

STROES-GASCOYNE, S. 1989. *The potential for microbial life in a Canadian high-level fuel waste disposal vault – A nutrient and energy source analysis*. Atomic Energy of Canada Ltd. report **TR-9574**, Whiteshell, Canada.

——, FRANCIS, A. J. & VILKS, P. 1994. Microbial research. *In:* CRAMER, J. J. & SMELLIE, J. A. T. (eds) *Final report of the AECL/SKB Cigar Lake analogue study*. Swedish Nuclear Fuel and Waste Management Co. (SKB), Technical Report series **TR 94-04**, Stockholm, Sweden.

SUMERLING, T. J. (ed.) 1992. *Dry run 3: A trial assessment of underground disposal of radioactive wastes based on probabilistic risk analysis: Overview*. UK DoE Report **DOE/HMIP/RR92.039** (and 9 supporting volumes), HMID, London.

THURY, M., GAUTSCHI, A., MAZUREK, M. *et al.* 1994. *Geology and hydrogeology of the crystalline basement of Northern Switzerland – Synthesis of regional investigations 1981–1993 within the Nagra radioactive waste disposal programme*. Nagra Technical Report Series **NTB 93-01**, Nagra, Wettingen, Switzerland.

VIENO, T., HAUTOJäRVI, A., KOSKINEN, L. & NORDMAN, H. 1992. *Safety analysis of spent fuel disposal*. Nuclear Fuel Waste Commission of Finnish Power Companies, Report, **YJT 92-33E**, Helsinki, Finland.

WANNER, H. 1985. Modelling radionuclide speciation and solubility limits in the near-field of a deep repository. *Scientific Basis for Nuclear Waste Management*, **IX**, 509–516.

WEST, J. M. & McKINLEY, I. G. 1985. The geomicrobiology of nuclear waste disposal. *Scientific Basis for Nuclear Waste Management*, **VII**, 487–494

——, —— & CHAPMAN, N. A. 1982. Microbes in deep geological systems and their possible influence on radioactive waste disposal. *Radioactive Waste Management in the Nuclear Fuel Cycle*, **3**, 1–15

——, —— & VIALTA, A. 1992. Microbiological analysis at the Pocos de Caldas natural analogue study sites. *Journal of Geochemical Exploration*, **45**, 439–449.

WITHERSPOON, P. A. (ed.) 1991. *Geological problems in radioactive waste isolation*. LBNL Report **29703**, Lawrence Berkeley National Laboratory, University of California.

—— (ed.) 1996. *Geological problems in radioactive waste disposal: second worldwide review*. LBNL Report **38915**, Lawrence Berkeley National Laboratory, University of California.

YU, J-W & NERETNIEKS, I. 1997. *Diffusion and sorption properties of radionuclides in compacted bentonite*. Swedish Nuclear Fuel and Waste Management Co. (SKB), Technical Report series **TR 97-12**, Stockholm, Sweden

YUSA, Y., KAMEI, G. & ARAI, T. 1991. Some aspects of natural analogue studies of long-term durability of engineered barrier materials – recent activities at PNC Tokai, Japan. *In:* COME, B. & CHAPMAN, N. (eds) *Natural analogues in radioactive waste disposal*. CEC Nuclear Science and Technology Report, **EUR11037EN**, CEC, Luxembourg, 153–164.

Characterizing the chemical containment properties of the deep geosphere: water–rock interactions in relation to fracture systems within deep crystalline rock in the Tono area, Japan

T. IWATSUKI & H. YOSHIDA

Japan Nuclear Cycle Development Institute (JNC), Tono Geoscience Center, 959–31, Jorinji, Izumi, Toki, Gifu, 509–51, Japan

Abstract: A water/rock interaction study has been integrated with interpretations of geological structures, in order to understand the geochemical conditions in deep granitic rock at the Tono study site, central Japan. Geological investigations show that fracture systems in the granitic rock can be classified into: intact zones; moderately fractured zones; and intensely fractured zones. This classification is based upon the frequency and width of fractures and fractured zones. Isotopic analyses of the groundwaters indicate that the groundwater at SL 80 m (SL: sea level) has recharged within the last 30 years. Petrological and mineralogical observations were undertaken, and coupled with theoretical calculations using the PHREEQE geochemical computer code. The results of this approach suggest that the chemical evolution of the groundwater is generally controlled by water–rock interactions involving plagioclase, clay minerals and ferric hydroxide in the intact zones and moderately fractured zones. The geochemical condition of the groundwater is correlated closely with the nature of the fracture systems, mineral compositions and water–rock interaction processes. Different fractures contain chemically distinct groundwaters, which have differing potential capacities to mobilize radionuclides. The investigation method, based on a comparison between geological structures and groundwater chemistry, can be applied to develop a realistic hydrogeochemical model for deep, fractured granitic rocks. Such an approach is necessary if the chemical containment potential of such lithologies is to be evaluated adequately. The work demonstrates that structural controls on chemical heterogeneities in groundwaters should be understood, before the disposal of redox-sensitive wastes can be undertaken safely in fractured crystalline rocks.

In any scenario for the disposal of radioactive waste, the waste will be contained within the geosphere partly by physical processes, and partly by chemical processes. If safe disposal options are to be designed for these wastes, the relative importance of the physical and chemical processes, and the inter-relationships between them, must be understood.

The chemical conditions of the groundwaters within the geosphere are an important control on the migration of radionuclides, and hence the ability of the geosphere to contain radioactive waste. However, these chemical conditions reflect the origins of the water, and the water–rock interactions that occur along groundwater flow paths. Therefore, the chemical conditions of the water are related to physical processes that control the geometries of these flow paths, and the rates of flow. Consequently, to evaluate the relative importance of physical and chemical containment processes, it is necessary to investigate the relationships between the physical hydrogeological characteristics of a groundwater system, and a groundwater's chemical evolution as it flows through the system. To develop this understanding, detailed hydrogeological and hydrogeochemical data are required.

In this paper, we describe various generic geochemical studies, aimed at developing techniques to acquire such detailed data. These studies have been conducted at the Tono research site in central Japan, as part of an on-going research and development programme aimed at understanding the deep geological environment. Specifically, the study aims to build a reliable conceptual model concerning geochemical conditions, by identifying relationships between the geological structure and groundwater chemistry.

From: METCALFE, R. & ROCHELLE, C. A. (eds) 1999. *Chemical Containment of Waste in the Geosphere.* Geological Society, London, Special Publications, **157**, 71–84. 1-86239-040-1/99/$15.00 © The Geological Society of London 1999.

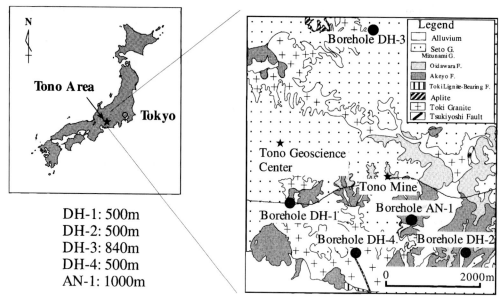

Fig. 1. Simplified maps illustrating the study area.

Geology of the study area

The study area is located in central Japan (Fig. 1), and occupies approximately 5 km × 5 km. In this area, Tertiary sedimentary rocks unconformably overlie Toki granitic basement rock (which has an age between 60 and 100 Ma; Katayama *et al.* 1974). The sedimentary rocks are composed of the Mizunami Group (15–20 Ma) and the Seto Group (0.7–5 Ma). The thickness of the sedimentary rocks is less than 200 m. The Toki granitic basement lithologically consists of medium- to coarse-grained biotite granite and medium-grained hornblende-biotite porphyry. Dyke rocks of quartz porphyry and aplite partly intrude into the granitic rock.

Previous studies of groundwater chemistry in the overlying sedimentary rocks

Previous studies have investigated the groundwater chemistry in the sedimentary rocks (e.g. Mizutani *et al.* 1992; Iwatsuki *et al.* 1995). The waters in these rocks have been classified according to the variation in Na^+, Ca^{2+}, and HCO_3^- concentrations. The dissolution of calcite and ion exchange between the clay minerals and groundwater have been identified as dominant reactions in the evolution of the groundwater (Iwatsuki *et al.* 1995). The isotopic data indicate

that the groundwater is of meteoric origin. The carbon-14 ages of the groundwater in the lower part of the sedimentary rocks range between 13 000 and 15 000 years (Mizutani *et al.* 1992).

Method

Geochemical investigations have been conducted in five boreholes (AN-1, DH-1, 2 3, 4). The locations of these boreholes were chosen to test the conceptual model that has been developed for the regional hydrogeological conditions, the structural geology and the groundwater chemistry. The geological structure, including the distribution of fractures in the rock mass, has been studied to aid an analysis of the groundwater flow path's geometry. Basically, the structural geology and fracture systems have been investigated by examining drill cores. Major structures within the granitic rocks were described from a geological and mineralogical point of view. The structural descriptions were used to choose groundwater sampling points located on fractures and fractured zones. An isotopic study of the groundwater has also been carried out to constrain the recharge time of the groundwater. Microscopic observations on rock samples, in combination with theoretical calculations, have been carried out to constrain the dominant water/rock interactions.

Analysis of geological structure in the Toki Granite

To predict the geochemical conditions of the groundwaters, it is important to characterize the geometry of the fracture system and to characterize the fracture-filling minerals. To understand the fracture systems of the Toki granite basement rocks in the Tono area, observations were made on core samples from the fractured and the intact zone, and geophysical logging was undertaken in all boreholes. A three-dimensional model for the distribution of fractures and fractured zones was then developed. Core logging, optical microscopy, X-ray diffractometry and SEM-EDS analysis were used to determine the characteristics of the fractures. The fracture parameters that were measured included: their lengths; their widths; the nature of wallrock alteration; and the identities of fracture filling minerals.

Geochemical investigation of groundwater from the Toki Granite

Groundwaters from two boreholes, DH-3 and DH-4, were sampled for analysis. These boreholes were drilled using fresh water as a drilling fluid, thereby minimizing the contamination of the groundwater samples that were collected. A multi-packer apparatus, incorporating a multi-piezometer monitoring system with several sampling ports, was used to sample groundwater from borehole DH-3. Eleven pairs of packers were set in the fractured and the intact zones, the packer spacing ranging from several metres to several tens of metres (Fig. 2). The drilling fluids in each sampling point were removed by pumping out fluid equivalent to several volumes of the borehole across each sampling interval. Sodium fluorescein (uranine) was added to the drilling fluids to allow the degree of contamination of the groundwater samples by drilling fluids to be determined quantitatively. In borehole DH-4, a double packer system was also used on a fractured rock zone (Fig. 2). These packer assemblies were able to collect the groundwater samples whilst maintaining the *in situ* water pressure (maximum pressure is several tens of kilogram f/cm^2) and minimizing perturbations to *in situ* chemical conditions.

Groundwater samples were collected from borehole DH-3 at 148 m above sea level (DH-3/148 m), 26 m above sea level (DH-3/26 m), 484 m below sea level (DH-3/−484 m). Groundwaters were sampled from borehole DH-4 at 80 m above sea level (DH-4/80 m). Chemical compositions

and physico-chemical parameters were determined, to characterize the chemistry of groundwaters in the fractures and fractured zones. The compositions were obtained by ion chromatography, atomic absorption spectrometry and inductive coupled plasma spectrometry. Isotopic compositions of groundwater samples (δD, $\delta^{18}O$, and tritium) were also determined to constrain the origin and age of the groundwater. The waters' saturation states with respect to various mineral phases, represented by saturation indices (SI), were then calculated using the computer code PHREEQE (Parkhurst *et al.* 1980).

Results

Geological structure

The observed frequency and width of fractures and fractured zones were used to classify the structure of the granitic rock into: intact zones; moderately fractured zones; and intensely fractured zones (Fig. 3).

The core from intact zones has several single fractures in a metre, and the wall rocks of each fracture contain altered plagioclase; the surrounding matrix is not altered. The fracture surfaces are sealed with a thin layer of clay minerals which replaces most of the plagioclase in these surfaces.

The rock quality designations (RQD; the total length of solid core pieces that have individual lengths more than 10 cm, measured axially between discontinuities) of both the moderately and the intensely fractured zones are nearly 0%. Each fracture and fractured zone has been characterized to determine its length, width and continuity and the nature of the filling minerals. These different zones have different types of wall rock alteration. The distributions of fractured zones are shown in Fig. 4.

The moderately fractured zones range from several metres to several tens of metres in width. The maximum thickness of an alteration zone is 50 m. The colour of the rock matrix is largely greenish grey, due to the occurrence of chlorite and montmorillonite. Some moderately fractured zones were seen in all boreholes (e.g. in boreholes DH-2, DH-3 and DH-4: at around SL −150 m). Several moderately fractured zones are associated with a pegmatitic texture.

The intensely fractured zones are associated with a strongly altered rock matrix. The maximum width of a zone is several metres. The intensely fractured zones are found at several points (e.g. DH-2/−280 m, DH-3/−200 m DH-4/−100, −200 m and AN-1/−270 m). The colour of

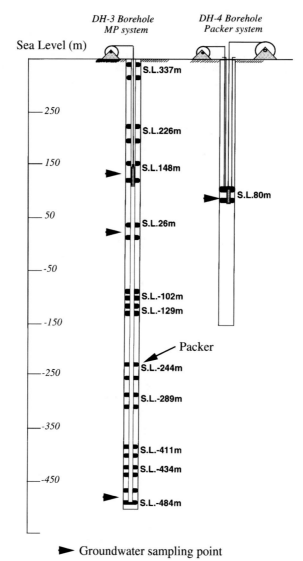

Fig. 2. Locations of the multi-packer (MP) systems and packer systems for groundwater sampling from boreholes DH-3 and DH-4.

the rock matrix is grey to white. The rock matrix shows some alteration to clay minerals such as kaolinite. The fracturing is commonly observed on both sides of these zones. The occurrence of the fractures suggests that the intensely fractured zones might be caused by faulting.

Mineralogy

The matrix of the granitic rock mainly consists of plagioclase, quartz, biotite and chlorite. Acces-

sory opaque minerals, pyrite and zircon also occur. Mineralogical data for fracture-filling minerals, as well as groundwater chemistry, can give information about the chemical conditions in the fractures and the fracture zones.

Iron hydroxides have precipitated in the microfractures of the minerals, notably quartz and plagioclase, in each borehole (Fig. 5). Some iron hydroxides are found around biotite and may have precipitated from groundwater containing Fe^{2+} dissolved from biotite. The maximum depth to which such iron hydroxide

A

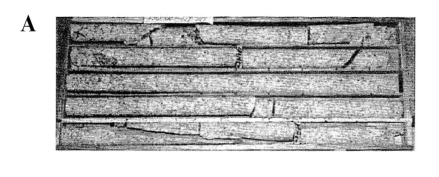

B

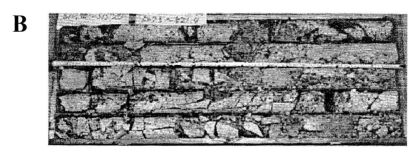

C

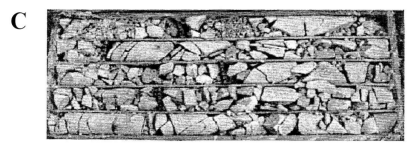

Fig. 3. Illustrations showing core in the three types of zone into which the granitic rock is classified on structural grounds. A, Intact zone (DH-3/226 m); B, moderately fractured zone associated with a pegmatitic texture (DH-3/−154 m); C, intensely fractured zone which might be caused by faulting (DH-4/−64 m).

precipitation is seen is 132 m below ground level.

The visual examination and analysis of the various minerals by SEM-EDS has revealed 'etch pits' on the mineral surfaces of pyrite crystals at DH-4/80 m, (Fig. 6) DH-3/−527 m. In contrast, the pyrite keeps its original form in the rock samples from DH-4/130 m (Fig. 6) and DH-3/−344 m. The etch pits have also been observed on silicate mineral surfaces in certain fractures and fractured zones (Fig. 6). Most of the silicate mineral surfaces are considered to be coated by stable clay minerals in the fractured zones. The predominant clay mineral of the fracture fillings in the moderately and

intensely fractured zone is montmorillo-nite, regardless of the depth.

Groundwater chemistry

Groundwater samples have been extracted from: moderately fractured zones at DH-3/148 m, DH-3/26 m and DH-4/80 m; and from an intact zone at DH-3/−484 m. The chemical composition of the groundwaters are summarized in Table 1. The waters from all the sampling points had low total dissolved solid (TDS) contents, up to 150 mg/l. The chemical composition of the groundwater in the moderately fractured zones

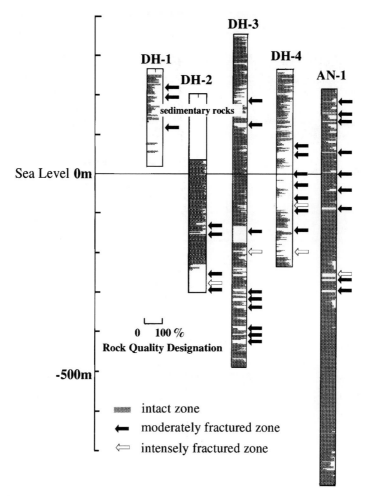

Fig. 4. Rock Quality Designation data (RQD; the total length of solid core pieces that have individual lengths more than 10 cm, measured axially between discontinuities) for each borehole.

Fig. 5. Photomicrographs under plane polarized light, illustrating the distribution of iron hydroxide (arrowed) in microfractures in minerals.

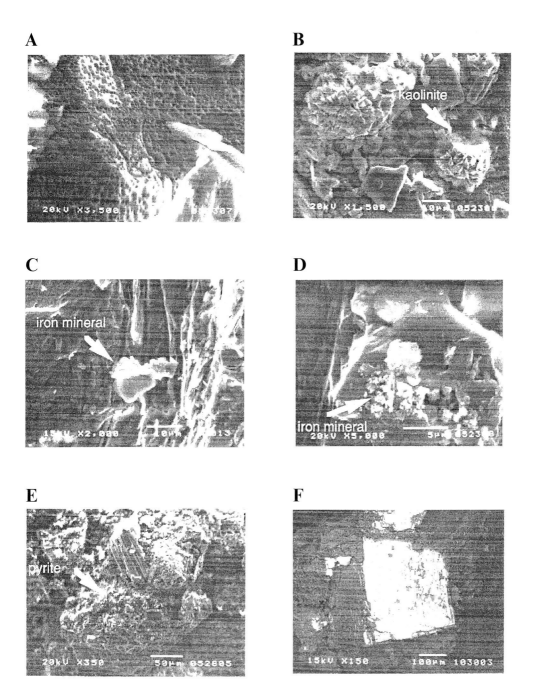

Fig. 6. Scanning Electron photomicrographs illustrating key mineral textures in core samples. A, Plagioclase surface showing dissolution textures (DH-3/−484 m); B, plagioclase surface coated by kaolinite (DH-3/−484 m); C and D, biotite and iron oxyhydroxide mineral (DH-3/−484 m); E, pyrite dissolution texture (DH-4/80 m); F, pyrite retaining its original form (DH-4/130 m).

Table 1. *Chemical and isotopic compositions of the groundwaters (mg/l except where stated)*

Sampling point	Surface	DH-4	DH-3	DH-3	DH-3
Depth (GL − m)	0	−186	−208	−330	−840
Sea level (SL m)	356	80	148	26	−484
Zone	–	MFZ	MFZ	MFZ	IZ
pH	6.2	6.8	9.7	8.9	9.3
Temperature (°C)	18	13	15	15	30
TDS (mg/l)	70	150	105	110	140
Si	5.96	5.6	8.23	13.8	5.53
Ti	<0.01	<0.01	<0.01	<0.01	<0.01
Al	0.16	<0.02	<0.01	<0.01	<0.01
Fe^{2+}	<0.05	9.34	<0.05	<0.05	<0.05
Fe^{3+}	<0.05	<0.05	<0.05	<0.05	<0.05
Σ Fe	0.20	9.34	<0.02	0.06	<0.02
Mn	0.76	0.77	<0.01	<0.01	<0.01
Mg^{2+}	0.18	1.88	0.14	0.12	0.05
Ca^{2+}	1.04	17.6	16.0	12.2	3.72
Sr^{2+}	<0.01	0.11	0.12	0.12	0.03
Na^+	12.3	13.3	8.2	11.6	39.5
K^+	10.1	6.13	3.45	0.93	0.80
F^-	0.17	4.99	1.99	4.09	9.73
Cl^-	1.43	2.74	2.30	3.69	3.11
NO_2^-	<0.02	<0.02	<0.02	<0.02	<0.02
PO_4^-	<0.02	<0.02	<0.02	<0.02	<0.02
Br^-	<0.02	0.08	<0.02	<0.02	0.04
NO_3^-	0.08	0.02	<0.02	<0.02	<0.02
SO_4^{2-}	1.37	0.09	2.27	6.66	6.21
CO_3^{2-}	<1*	<1†	12.1*	1.8*	6.2*
HCO_3^-	42.8*	88.5†	48.8*	45.1*	62.4*
TC	47.0	nm	14.0	11.8	14.0
IC	8.4	nm	12.0	9.2	13.5
TOC	38.6	nm	1.99	2.61	0.5
δD (‰ SMOW)	−51.7	−52.5	−53.7	−53.6	−53.2
δO-18 (‰ SMOW)	−8.1	−8	−8.2	−8.3	−8.0
Tritium (TU)	nm	4.6	nm	nm	nm

nm, not measured
MFZ, moderately fractured zone.
IZ, intact zone.
* Calculated from pH and IC.
† Measured by high temperatue–nondispersive infrared gas analysis.

are dominated by Na^+, Ca^{2+} and HCO_3^- at DH-3/148 m and DH-3/26 m. In contrast, the waters are dominated by Na^+, Ca^{2+}, Fe^{2+} and HCO_3^- at DH-4/80 m. The deepest groundwater sampled, from DH-3/−484 m was enriched in Na^+ and HCO_3^- relative to the other waters.

The saturation states of these waters with respect to a range of minerals were calculated using the computer code PHREEQE. This approach indicated that the groundwaters are approximately in equilibrium with respect to quartz, calcite, kaolinite, fluorite, ferrosilite, ferric hydroxide and siderite (Table 2). The water/rock interactions deduced from the SEM-EDS analyses and the theoretical calculations using PHREEQE, are summarized in Table 3.

Isotopic composition of groundwater in the Toki granite

The stable isotopic compositions of these waters (δD SMOW and $\delta^{18}O$ SMOW) indicates that all the groundwaters are meteoric in origin (Fig. 7). Water sampled from the granite at DH-4/80 m contained detectable tritium (4.6 TU).

Discussion

Timing of recharge of waters in the Toki Granite

The isotopic composition of the groundwater in the granitic rock plots near to the composition

Table 2. *Saturation indices of constituent minerals at each sampling point*

Borehole	Surface	DH-4	DH-3	DH-3	DH-3
Depth (GL − m)	0	−186	−208	−330	−840
Sea level (SL m)	356	80	148	26	−484
pH	6.2	6.8	9.7	8.9	9.3
Eh (mV)*	240	−200–0	−200–0	−200–0	−200–0
Quartz	−0.40	0.15	0.15	0.57	−0.16
Albite	−5.07	−3.10	−0.73	0.63	−1.77
Anorthite	−10.68	−7.99	−2.55	−1.87	−5.64
Microcline	−3.78	−2.06	0.10	0.75	−1.34
Mica (Muscovite)	−0.53	0.53	1.98	4.09	2.10
Calcite	−3.80	−0.91	1.00	0.14	−0.12
Fluorite	−2.55	0.28	−0.47	0.09	0.15
Kaolinite	−1.42	−0.96	−1.74	0.56	−0.64
Gibbsite	−0.72	−1.05	−1.40	−0.67	−0.64
Pyrite	−82.79	8.43 ~ −37.47	−35.61 ~ −85.37	−21.08 ~ −68.4	−19.16 ~ −65.87
Ferrosilite	−5.30	−1.41	0.96 ~ −2.17	0.93 ~ −0.17	−0.64 ~ −0.73
Goethite	4.97	1.57 ~ 4.90	6.60 ~ 6.79	5.36 ~ 7.93	4.77 ~ 8.01
Siderite	−3.26	0.72	−0.65 ~ −3.78	−0.36 ~ −1.11	−0.44 ~ −0.55
Fe(OH)₃	−1.47	−4.87 ~ −1.54	0.34 ~ 0.53	−0.89 ~ 1.67	−2.03 ~ 1.20

The saturation index equals log (IAP/KT), where IAP is the ion activity product and KT is the temperature-corrected equilibrium constant.
* *Eh* values of the groundwater are assumed between −200 mV and 0 mV.

of modern recharge water (Fig. 7). This relationship contrasts with that between the data for deeper groundwaters from the sediments, and present surface waters. These groundwaters in the sediments are significantly depleted in heavy isotopes compared to present recharge waters, and are considered to have been recharged between about 13 000 and 15 000 years ago, under colder climatic conditions than those prevailing at the present (Mizutani *et al.* 1992). These findings suggest the possibility that the groundwater in the granitic rock recharged more recently than the water in the sedimentary rocks, due to the existence of a pathway for enhanced groundwater flow within the fracture system.

A possible objection to this hypothesis is that the similar stable isotopic compositions of waters from all three fracture zones (Fig. 7), do not preclude the possibility that groundwaters in different parts of the granite were recharged at different times, possibly more than 10^4 years ago. This objection arises because stable isotopic compositions are imprecise indicators of absolute groundwater 'age', and reflect the conditions (temperature, altitude, atmospheric moisture content etc.) during recharge, rather than the absolute timing of recharge (e.g. Craig *et al.* 1963; Stewart 1975; Gat 1981; Yurtsever & Gat 1981). Consequently, the waters in the granite could have been recharged at different times, but under similar conditions. However, tritium was detected in the groundwater

sampled from the granite at DH-4/80 m. This finding implies that at least some of the groundwater in the granite recharged from the surface within the last 30 years or so. Thus, it appears that the depth at which a water occurs in this

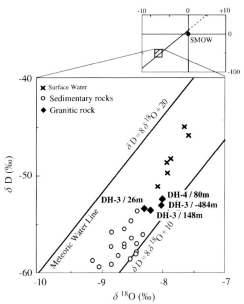

Fig. 7. Stable isotopic compositions (δD versus $\delta^{18}O$, both in per mil relative to Standard Mean Ocean Water, SMOW) of groundwaters from the Tono area.

Table. 3. *Water–rock interactions deduced from SEM-EDS analysis and PHREEQE calculations*

Point	Water type	SEM observation	Reaction	Calculation*
DH-4 SL 80 m	Na^+–Ca^{2+}–Fe^{2+}–HCO_3	Quartz dissolution	$SiO_2 + 2H_2O \rightarrow Si(OH)_4$	Saturated
		Plagioclase $\rightarrow$ Kaolinite	$Albite + \frac{9}{2}H_2O \rightarrow \frac{1}{2}Kaolinite + Na^+ + 2Si(OH)_4$	Under-saturated
			$Anorthite + 2H^+ + H_2O \rightarrow Kaolinite + Ca^{2+}$	Under-saturated
		Biotite $\rightarrow$ Kaolinite	$Biotite + 7H^+ + \frac{1}{2}H_2O \rightarrow Kaolinite + K^+ + 3Fe^{2+} + 2Si(OH)_4$	Saturated
		Calcite dissolution	$CaCO_3 \rightarrow Ca^{2+} + CO_3^{2-}$	Saturated
		Pyrite dissolution	$Pyrite + 8H_2O \rightarrow 2SO_4^{2-} + Fe^{2+} + 16H^+ + 14e^-$	Under-saturated
DH-3 SL 148 m SL 26 m	Na^+–Ca^{2+}–HCO_3	Quartz dissolution	$SiO_2 + 2H_2O \rightarrow Si(OH)_4$	Saturated
		Plagioclase $\rightarrow$ Kaolinite	$Albite + \frac{9}{2}H_2O \rightarrow \frac{1}{2}Kaolinite + Na^+ + 2Si(OH)_4$	Saturated
			$Anorthite + 2H^+ + H_2O \rightarrow Kaolinite + Ca^{2+}$	Under-saturated
		Biotite $\rightarrow$ Kaolinite	$Biotite + 7H + \frac{1}{2}H_2O \rightarrow Kaolinite + K^+ + 3Fe^{2+} + 2Si(OH)_4$	Supersaturated
		Fe^{2+} oxidation	$Fe^{2+} + 3H_2O \rightarrow Fe(OH)_3 + 3H^+ + e^-$	Saturated
		Calcite dissolution	$CaCO_3 \rightarrow Ca^{2+} + CO_3^{2-}$	Saturated
DH-3 SI −484 m	Na^+–HCO_3	Quartz dissolution	$SiO_2 + 2H_2O \rightarrow Si(OH)_4$	Saturated
		Plagioclase $\rightarrow$ Kaolinite	$Albite + \frac{9}{2}H_2O + H^+ \rightarrow \frac{1}{2}Kaolinite + Na^+ + 2Si(OH)_4$	Under-saturated
			$Anorthite + 2H^+ + H_2O \rightarrow Kaolinite + Ca^{2+}$	Under-saturated
		Biotite $\rightarrow$ Kaolinite	$Biotite + 7H^+ + \frac{1}{2}H_2O \rightarrow Kaolinite + K^+ + 3Fe^{2+} + 2Si(OH)_4$	Supersaturated
		Fe^{2+} oxidation	$Fe^{2+} + 3H_2O \rightarrow Fe(OH)_3 + 3H^+ + e^-$	Under-saturated
		Calcite dissolution	$CaCO_3 \rightarrow Ca^{2+} + CO_3^{2-}$	Saturated

*Saturated; SI $= -1 \sim +1$, Under-saturated; SI < -1, Supersaturated; SI $> +1$.

system does not necessarily reflect its age; waters in the granite can be 'younger' than shallower waters in the overlying sediments.

Water–rock interactions in the fractures and fracture zones

Structural zones within the granitic rock, that are defined by structural data (intact zones, moderately fractured zones, and intensely fractured zones) are accompanied by variations in groundwater chemistry, and variations in fracture-filling minerals.

The chemical composition of the water in the intact zone that was sampled at DH-3/−484 m is dominated by Na^+ and HCO_3^-. In this zone, the surface of plagioclase in fractures is coated by kaolinite. Calculated saturation indices indicate that the plagioclase is under-saturated, while the secondary mineral phases, kaolinite and gibbsite, are saturated. These results suggest that the reactions between groundwater and plagioclase have not reached equilibrium, but as they move towards an equilibrium state, kaolinite has been produced. However, the kaolinite has sealed the fracture surface and has thus decreased the surface area available for the water–rock interactions. The reaction of plagioclase to form kaolinite is considered to be the dominant reaction in single fractures of the intact zone.

The groundwaters in the moderately fractured zones have proportionately higher concentrations of Ca^{2+} than the waters in the intact rock. These waters are of $Na^+–Ca^{2+}–HCO_3^-$ type in the moderately fractured zones in shallower part at DH-3/148 m and DH-3/26 m. In contrast, the water from the fractured zone at DH-4/80 m has a proportionately higher Fe^{2+} content, and is of $Na^+–Ca^{2+}–Fe^{2+}–HCO_3^-$ type. In these fractured zones, the surfaces of plagioclase crystals alter to kaolinite and montmorillonite, according to a similar reaction to that deduced to occur in the intact zone. However, the intensity of the alteration differs in different fractured zones. The rock sampled from the zones at DH-3/148 m and DH-3/26 m have been altered more strongly than that sampled from the zone at DH-4/80 m. In the latter zone, only the surfaces of plagioclase crystals are generally altered, while in the other zones, the centres of the crystals are altered. The chemical calculations indicate that the waters sampled from the fractured zones at DH-3/148 m and DH-3/26 m are saturated with respect to plagioclase (albite and microcline). In contrast, the water from the zone at DH-4/80 m is under-saturated with respect to plagioclase.

At low temperatures ($<c. 50°C$), the reactions between groundwaters and minerals tend to be slow. Secondary mineral formation on the surfaces of primary minerals can slow the rates of reaction still further. Consequently, equilibrium between waters and silicates is attained only after a groundwater has resided in a rock for at least 10^4 to 10^6 years (e.g. Brantley 1992). Hence, the observation that groundwaters from the DH-3/148 m and DH-3/26 m are saturated with respect to albite and microcline, implies that the waters have been resident in the fractures for a considerable period of time. Conversely, the fact that groundwater from the fractured zone at DH-4/80 m is under-saturated with respect to plagioclase, implies that the water here has not resided in this fracture for 10^4 to 10^6 years. These findings imply that different fractures show different degrees of interconnectivity with one another. A corollary of this conclusion is that different parts of the fracture network have different permeability.

An alternative explanation for the differing mineral saturation states shown by groundwaters from different fracture zones is that the waters were recharged at the same time, but have reacted with minerals at different rates. The different rates might be explained by different fracture zones having different degrees of fracturing, and different fractures having different reactive surface areas. In this scenario, the differing degrees of fracturing make different mineral surface areas available for reaction.

A third explanation is that the different rates could reflect differences in the pH of the waters. The rate of feldspar dissolution is enhanced at $pH > 7$ (e.g. Knauss & Wolery 1986). This fact might explain the observation that the alteration is more intense at DH-3/26 m where the sampled water had a pH of 8.9–9.3, than at DH-4/80 m where the water had a pH of 6.8. Knauss & Wolery (1986) found a rate of albite dissolution some five times faster at the higher pH levels. However, it follows from this hypothesis that waters in different fractures have acquired distinct pH values. Once again, this observation may reflect the differing natures of water/rock interactions in different fractures, which may in turn be linked to variable hydraulic properties of different fractures.

Redox conditions in the granitic rock

The redox conditions of a groundwater will tend to reflect the minerals present in the host rock, even when chemical equilibrium has not been attained between all the solid mineral phases and all the dissolved species present in the ground-water (e.g. Tullborg 1986; Wikberg 1987). Iron minerals such as pyrite, siderite and ferric hydroxide often react sufficiently rapidly to approach equilibrium with a groundwater. Such minerals comprise only a small proportion of the rock mass of the granitic basement at Tono. However, because of their reactivity, many features of the groundwater chemistry can be explained by reactions involving these minerals.

The water sampled from the fracture zone at DH-4/80 m was calculated to be either super-saturated or under-saturated with respect to pyrite, depending upon the redox potential (Eh) used in the calculations. If the water is under-saturated with respect to pyrite, then this mineral will dissolve. The etch pits seen on surfaces of pyrite crystals indicate that such dissolution could be occurring. However, the amount of SO_4^{2-} in the groundwater is insuffi-cient to balance the amount of dissolved Fe^{2+} present. Therefore, pyrite dissolution cannot contribute all the Fe^{2+} dissolved in the water.

Ferrous iron can also be dissolved from easily weathered minerals, such as biotite. It is thought that the excess Fe^{2+} in the water from the fractured zones at Tono originates in weathering reaction of biotite as follows. In the presence of dissolved oxygen, this Fe^{2+} can be oxidized and precipitated as ferric oxyhydroxide (Hem 1961; Siever & Woodford 1979; Grenthe et al. 1992), according to:

$$2KFe_3AlSi_3O_{10}(OH)_2 + 2H^+$$
$$+ 3.5O_2 + 8H_2O$$
$$= 6Fe(OH)_3 + Al_2Si_2O_5(OH)_4$$
$$+ 2K^+ + 6SiO_2 \qquad (1)$$

Therefore, it can be assumed that the Eh con-dition is controlled by the $Fe(OH)_3/Fe^{2+}$ redox couple. The thermodynamic stability of iron minerals and aqueous iron species is represented on a plot of Eh versus pH in Fig. 8. Assuming that the redox-controlling iron mineral is ferric hydroxide, the Eh value best describing redox conditions in the deep Tono groundwaters lies around -200 mV, for a pH in the range 9 to 10.

The distributions of these various iron miner-als can yield information concerning the varia-bility of the redox conditions in the fractures and fractured zones. The heterogeneous distribution of ferric oxyhydroxides, and pyrite in the granite at Tono, suggest the possibility that each frac-ture and fractured zone has distinctive Eh–pH conditions. In the Tono area, iron hydroxides have been observed in the micro-fractures of minerals in each borehole. This finding indicates the depth to which oxidizing meteoric water has recharged into the rock mass. The maximum depth at which these minerals are found is 132 m below the ground level.

There are two possible explanations for the waters remaining oxidizing at such depths. Potentially, the flow rate has been sufficiently fast that there has not been enough time for water–rock interactions higher along the flow path to reduce the concentration of dissolved oxygen. Alternatively, higher along the flow path, the capacity of the rock to reduce oxygen could have been exhausted by previous water–rock interactions during the infiltration of oxidizing waters (Iwatsuki & Yoshida 1999).

Implications for chemical containment of wastes

The findings from the Tono research site indi-cate that complex age relationships and chemical relationships exist between groundwaters in dif-ferent parts of a fractured crystalline rock mass. In particular, the intensity of fracturing, and the degree to which fractures are interconnected, controls the subsurface penetration and disper-sion of relatively oxygenated recharge waters. A consequence is that, in such lithologies, varia-tions in chemical conditions are not related simply to depth. Potentially, relatively deep waters could be 'younger' and more oxidizing than more shallow waters at the same locality.

This structural control on the distribution of chemically distinct groundwater conditions is important, because redox-sensitive reactions control the migration of many radionuclides, such as those that might originate in radioactive wastes. This redox control can be illustrated by considering the redox conditions under which uranium is immobile (Fig.8). Under reducing conditions ($Eh < c. -200$ mV at pH 9 to 10), uranium will be immobilized in solid mineral phases such as uraninite. However, under more oxidizing conditions, uranium could be mobi-lized. By comparing the redox-dependent specia-tion of uranium with that of iron (Fig. 8), it is apparent that textural evidence for pyrite disso-lution, accompanied by evidence for iron hydro-xide precipitation, might indicate conditions that are inappropriate for the chemical containment of uranium-bearing wastes (although of course other factors, such as the ability of the rock to retard radionuclides by sorption mechanisms, would also need to be evaluated before the per-formance of the geosphere as a chemical barrier can be assessed thoroughly).

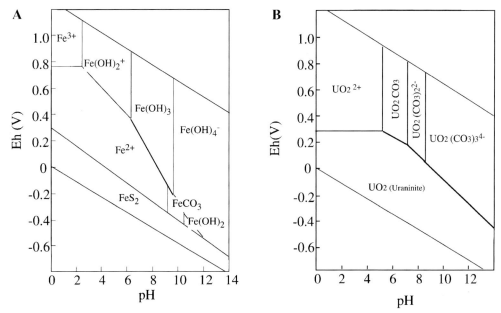

Fig. 8. *Eh*–pH diagrams showing predominance fields of aqueous species and minerals at 25°C. Iron species and minerals are shown in (**A**); uranium species and minerals are illustrated by (**B**). The boundaries of the fields were drawn for a total dissolved Fe concentration of 0.01 ppm, a total dissolved carbon concentration of 10 ppm, a total sulfur concentration of 10 ppm, and a total dissolved uranium concentration of 0.001 ppm.

Thus, the findings from Tono illustrate the importance of detailed structural and hydro-geochemical investigations when evaluating the potential of a given fractured crystalline rock mass to retain redox-sensitive radionuclides. The chemically heterogeneous character of ground-waters from structurally distinct parts of a rock mass must also be considered if it is proposed to dispose of any other type of redox-sensitive waste within a crystalline rock.

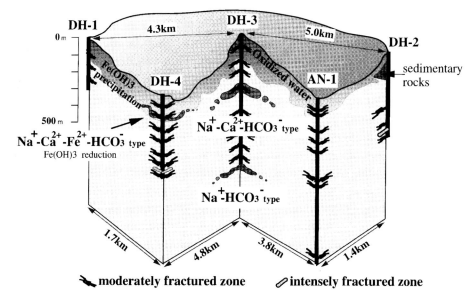

Fig. 9. Conceptual three-dimensional diagram summarizing a hydrogeochemical model of the granitic rock in the Tono area.

Conclusions

The results presented in this paper are summarized in Fig. 9. The main findings are: (1) The structure of the granitic rock in the Tono area can be classified into three zones (intact zones, moderately fractured zones and intensely fractured zones) according to the frequency and width of the fractures and fractured zones and their mineral contents. These zones can be seen at several levels in all the boreholes within the deep granitic rock in the Tono area. (2) The chemical evolution of groundwater is generally controlled by water–rock interactions between plagioclase, ferric hydroxide and groundwater. (3) Redox conditions in the granitic rock are controlled by water–rock interactions involving iron minerals such as ferric hydroxide and pyrite. (4) The chemistry of the groundwaters is heterogeneous, and different fracture systems contain chemically distinct groundwaters. Correlations between geological structures and groundwater chemistry can be applied to understand the geochemical environment in the deep granitic rock and to develop a realistic hydrogeochemical model for such fractured rocks. Such an approach is necessary in order to characterize the chemical heterogeneity of waters in fractured crystalline rocks, and to develop safe disposal options for redox-sensitive wastes, such as radioactive wastes.

We thank Dr R. Metcalfe of British Geological Survey and the staff members of Tono Geoscience Center for offering numerous suggestions and improving the manuscript.

References

BRANTLEY, S. L. 1992. Kinetics of dissolution and precipitation – experimental and field results. *In*: KHARAKA, Y. K. & MAEST, A. S. (eds) *Proceedings of the 7th International Symposium on Water/rock Interaction (WRI 7)*. Balkema, Rotterdam, 3–6.

CRAIG, H., GORDON, L. I. & HORIBE, Y. 1963. Isotopic exchange effects in the evaporation of water. *Journal of Geophysical Research*, **68**, 5079–5087.

GAT, J. R. 1981. Groundwater. *In*: GAT, J. R. & GONFIANTINI, R. (eds) *Stable Isotope Hydrology: Deuterium and oxygen-18 in the water cycle 210*. International Atomic Energy Agency, Vienna 223–240.

GRENTHE, I., STUMM, W., LAAKSUHARJU, M., NILSSON, A.-C. & WIKBERG, P. 1992. Redox potentials and redox reactions in deep groundwater systems. *Chemical Geology*, **98**, 131–150.

HEM, J. D. 1961. Stability field diagrams as aids in iron chemistry studies. *Journal of the American Water Works Association*, **53**, 211–229.

IWATSUKI, T. & YOSHIDA, H. 1999. Groundwater chemistry and fracture mineralogy in the basement granitic rock in the Tono uranium mine area, Gifu Prefecture, Japan – Groundwater composition, Eh evolution analysis by fracture filling minerals. *Geochemical Journal*, **33**, 19–32.

——, SATO, K., SEO, T. & HAMA, K. 1995. Hydrogeochemical investigation of groundwater in the Tono area, Japan. *Materials Research Society Symposium Proceedings*, **353**, 1251–1257.

KATAYAMA, N., KUBO, K. & HIRONO, S. 1974. Genesis of uranium deposits of the Tono mine, Japan. *Proceedings of a Symposium on the Formation of Uranium Ore Deposits, Athens*. IAEA-SM-183/11, 437–452.

KNAUSS, K. G. & WOLERY, T. J. 1986. Dependence of albite dissolution kinetics on pH and time at 25°C. *Geochimica et Cosmochimica Acta*, **50**, 2481–2497.

MIZUTANI, Y., SEO, T., OTA, K., NAKAMURA, N. & MURAI, Y. 1992. *Proceedings of a Symposium on Accelerator Mass Spectrometry and Interdisciplinary Application of Carbon Isotopes*, 159–168.

PARKHURST, D. L., THORSTENSON, D. C. & PLUMMER, L. N. 1980. *PHREEQE – A computer program for geochemical calculations*. United States Geological Survey Water Resource Investigations Paper, 80–96.

SIEVER, R. & WOODFORD, N. 1979. Dissolution kinetics and weathering of mafic minerals. *Geochimica et Cosmochimica Acta*, **43**, 717.

STEWART, M. K. 1975. Stable isotope fractionation due to evaporation and isotopic exchange of falling water drops: applications to atmospheric processes and evaporation of Lakes. *Journal of Geophysical Research*, **80**, 1133–1146.

TULLBORG, E-L. 1986. *Fissure fillings from the Kliperås study site*. SKB Technical Report **86-10**. Stockholm.

WIKBERG, P. 1987. *The chemistry of deep groundwaters in crystalline rocks: development of inorganic chemistry*. The Royal Institute of Technology, Stockholm.

YURTSEVER, Y. & GAT, J. R. 1981. Atmospheric waters. *In*: GAT, J. R. & GONFIANTINI, R. (eds) *Stable Isotope Hydrology: Deuterium and oxygen-18 in the water cycle 210*. International Atomic Energy Agency, Vienna, 103–142.

Protecting the redox stability of a deep repository: concepts, results and experience from the Äspö hard rock laboratory

STEVEN A. BANWART,[1] PETER WIKBERG[2]
& IGNASI PUIGDOMENECH[3]

*Department of Civil and Structural Engineering, University of Sheffield,
Mappin Street, Sheffield S1 3JD, UK
The Swedish Nuclear Fuel and Waste Management Co., Box 5864,
102 40 Stockholm Sweden
Department of Inorganic Chemistry, The Royal Institute of Technology,
100 44 Stockholm, Sweden*

Abstract: Redox potential and net reducing capacity are rigorously defined geochemical parameters that provide a framework for assessing the redox-related performance of waste containment in the sub-surface. These concepts are illustrated by performance assessment issues in the geological disposal of spent nuclear fuel, and also by consideration of chromium and tetrachloromethane as illustrative compounds representing other types of hazardous waste. Recent and ongoing research at the Äspö Hard Rock Laboratory, a full-scale prototype repository, provides a case study of the redox-related performance of the geological barrier and its function to protect a deep repository from intrusion of dissolved oxygen. Enhanced recharge during construction of the underground laboratory at Äspö strongly influenced redox conditions in the shallow bedrock. The rapid recharge, caused by tunnel construction through a vertical fracture zone, mobilized organic carbon into the groundwater. Microbial processes were critical to the performance of the shallow bedrock, as a barrier to oxygen intrusion during enhanced recharge, by efficiently scavenging oxidants through intense respiration of the added carbon. Because the study site was mainly exposed bedrock with very little soil cover, this effect may be even more pronounced at potential repository sites with thicker overburden.

Crucial to national strategies for deep geological disposal of spent nuclear fuel is a need for reliable information about the design and performance of underground repositories. The Äspö Hard Rock Laboratory near Oskarsham in southeastern Sweden addresses this need with a prototype repository research programme. Although the laboratory is not an actual repository (nuclear waste will never be put in the ground at this location) the facility comes as close as possible to study simulated full scale emplacement of nuclear materials in a bedrock environment that is characteristic of future underground facilities. Banwart *et al.* (1997) have recently reviewed key performance assessment issues for the Swedish repository concept and the role of the laboratory in the Swedish nuclear waste programme and provide a summary of current and planned activities at the facility.

The Swedish concept for disposal of spent nuclear fuel includes isolation of the fuel rods in steel-lined copper canisters buried approximately 0.5 km deep in granite bedrock. Repository safety depends on the design of the engineered and geological barriers: spent fuel matrix, canisters, buffer material filling the deposition hole, backfill material sealing the tunnel and shafts and the bedrock isolating the repository from the biosphere.

The primary engineered barrier is the spent fuel itself. It is a sparingly soluble oxide ($UO_2(s)$) which dissolves extremely slowly under anoxic conditions, but can corrode to more soluble forms in the presence of dissolved oxygen (O_2) (summarized in Miller *et al.* 1994, Section 4.2). This is also the case for the copper canister, and the steel lining of the canister, should it be breached. The copper canister can also undergo sulphidic corrosion (described below) under reducing conditions. The buffer material is intended to be compacted bentonite clay, with a mixture of bentonite and sand to be used as fill material for the tunnel and shafts. The compacted bentonite buffer is designed to maintain low hydraulic conductivity in order

From: METCALFE, R. & ROCHELLE, C. A. (eds) 1999. *Chemical Containment of Waste in the Geosphere.* Geological Society, London, Special Publications, **157**, 85–99. 1-86239-040-1/99/$15.00 © The Geological Society of London 1999.

to keep groundwater flow rates low, and to provide a large sorptive surface area to retain soluble radionuclides, if they are released by dissolution of spent fuel.

Within the Source–Pathway–Receptor framework for risk assessment, repository safety depends in part on maintaining the integrity of the engineered barriers in order to eliminate the source for as long as possible. However, performance assessment of the complete disposal system assumes that the engineered barriers will eventually fail. In this scenario, a lengthy pathway for groundwater flow between repository and surface ensures long time periods for radioactive decay before soluble radionuclides reach the biosphere. Stagnant or very slow groundwater flow prevents corrosive surface water that can degrade engineered barriers from reaching repository depths. If radionuclides are released, slow groundwater flow acts along with sorption processes to retard their migration to the surface. Dilution by groundwater at shallower depths reduces any residual radioactivity dose to the biosphere.

Because of the possible impact of corrosion on the performance of the engineered barriers, and the redox-dependent solubility of some radionuclides, the performance of both the engineered and geological barriers depends critically on redox conditions in the repository and surrounding groundwater environment. Of special concern are long-lived isotopes including neptunium (Np), technetium (Tc), iodine (I) and caesium (Cs). For reducing conditions, Np and Tc form sparingly soluble solids. Under oxic conditions they can exist in higher valent states which are more soluble, particularly Tc, and thus mobile with groundwater flow.

Performance assessment issues regarding groundwater redox chemistry are listed below.

(a) The persistence of O_2, trapped in the repository at the time of closure, could lead to corrosion of the canister and subsequently the spent fuel, releasing radionuclides into solution.

(b) Production of hydrogen sulphide, resulting from biogenic sulphate reduction, could lead to sulphidic corrosion of the copper canister.

(c) Dissolved oxygen in the hydraulically resaturated buffer material, backfill or in fracture zones would favour the solubility and thus mobility of some radionuclides such as Tc.

For both the integrity of the engineered barriers, and the optimal performance of the geological barrier, post-closure exposure of the repository to O_2 must be minimized or otherwise controlled. During construction and operation of the repository, the underground environment will be open to atmospheric conditions. There will also be enhanced recharge of oxygenated surface waters into vertical and subvertical fracture zones in the rock mass around the repository.

In the following sections a rigorous geochemical framework is developed within which redox-related performance issues can be interpreted. A recent field study at the Äspö Hard Rock Laboratory (Oskarsham, southeastern Sweden) studied the performance of the geological barrier to protect the repository from intrusion of dissolved oxygen from the surface during construction of a repository. The performance of the shallow bedrock, primarily as a barrier to atmospheric O_2, is reviewed in light of results from that study. Finally, the performance assessment concepts presented here are extended to containment of other types of wastes in the geosphere.

Technical background

The redox intensity of groundwater

Figure 1 shows a redox scale relating redox potentials to chemical equilibria for some major elements of interest in groundwater. Oxidizing species or solid phases are written on the left hand side of the scale, and reducing species or phases for the same redox-reactive element (O, N, Mn, Fe, S, H, C) are written on the right hand side. Any reductant on the right can react with any oxidant above it, on the left of the scale. For example, organic carbon written here as CH_2O, can react with (i.e. reduce) all oxidants other than CO_2 on the scale; H_2O, SO_4^{2-}, $Fe(OH)_3(s)$, $MnO_2(s)$, NO_3^-, $O_2(g)$. On the other hand, Mn^{2+} can only reduce nitrate ion (NO_3^-) and $O_2(g)$. It does not have sufficient reducing intensity to react with $Fe(OH)_3(s)$, SO_4^{-2}, CO_2 and H_2O.

In Fig. 1 the redox potentials are calculated for unit activities of reactants and products except protons which are assigned a value corresponding to pH 7. Redox potentials depend on the activities of both reactants and products, and will change as the aqueous composition of the groundwater changes. The reader is referred to a general text such as Stumm & Morgan (1996) to review how redox potentials are calculated and how they depend on solution composition and aqueous speciation.

Each redox couple, O_2/H_2O, $NO_3^-/N_2(g)$, etc., is associated with a unique position on the scale,

Eh (pH = 7), Volts

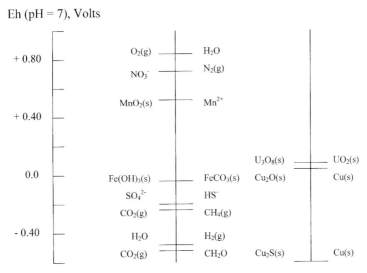

Fig. 1. Redox scales for important components of groundwater and engineered barriers in the Swedish repository design concept. At left an arithmetic scale lists redox potentials from -0.6 V up to $+1.0$ V. The scales are qualitative, but are based on unit activities of all reactants and products except protons (pH = 7) at 25°C. The centre scale lists some major redox couples that commonly occur in groundwater. The oxidized form of each element is listed on the left, and the reduced form on the right. Oxidized forms (the oxidants) can react with any reduced forms (reductants) that are lower on the scale. The scale at right lists important redox couples that correspond to important engineered barriers in the Swedish repository concept. The possibility of copper canister corrosion and spent fuel oxidation is of concern to repository performance.

reflecting the redox potential of that couple. The value for the redox potential associated with each couple is given on the arithmetic scale at the left of Fig. 1. For example, the O_2/H_2O couple has a redox potential near 0.8 V; the SO_4^{2-}/HS^- couple has a potential near -0.2 V. The relative value of the redox potential reflects the relative reducing or oxidizing intensity associated with each redox couple. This intensity parameter (Eh V) determines whether a specific redox reaction is thermodynamically favoured or not; it is a measure of the tendency for that couple to react as an oxidant or reductant with other couples. If two couples occupy the same position on the scale, thus having the same redox intensity, then they are at equilibrium and will not react further.

For example, in the case of O_2/H_2O and SO_4^{2-}/HS^-, monohydrogen sulphide (HS^-) is thermodynamically favoured to reduce O_2 to water, producing sulphate in the reaction. Alternatively, comparing the redox potentials for SO_4^{2-}/HS^- and $CO_2(g)/CH_2O$ indicates that HS^- is not a strong enough reductant to reduce carbon dioxide to organic carbon. Rather, organic carbon will instead react with sulphate, forming dissolved sulphide and carbon dioxide gas in the process. Comparing redox potentials for various redox couples (O_2/H_2O, $N_3^-/N_2(g)$,

etc.), provides a prediction of the tendency for certain reactions to occur. Because many of these reactions are microbially mediated in aquifers, soils and sediments, they are often referred to as biogeochemical processes.

Figure 1 also compares redox couples that are of importance to the performance and safety of a deep geological repository for spent nuclear fuel. They reflect the performance assessment issues raised in the introduction to this paper, and have been reviewed by Banwart et al. (1997). The couples $Cu_2O(s)/Cu$ and Cu_2S/Cu reflect corrosion processes that can destabilize the copper canister in which spent fuel rods will be placed. These couples represent oxic and sulphidic corrosion of copper, respectively.

Di-copper oxide $Cu_2O(s)$ is a corrosion product from reaction of O_2 with copper metal. The oxide is more soluble, and thus less stable, with respect to maintaining the integrity of the copper canister. Dissolved oxygen, NO_3^- and $MnO_2(s)$ have sufficient oxidizing intensity to corrode copper, forming Cu_2O. Nitrate ion and manganese dioxide might not be considered, if they are not relatively abundant, in the repository environment. However, dissolved oxygen will be present while the repository is open during construction and operation. Any remaining O_2, after closure of the repository, could lead to canister corrosion. Although copper is chosen as the canister material because its corrosion is an extremely slow process (see next paragraph), the redox scale in Fig. 1 does show that there is a tendency for O_2 and Cu to react.

Formation of $Cu_2S(s)$ is a more complex process. Comparing the redox scales in Fig. 1 shows that the groundwater itself ($H_2O/H_2(g)$) has sufficient oxidizing intensity to react with Cu, producing hydrogen gas. At these low redox potentials; i.e. below -0.2 V according to Fig. 1, sulphur is stable in its reduced form, and can react with the oxidized copper ions. Under these conditions the stable oxidation product of Cu is $Cu_2S(s)$. The process of sulphidic corrosion is substantially slower than oxic corrosion of copper; a conservative estimate of the composite corrosion rate (oxic and sulphidic) is 5 millimetres in 10^5 years (SKB 1995). A number of natural and archeaological analogue studies of copper corrosion rates are presented by Miller *et al.* (1994, Section 4.3).

Figure 1(b) also shows the redox potential for the $U_3O_8(g)/UO_2(s)$ redox couple. Uranium dioxide ($UO_2(s)$) is a sparingly soluble mineral phase, and is the predominant constituent in spent fuel rods. Upon corrosive attack by dissolved oxygen, it is converted to the solid phase $U_3O_8(s)$, which is also relatively insoluble. If oxic conditions persist, $U_3O_8(s)$ can be further oxidized to hexavalent uranium which is soluble and can be transported with groundwater flow. Groundwater entering the repository after closure will undergo radiolysis, whereby exposure to radiation splits water into O_2 and H_2, (also producing H_2O_2). If a canister is breached, then the O_2 that is produced can lead to dissolution of the spent fuel. Although the spent fuel is primarily composed of $UO_2(s)$, its dissolution is associated with the release of other radionuclides present in the solid matrix. Subsequent adsorption of released radionuclides on clay backfill around the repository, and on fracture minerals in contact with groundwater flow paths, will help prevent radionuclide migration from the repository. However, anoxic conditions and the resulting low solubility and slow dissolution of $UO_2(s)$ is critical for the intended performance and safety of the repository (see review by Miller *et al.* l994, Section 4.2).

The redox capacity of groundwater

The reducing capacity of a groundwater is the net amount of reductants available to maintain the redox intensity below a fixed threshold if oxidants, such as O_2, are introduced. Repository safety requires that even if O_2 is introduced

during operation, the amount of reductants available for reaction after closure should be overwhelming by comparison. If this is the case, the O_2 will be completely consumed, and the groundwater will return to the reducing conditions that are favourable for repository performance.

For example, if the amount of ferrous iron minerals present in repository backfill and fracture minerals (represented by $FeCO_3(s)$ in Fig. l(a)) is much greater than the amount of O_2 remaining after closure, then with time, all O_2 will be reduced to H_2O by these minerals, producing iron hydroxide in the process. This would ensure that the reducing intensity would return to values at least as low as the redox potential of the $Fe(OH)_3(s)/FeCO_3(s)$ couple (near -0.05 V). This is below the threshold for corrosion of either copper or uranium oxide by O_2. It is also slightly above the threshold for sulphide production by sulphate reduction (-0.2 V). The presence of ferrous minerals thus buffers the redox intensity of the repository to conditions that are favourable for repository performance.

The reducing capacity provided by ferrous carbonate can be extended to include all reductants in the groundwater. The total reducing capacity, with respect to oxic corrosion of copper is defined by the total amount of all reductants, below the $Cu_2O(s)/Cu(s)$ redox potential, minus the total amount of oxidants above it. The amount in moles per litre of groundwater, of each oxidant or reductant, must also be weighted by the mole equivalents of electrons transferred when 1 mole of the couple reacts. The following equation defines the net reducing capacity (RDC) of the repository, with respect to a specified redox intensity. The example chosen here is the redox potential for copper corrosion by O_2. All reductants below this potential contribute to reducing capacity, while all oxidants above this potential contribute to oxidizing capacity. The net RDC is the sum of reducing capacity with the oxidizing capacity subtracted from it, where [X] is moles of a species (or solid phase) X per litre of groundwater.

$$RDC = [FeCO_3(s)] + 14[FeS_2(s)] + 8[CH_4]$$

$$+ 2[H_2] + 4[CH_2O] - 2[MnO_2(s)]$$

$$- 5[NO_3^-] - 4[O_2] \tag{1}$$

Although RDC provides a rigorous definition of the buffer capacity required to maintain the redox intensity below that associated with copper corrosion, engineering application to real systems is difficult due to the large uncertainty in the site information that is needed to characterize RDC. Above all, the amount of reactive solid phases present (i.e. '[FeCO$_3$(s)]', mole l^{-1}) in the repository or in contact with groundwater flow is difficult to quantify. For groundwaters (hydraulically saturated), the molar concentration of a mineral depends on the mass fraction of reactive mineral (m, kg mineral/kg rock), the dry bulk density of the geological medium (ρ_b, kg l^{-1}), the porosity (ε, 1 fluid 1 bulk medium) and the molecular formula weight of the mineral (F_w, kg mol^{-1}). A further assumption is that the porosity is sufficiently small that the mass of fluid is negligible compared the mass of rock in the bulk geological medium. The molar concentration of mineral in contact with the groundwater is therefore defined in units of moles of mineral per litre of groundwater as:

$$[\text{mineral}] = \frac{m\rho_b}{F_w\varepsilon}. \qquad (2)$$

The factors that control the availability of the mineral reductants to react with dissolved O$_2$ are not well understood, and can be difficult to quantify (Malmstrom & Banwart 1996). The extent of reaction depends on redox reactions that occur on mineral surfaces and on dissolution reactions that release the structural reductants such as Fe(II) to solution. The rate of reaction depends on the amount of surface area in contact with flowing groundwater and on slow dissolution and diffusion processes within the rock matrix. The reactions may be microbially mediated. In the absence of verified models for these processes, and lacking field methods to determine relevent parameters, empirical approaches to measuring RDC are useful.

Volumetric titrations are often used to determine capacity parameters. A conceptual titration to determine RDC might be as follows. A sample of material would have to be titrated by standard additions of an oxidant such as dissolved O$_2$. After each addition of oxidant is consumed, another dose is added. When an addition of oxidant fails to be consumed, the reaction is assumed to have reached completion. The total equivalents of oxidant added provides a measure of RDC.

Research and development associated with the Äspö Hard Rock Laboratory (Swedish Nuclear Fuel and Waste Management Co. 1995, p. 171) aims to develop and verify lab and field methods to assess the extent (RDC) and rate (half-life) of O$_2$ consumption in fracture zones. These parameters relate directly to the redox-related performance of the geological barrier function to prevent O$_2$ from entering from the surface, and

to maintain reducing conditions in the repository and surrounding bedrock. The research is currently in progress. Key issues that are being addressed include the role of surface geochemical reactions in O$_2$ reduction by ferrous and sulphide minerals and the impact of microbial populations on such processes.

The Äspö groundwater redox chemistry project

The project was designed to demonstrate and quantify the reducing capacity provided by a vertical conductive fracture zone down which oxygenated surface waters were permeating. The proposed methodology was to observe the amount of time required for oxic conditions to develop after a fracture zone was intersected at a depth of 70 m during construction of the underground laboratory at Äspö. Enhanced recharge was anticipated due to the large hydraulic gradient created by tunnel construction. The rapid influx of surface waters was predicted to add dissolved oxygen and 'titrate' the fracture until all reducing capacity had been consumed.

Initial contributions to the redox capacity were assumed to be dissolved organic carbon, dissolved iron and manganese, and Fe(II)- and Mn(II)-bearing fracture minerals. An unexpected result was the massive mobilization of organic carbon that entered the fracture zone with the recharging surface waters, indicated by an increase by up to 20 mg l^{-1} of DOC in the groundwater draining into the tunnel. Rather than titrating the fracture with O$_2$ thereby decreasing the RDC of the rock and groundwater, the enhanced recharge actually provided additional RDC to the bedrock environment.

Site description

The field experiment was carried out within a vertical fracture zone on the island of Hålö, located in the Baltic Archipelago of southeastern Sweden (Fig. 2A). The inclined entrance tunnel to the Äspö Hard Rock Laboratory which begins at the mainland, passes beneath the island, to the island of Äspö, where the underground laboratory is located. The island of Hålö comprises a slightly undulating topography of well exposed rock 5–10 m above sea level. The geology is characterized by a red to grey porphyritic granite-granodiorite known locally as 'Småland' granite, belonging to the vast Transscandinavian Granite-Porphyry Belt (Gàal & Gorbetschev 1987) with U-Pb intrusion ages between 1760

A.

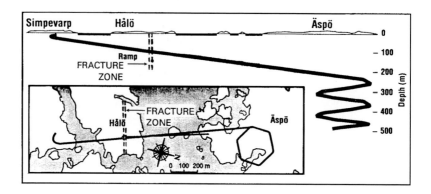

B.

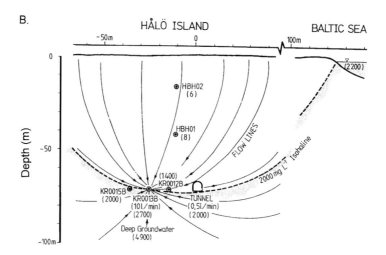

Fig. 2. Site plan and identification of sampling locations. A. The cross-section view of the islands of Hålö and Äspö, and the corresponding inset, showing the location of the entrance tunnel as a heavy black line extending from Simpevarp on the Swedish coast to Äspö. B. The view of the fracture zone facing Åpsö showing the borehole locations. The horizontal scale has a value of zero at the centreline of the tunnel. The chloride concentrations of sampled waters are shown adjacent to each borehole. The large salinity gradient between the shallow ($[Cl^-] = 10\,mg\,l^{-1}$) and the undisturbed, deep ($[Cl^-] = 4890\,mg\,l^{-1}$) groundwater allows use of chloride as a natural tracer for shallow water inflow. The dashed line represents the position of the interface between these two water types. The $2000\,mg\,l^{-1}$ isohaline was estimated qualitatively from measured chloride concentrations, sampled $7\frac{1}{2}$ months after the start of the experiment, in the inflow to the tunnel and in boreholes HBH01, KR0012B, KR0013B and KR0015B.

and 1840 Ma (Johansson 1988), i.e. late- to post-orogenic in relation to the Svecofennian origin (1800–1850 Ma). The major fractures and fracture zones control recharge, discharge and groundwater flow through the island(s) (Smellie *et al.* 1995).

Figure 2B is a representation of the cross-section of the fracture zone studied. It transects the island of Hålö and appears from the surface as a small depression in the exposed granite. The depression is 2–5 m wide, 2–3 m deep, and extends 220 m laterally across the northern tip of the island. The access tunnel to the Hard Rock Laboratory slopes downwards toward Äspö, intersecting the fracture zone at a depth of 70 m.

The topography of the island along the fracture zone defines a catchment area, on Hålö itself, of $10\,000\,m^2$. Hydrometeorological data from nearby stations (Swedish Meteorological and Hydrological Institute, Norköping) reported in Banwart et al. (1994, and included references) show values for average annual precipitation to be in the range 550–675 mm) and the annual average evapotranspiration to be 500 mm. This leaves 50–175 mm each year for runoff and groundwater recharge, corresponding to a maximum recharge of $1.0–3.3\,l\,min^{-1}$ from Hålö alone. Banwart & Gustafsson (1997) estimate the total recharge to be near $8\,l\,min^{-1}$ indicating additional recharge from areas other than the island of Hålö. Prior to intersection of the fracture zone by tunnel construction, the water table above the tunnel was approximately 0.5 m below the ground surface and approximately 1.5 m above the surrounding level of the Baltic Sea. There was only a slight drawdown in the water table during the experiment, i.e. the fracture zone remained hydraulically saturated.

Within the depression, directly above the tunnel, there is a 0.2 m deep layer of organic-rich soil. This soil overlies a zone of sand and gravel that is substantially reworked, presumably by wave action during isostatic uplift of the island from the Baltic Sea, and extending to 0.5 m depth. Below this a layer of glacial clay extends to at least 1 m depth. A percussion borehole drilled from the surface showed a granite base at a depth of 5 m. A moraine layer exists between the glacial clay and the granite, as indicated by loose moraine fragments that were observed during attempts to drill a borehole into the fracture zone at a depth of only 3 m.

The plane of fracture is approximately vertical and is clearly visible from the interior of the access tunnel as a band of water-bearing fractured rock with a nominal width of 1 m. The tunnel intersects the vertical plane of the fracture perpendicularly at a distance 513 m from the tunnel mouth.

On 13 March 1991 construction of the access tunnel to the Äspö Hard Rock Laboratory (HRL) intersected the vertical fracture zone at a depth of 70 m. The fracture zone hydrochemistry was then monitored through time. Groundwater sampling and analysis before intersection by tunnel construction, and examination of drillcores taken from the fracture zone early in the experiment, provided a reference state against which to compare subsequent evolution of groundwater conditions. Figure 2B identifies borehole locations where sampling occurred.

Table 1 shows that the shallow groundwater at the site is dilute while the older and deeper groundwater is relatively saline. Dilution of the chloride ion is therefore a good indicator of surface water intrusion. Details of the experimental methodology and site characterization are reported by Banwart et al. (1994, 1996).

The objectives of the experiment were:

(a) to determine the extent of surface water intrusion induced by opening the fracture zone by tunnel construction at a depth of 70 m;

(b) to observe if O_2 transport from the surface or from the tunnel could create oxic conditions in the fracture zone;

(c) to identify and quantify dominating transport and reaction processes.

Main results

Key results that concern objectives (a) and (b) are reported here. Initial development of a site model (objective (c)) is reported by Banwart (1995). Figure 3 compares changes in concentration of chloride ion, bicarbonate ion, dissolved Fe(II) and pH during the first 2 years of the field experiment. A sharp dilution front arrived in the access tunnel 21 days after intersection of the fracture zone. The decrease in chloride concentration is consistent with approximately 80% dilution of the undisturbed, saline groundwater by the dilute shallow groundwater.

Input of marine water is ruled out. Inspection of Table 1 shows that the $[Mg^{2+}]/[Cl^-]$ and $[K^+]/[Cl^-]$ ratios differ greatly between the undisturbed, saline groundwater and the Baltic Sea water. Pitkänen et al. (1994) showed that these ratios vary during the experiment only within a narrow range that corresponds closely with the ratios observed for the undisturbed, saline groundwater, and not the Baltic Sea water.

Because the shallow groundwater has a much higher bicarbonate ion concentration (Table 1), an increase in concentration is also expected in the tunnel during breakthrough of the shallow water. This is clearly seen between days 0 ($40\,mg\,l^{-1}$) and 20 ($100\,mg\,l^{-1}$) in Fig. 3. The measured concentration in the tunnel is slightly less than that expected from conservative mixing of shallow and native groundwater (near $120\,mg\,l^{-1}$) according to the mixing ratio predicted from chloride dilution. However, after day 50, bicarbonate ion concentrations increased steadily to values over $300\,mg\,l^{-1}$.

An oxidizing disturbance occurred between days 25 and 50, as evidenced by the disappearance of dissolved iron in the inflow to the access tunnel. At that time, the results were in accordance with what was expected. After this

Table 1. *Chemical compositions of groundwater and Baltic water at Hälö Island*

Location i.d.	Undisturbed saline groundwater	HBH05	HBH02	HBH01	KR0012B	KR0013B	KR0015B	Baltic Sea water
depth	70 m tunnel centreline	5 m	15 m	45 m	70 m 10 m l.o.t.*	70 m 20 m l.o.t.*	70 m 30 m l.o.t.*	
date	14 Mar. 1991	28 Oct. 1992	23 Apr. 1992	26 Jun. 1992	6 May 1992	6 May 1992	6 May 1992	28 Aug. 1992
day	−1	546	407	501	420	420	420	546
pH	7.5	7.6	6.6	7.4	7.8	7.7	7.8	8.2
Na (mg l)	1480	15.4	7.5	441	597	926	641	1960
K (mg l)	9.1	2.6	1.3	5.0	5.1	4.5	3.7	95
Ca (mg l)	1250	38	21	180	255	502	296	94
Mg (mg l)	132	4	3	30	37	70	40	234
Cl (mg l)	4890	11	11	932	1290	2340	1475	3758
HCO$_3$ (mg l)	42	137	63	256	248	245	327	90
SO$_4$ (mg l)	60	23	18	130	132	144	134	504
DOC (mg l)	†	10	18	14	13	15	16	14
δ^{18}O‰ SMOW	−11.3	−9.5	−10.2	−10.7	−9.9	−10.1	−10.4	−5.9

*l.o.t. refers to 'left of tunnel' when facing in the direction of tunnel construction, i.e. towards Äspö.
† No DOC data is available for the native groundwater. Comparable saline groundwaters in the area of Äspö exhibit DOC values near or below detection limits.

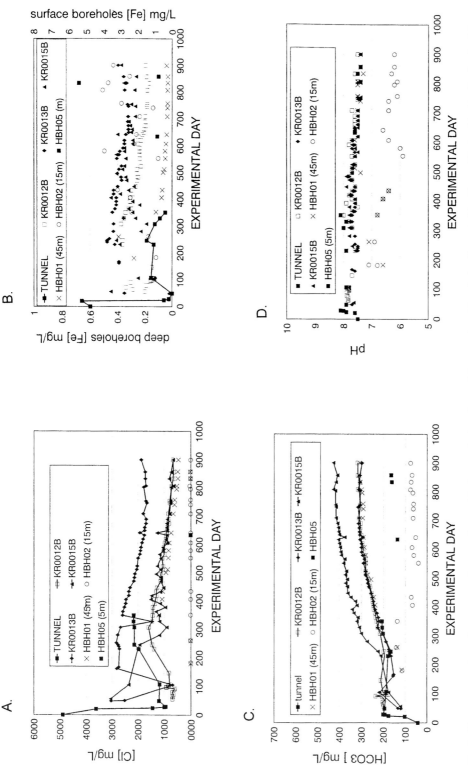

Fig. 3. The concentration of some groundwater solutes with time at the locations of the tunnel and boreholes illustrated in Figure 2B; A. chloride, B. dissolved iron, C. alkalinity, and D. pH. The sharp dilution front observed in the entrance tunnel at the beginning of the experiment is seen in the plot for chloride. The associated decrease in dissolved Fe was brief and the fracture zone remained anoxic after day 50, as indicated by significant (>0.05 mg l^{-1}) dissolved iron concentrations throughout the zone.

Table 2. *Analysis of dissolved gases*

Borehole	Vol% gas	$H_2 + He$	O_2	N_2	CH_4	CO	CO_2	C_2H_4	C_2H_6	C_2H_2
KR0012B	2.9	40	180	22 000	1030	0.1	6050	<0.1	<0.1	<0.1
KR0013B	3.7	110	310	25 000	1970	0.2	9 640	<0.1	<0.1	<0.1
KR0015B	4.0	64	130	22 000	4070	0.1	15 037	<0 1	<0.1	<0.1

μL gas/L water, sampled August, 1993.
From Banwart *et al.* 1996.

time, however, dissolved iron(II) concentrations began to increase simultaneously with the further increase in bicarbonate concentrations. These results suggested a return to anaerobic conditions with intensive respiration of organic carbon.

Speciation calculations for the carbonate system using the alkalinity data and pH values after day 50 and plotted in Fig. 3, are consistent with $CO_2(g)$ partial pressures ranging between 10^{-3} and 10^{-2} atm. These values agree well with the values for carbon dioxide determined by gas chromatography of dissolved gases collected from water samples in boreholes at the site (Table 2). There is clearly a dominant source of carbon dioxide to the groundwater at the site, presumably microbial respiration or fermentation of organic carbon.

During the first 100 days of the experiment, when bicarbonate ion concentration increased dramatically, the stable isotope composition of dissolved inorganic carbon remained remarkably constant with values near $\delta^{13}C = -16‰$ PDB (Table 3). Fractionation equilibrium between bicarbonate ion with this signature, and a hypothetical source of carbon dioxide gas, shows that the $CO_2(g)$ would have an isotopic signature of -28 to $-26‰$ PDB (Banwart *et al.* 1994). This value is consistent with typical values associated with soil-derived biogenic $CO_2(g)$ (Pearson & Hanshaw 1970; Dienes 1980), and is indicative of microbial production of CO_2 during decomposition of natural organic matter in the groundwater.

The dissolved organic carbon (DOC) concentration is at or near detection limits in saline groundwaters at Äspö (Smellie *et al.* 1995). The undisturbed, deeper groundwater at this site has a composition which is characteristic of the saline groundwater, and is expected to likewise be at or below detection limits for DOC. However, concentrations in groundwater flowing into the tunnel after the start of the experiment showed much higher values; $10 < DOC < 20 \text{ mg l}^{-1}$. An increase in DOC is expected due to displacement and mixing of the undisturbed, saline groundwater with shallow groundwater which had measured DOC concentrations of 10 and 18 mg l^{-1} respectively at depths of 5 and 15 m.

An alternative explanation to anaerobic microbial processes, for the source of inorganic carbon, is aerobic oxidation of organic carbon and subsequent dissolution of fracture calcites. This explanation was ruled out early in the experiment. Calcites sampled from drillcores at the site (Banwart *et al.* 1994) exhibited relatively high $\delta^{13}C$ values and a very wide range in values $(-16‰ < \delta^{13}C < -6‰)$, compared to the values determined for dissolved inorganic carbon in the groundwater $(-17.4 < \delta^{13}C < -15.8)$.

Although drillcore investigations only probe a limited volume of geological solids at a site, the 10 calcite samples analysed from various drillcores at this site are not consistent with the hypothesis that they are the source of the $\delta^{13}C$ signature observed for the groundwater sampled during the experiment. Additional evidence

Table 3. $\delta^{13}C$ *and* $\delta^{18}O$ *values for total dissolved inorganic carbon*

Borehole	Date 1991	Day	$\delta^{13}C‰$	$\delta^{18}O‰$
KR0013B	May 1	49	−16.2	19.9
KR0013B	May 1	49	−17.4	19.7
KR0012B	Jun. 19	98	−15.8	19.4
KR0012B	Jun. 19	98	−15.8	21.4
KR0012B	Jun. 19	98	−15.9	21.3

Data from Banwart *et al.* (1994).

Table 4. *Carbon-14 dating analyses for fulvic acid and bicarbonate*

	Carbon-14 age (a)	Per cent modern carbon
Dec. 1991* borehole KR0013B		
Fulvic acid	640	91.8
Bicarbonate	3880	61.4
May 1993 borehole KR0013B		
Fulvic acid	115 ± 85	98.1
Bicarbonate	1675 ± 80	80.8
borehole KR0012B		
Fulvic acid	140 ± 85	97.8
Bicarbonate	1565 ± 130	81.8

From Banwart *et al.* 1996.
* Values from Pettersson 1992.

against oxic conditions, and thus significant aerobic respiration, within the fracture zone are the very stable and reducing redox potentials ($-120 \, \text{mV}$) measured continuously at the site (Banwart *et al.* 1994).

The significant increase in modern carbon content over time for both dissolved organic and inorganic carbon species sampled from the deeper boreholes (Table 4) shows conclusively that increasingly younger organic carbon is being transported downward and oxidized. This strongly supports our hypothesis that the input of shallow groundwater adds reducing capacity in the form of reactive dissolved organic carbon to the deeper reaches of the fracture zone. Considering the accompanying increase in alkalinity shown in Fig. 3, the excess bicarbonate must be very young, i.e. with a ^{14}C age on the order of decades. Tullborg & Gustafsson (1997) show that if the bicarbonate content of the undisturbed, saline groundwater, and its likely carbon-14 activity is accounted for, the freshwater component of the fracture discharge has a carbon-14 activity that approaches 100% modern carbon, essentially the same as that of the organic carbon in the discharge. The likely source of the young organic carbon is ongoing degradation of biomass within the recharge areas where the fracture zone intersects the land surface.

Conclusions

Site specific conclusions

The main conclusion from the experiment at Äspö was that rather than adding dissolved

oxygen that would titrate the reducing capacity in the fracture zone, the enhanced recharge caused by tunnel construction predominantly added reducing capacity in the form of organic carbon. This carbon was efficiently used to scavenge oxidants such as O_2 from the groundwater, thus helping to maintain reducing conditions in the upper bedrock.

With pH remaining nearly constant, the large increase in alkalinity observed during the experiment corresponds to a large source of inorganic carbon and to continually increasing values for P_{CO_2} in the groundwater. The observed values for bicarbonate concentration in the discharge are much higher than those observed for shallow or deeper groundwater or Baltic Sea water. This provides strong evidence for microbial respiration as a source of dissolved inorganic carbon. Additional evidence for microbial respiration includes carbon-13 signatures for dissolved inorganic carbon that indicate an organic source for $H^{13}CO_3^-$ and increases in the carbon-14 activity of organic carbon to near 100% modern carbon with a corresponding increase in the modern carbon content of the dissolved inorganic carbon.

Microbial respiration was critical to the performance of the geological barrier to scavenge oxidants, by catalysing the oxidation of the organic carbon that entered with the recharge. The initial concept for this field study presumed that dissolved organic carbon would contribute to the initial reducing capacity of the fracture zone. This turned out to be correct. However, the initial concept incorrectly assumed that this supply of carbon would be finite. This scenario turned out to be very pessimistic. In fact, the supply of O_2 appeared to be finite, with no evidence of oxic conditions after day 50. On the other hand, organic carbon was continuously added with recharge during the experiment, as evidenced by the DOC, carbon-13 and carbon-14 determinations. This result is particularly significant because this experimental site was specifically selected in order to favour development of oxic conditions; i.e. rapid recharge from a catchment of exposed bedrock with almost no soil.

Conclusions on the performance of the geological barrier under increased recharge

In this experiment there are two dominant effects from increased groundwater recharge under the disturbed hydraulic conditions created by tunnel construction.

(a) *Increased groundwater recharge adds reducing capacity* in the form of organic carbon to the groundwater. Because the fracture zone remains persistently anoxic, we conclude that dissolved oxygen is consumed within its topmost regions. The increased inflow of surface water thus adds reducing capacity to the deeper regions of the fracture zone in the form of organic carbon, rather than oxidizing capacity in the form of dissolved oxygen. This additional reducing capacity would help to maintain anoxia deep in the fracture zones while a repository remains open.

(b) *Increased groundwater recharge causes dilution* of the undisturbed, saline groundwater. Groundwater salinity affects aqueous speciation and colloid stability through ionic strength effects. These have an impact on the migration behaviour of radionuclides in several ways. Changes in aqueous speciation affect the solubility of mineral phases and the sorption behaviour of trace elements. Decreasing ionic strength favours colloid stability and thus transport of colloids and sorbed trace elements. Ion exchange surfaces are increasingly selective for divalent cations upon increasing dilution. Groundwater dilution and reaction with ion-exchange minerals may therefore result in lower calcium/sodium ratios in solution. This also favours colloid stability and can affect calcite solubility.

Conclusions on the generality of the results from the Äspö experiment

Figure 4 represents several generalized geological settings, particularly relevent to the Fennoscandian Shield, that influence recharge to groundwater. As noted in the figure, an island setting such as the field site for this experiment, favours recharge. Given the glacial history of the site, and the uplift above sea level during the past two to three thousand years, there has been little opportunity for development of soil and the associated accumulation of organic carbon. In spite of being soil-poor, the fracture zone recharge area provided a significant input of organic carbon, rather than oxygen, to the groundwater.

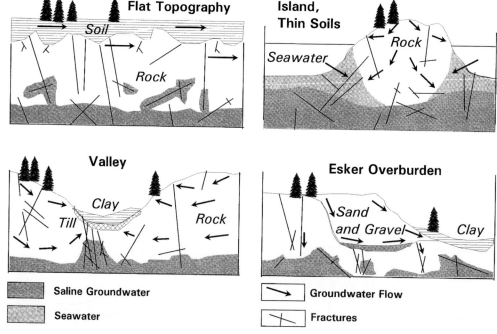

Fig. 4. A representation of hydrogeological environments where saline groundwater interfaces with recharge waters (after Olofsson 1994). Repository construction will introduce large amounts of recharge water below the salinity interface. The Äspö Groundwater Redox Chemistry Experiment showed that even a very soil-poor site like the island of Hålö can provide organic carbon to the groundwater during repository construction. This result is expected for any repository site, and would be even more likely at sites with thicker soil cover, due to the greater pool of available carbon.

This result is certainly expected at other sites, particularly those with thicker soils. Such sites would have an even larger soil carbon reservoir available for mobilization into the groundwater. Such sites would also have a longer travel time for recharge water through the thicker overburden, thus allowing more time for oxygen to be consumed, and more time for additional organic carbon to be solubilized during infiltration. This effect is assumed to occur independently of the geological setting for a potential repository, since hydraulic gradients induced by repository construction would probably overide any natural hydraulic gradients existing during undisturbed conditions.

Other types of waste in the geosphere

Other types of wastes and waste repositories have redox-related performance issues. Figure 5 shows redox potentials associated with (1) reductive precipitation of chromate ion and (2) the progressive reductive dechlorination of tetrachloromethane to methane. These two contaminants are selected as illustrative compounds to represent components of landfill leachate and hazardous industrial waste.

Chromate ion is a highly toxic mutagen. It is very soluble under oxic conditions and does not adsorb strongly above neutral pH. Under these conditions it is mobile and can pose chemical risk as a soluble ion in groundwater. Under anoxic conditions, chromium reacts with reductants such as ferrous iron, sulphide or organic carbon, and is reduced to form sparingly soluble chromium hydroxide mineral.

The redox potential for reduction of chromate therefore represents a critical threshold, below which chromium is contained, and above which chromium is mobilized. Sufficient reducing capacity to maintain the redox intensity below this threshold could be provided, for example, by including zero-valent iron or organic compost in reactive engineered barriers. This is not necessarily suggested as an engineering solution for Cr containment in particular; each contaminant and each site need full and detailed consideration. It is only used here to illustrate the point that redox transformations have a potentially important role in the design and performance of reactive barriers for containment of wastes other than radionuclides.

Chlorinated solvents are a common organic component in landfill leachate, and in industrial waste. Carbon atoms in these molecules are relatively oxidized, compared to the dechlorinated daughter compounds. This means that there is relatively little energy that can be gained by micro-organisms that seek to use them for an energy source. Because of the poor energy yield, these compounds are relatively recalcitrant with respect to biodegradation (see review by Vogel et al. 1987).

Eh (pH = 7), Volts

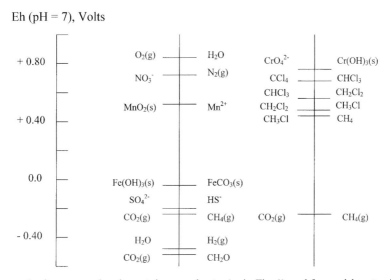

Fig. 5. Redox scales for some major elements in groundwater (as in Fig. 1), and for model contaminants representing forms of hazardous waste other than radionuclides. The scale at right shows the potentials associated with the mobility of chromium, and the reductive dechlorination of tetrachloromethane to methane (Vogel et al. 1987) and oxidation of methane to $CO_2(g)$.

Reaction with reductants such as ferrous iron is associated with loss of chlorine atoms from the molecule, a reaction process termed reductive dechlorination. Figure 5 shows the redox potentials associated with stepwise removal of chlorine atoms from tetra-chloromethane during a series of reductive dechlorination reactions that convert CCl_4 to CH_4. Methane, unlike the chlorinated parent compound, can be relatively easily oxidized to CO_2 in a separate step. Thus, a reducing environment is required to initiate the reaction sequence leading to complete biodegradation.

Again, depending on individual contaminants and site-specific considerations, redox properties can be considered in the design and performance of reactive barriers. Zero-valent (i.e. metallic) iron is now being considered as a possible component in reactive barriers that help promote rapid reductive dechlorination.

Summary

Using these examples, and the case study from the Äspö Hard Rock Laboratory, we summarize two linked principles for the redox-related performance of sub-surface repositories and the risk posed by wastes in the geosphere. *Redox potentials* need to be considered as critical thresholds for many controlling processes: corrosion of engineered barriers, solubility and sorption of metals and attenuation of organic contaminants through biodegradation. *Redox capacities*, on the other hand, indicate the total amount of reductants or oxidants required to maintain redox conditions above or below these critical thresholds for performance and risk.

These principles currently provide a rigorous geochemical framework within which to design experiments and interpret field observations on the redox-related performance of a deep geological repository for spent nuclear fuel. They can be extended to consider other types of hazardous waste and containment barriers in the geosphere. Of particular interest may be active containment barriers where redox-transformations of mobile contaminants are desirable.

The authors wish to acknowledge the investigators who participated in the Äspö Redox Experiment; M. Laaksoharju, Intera KB (Stockholm); E.-L. Tullborg, Terralogica AB (Gräbo); B. Wallin, Geokema AB (Lidingo); E. Gustafsson, GEOSIGMA (Uppsala); K. Pedersen, University of Gothenburg and A.-K. Nilsson, The Royal Institute of Technology (Stockholm), M. Snellman, I. Voima Oy (Vantaa, Finland) and P. Pitkanän and H. Leino-Forsman, Technical Research Centre of Finland (Espoo).

References

BANWART, S. (ed.) 1995. *The Äspö Redox Investigations in Block Scale. Project Summary and Implications for Repository Performance Assessment*. Technical Report **95-26**. The Swedish Nuclear Fuel and Waste Management Co., Stockholm.

—— & GUSTAFSSON, E. 1997. Natural tracers for shallow groundwater recharge and redox status. *In*: LAAKSOHARJU, M. & WALLIN, B. (eds). *Evolution of the Groundwater Chemistry at the Äspö Hard Rock Laboratory, Proceedings of the second Äspö International Geochemistry Workshop, June 6–7, 1995*, International Cooperation Report **97-04**, The Swedish Nuclear Fuel and Waste Management Co., Stockholm, 1-1–1-15.

——, WIKBERG, P. & OLSSON, O. 1997. A testbed for underground nuclear repository design. *Environmental Science and Technology*, **31**(11), 510A–514A.

——, GUSTAFSSON, E., LAASOHARJU, M., NILSSON, A.-C., TULLBORG, E.-L. & WALLIN, B. 1994. Large-scale instrusion of shallow water into a granite aquifer – Initial perturbation of a vertical fracture zone during tunnel construction at the Äspö Hard Rock Laboratory. *Water Resources Research*, **30**(6), 1747–1763.

——, TULLBORG, E.-L., PEDERSEN, K. *et al.* 1996. Organic carbon oxidation induced by large-scale shallow water intrusion into a vertical fracture zone at the Äspö Hard Rock Laboratory. *Journal of Contaminant Hydrology*, **21**, 115–125.

DIENES, P. 1980. The isotopic composition of reduced organic carbon. *In: The Terrestrial Environment, Handbook of Environmental Isotope Geochemistry. Volume 1*. Elsevier, Amsterdam, 329–406.

GAAL, G. & GORBATSCHEV, R. 1987. An outline of the Pre-cambrian evolution of the Baltic Shield. *Precambrian Research*, **35**, 15–52.

JOHANSSON, Å. 1988. The age and geotectonic setting of the Småland–Värmland granite-poryphry belt. *Geologiska Föreningens i Stockholm Förhandlingar*, **110**, 105–110.

MALMSTRÖM, M. & BANWART, S. 1996. Biotite dissolution at 25°C: the pH dependence of dissolution rate and stoichiometry. *Geochimica et Cosmochimica Acta*, **61**, 2779–2799.

MILLER, W., ALEXANDER, R., CHAPMAN, N., MCKINLEY, I. & SMELLIE, J. 1994. *Natural Analogue Studies in the Geological Disposal of Radioactive Wastes*. Elsevier Amsterdam.

OLOFSSON, B. 1994. Salt groundwater in Sweden. *In: Salt Groundwater in the Nordic Countries, Proceedings of a workshop, Saltsjöbaden, Sweden, Sept. 30–Oct. 1, 1992*. Nordic Hydrological Programme (NHP) Report No. **35**. The Research Council of Norway, Oslo.

PEARSON, F. J. JR & HANSHAW, B. B. 1970. Sources of dissolved carbonate species in groundwater and their effect on carbon-14. *In: Isotope Hydrology, Symposium proceedings of the International Atomic Energy Agency*, March 9–13, 1970, Vienna, 217–286.

PETTERSSON, C. 1992. *Properties of humic substances from groundwater and surface waters.* PhD Thesis, Linköping Studies in Arts and Sciences, University of Linköping, Linköping, Sweden.

PITKÄNEN, P., SNELLMAN, M., BANWART, S., LAAKSOHARJU, M. & LEINO-FORSMAN, H. 1994. The Äspö redox experiment in block scale-testing endmembers for mixing models. *In*: BANWART, S. (ed.), *Proceedings of the Äspö International Geochemistry Workshop, June 2–3 1994.* Äspö Hard Rock Laboratory, **B67-77**. The Swedish Nuclear Fuel and Waste Management Co., Stockholm.

SMELLIE, J. A. T., LAAKSOHARJU, M. & WIKBERG, P. 1995. Äspö, SE Sweden – A natural groundwater flow model derived from hydrogeochemical observations. *Journal of Hydrology*, **172**, 1–5, 147–169.

SWEDISH NUCLEAR FUEL AND WASTE MANAGEMENT CO. 1995. *RD&D Programme 95 – Treatment and Final Disposal of Nuclear Waste. Programme for Encapsulation, Deep Geological Disposal, and Research Development and Demonstration.* The Swedish Nuclear Fuel and Waste Management Co., Stockholm.

STUMM, W. & MORGAN, J. J. 1996. *Aquatic Chemistry.* Wiley, New York.

TULLBORG, E.-L. & GUSTAFSSON, E. 1997. Carbon-14 in bicarbonate and fulvic acid – a useful tracer? *In*: LAAKSOHARJU, M. & WALLIN, B. (eds), *Evolution of the Groundwater Chemistry at the Äspö Hard Rock Laboratory. Proceedings of the second Äspö International Geochemistry Workshop, June 6–7, 1995,* International Cooperation Report **97-04**, The Swedish Nuclear Fuel and Waste Management Co., Stockholm, 9-1–9-10.

VOGEL, T. M., CRIDDLE, C. P. & MCCARTY, P. L. 1987. Transformation of halogenated aliphatic compounds. *Environmental Science and Technology*, **21**(8), 722–735.

WALLIN, B. 1993. *Organic carbon input in shallow groundwater at Äspö, southeastern Sweden, paper presented at Conference on High Level Waste Management.* Am. Nucl. Soc., LaGrange Park, IL.

The role of sorption onto rocks of the Borrowdale Volcanic Group in providing chemical containment for a potential repository at Sellafield

J. A. BERRY, A. J. BAKER, K. A. BOND, M. M. COWPER, N. L. JEFFERIES & C. M. LINKLATER

AEA Technology plc, 220 Harwell, Didcot, Oxfordshire, OX11 0RA, UK

Abstract: The current Nirex design concept for the disposal of intermediate-level and certain low-level radioactive wastes in the UK involves emplacement in a cementitious repository deep underground. A site near Sellafield has been investigated to see if it is suitable. The Nirex Safety Assessment Research Programme includes a study of the sorption of radionuclides on the potential host rock formation, as sorption processes are important mechanisms for retarding radionuclides emanating from a repository. The potential host rock is the Borrowdale Volcanic Group (BVG) which is comprised of tuffs characterized by low porosity and permeability. This paper gives an overview of studies of the sorption of radionuclides, in particular actinides, on BVG rock and includes a description of experimental techniques in use at AEA Technology, the difficulties inherent in working with intact-rock samples of such low porosity and permeability and the use of data from both crushed and intact-rock experiments in performance assessment studies. The main sorbing minerals in the BVG are described, and the study of perturbing influences such as the presence of degradation products from organic materials in the wastes and highly alkaline porewater from the cementitious backfill are also discussed. Illustrative results are presented.

The current design concept for the disposal of intermediate-level and certain low-level radioactive waste in the UK involves emplacement in a deep geological repository. United Kingdom Nirex Limited (Nirex) has undertaken investigations to determine whether the Longlands Farm site, near Sellafield, is suitable for the siting of such a repository. In common with other radioactive waste management agencies, Nirex has adopted a multi-barrier containment concept for the potential repository. The waste packages are designed to provide physical containment of the waste. The other component of the engineered barrier is a cementitious backfill, which is used to fill the space between the waste containers. Dissolution of calcium hydroxide from the backfill and corrosion of steel containers generate conditions in the repository porewater which are both highly alkaline and chemically reducing. Such conditions limit the solubility of key radioelements such as uranium and plutonium (Chambers *et al.* 1995). The timescale over which this near-field chemical barrier is effective mainly depends on the mass of backfill in the repository and the groundwater flux through the repository after closure.

Although most of the radioactivity will decay in the repository, some residual radionuclides will be transported by groundwater into the rocks that surround the repository. Ultimately, some of this radioactivity will discharge into the biosphere, where it may come into contact with people. The geosphere barrier has several functions, including: to restrict the groundwater flow through the repository (and hence limit the rate at which radionuclides leach into the groundwater system), to dilute the contaminated water that leaves the repository (through mixing with larger volumes of groundwater along the flowpath) and to delay the return of radionuclides to the biosphere (giving more time for radioactive decay to reduce further the radionuclide concentrations). For some radionuclides, the geosphere is the most important component of the multi-barrier containment. For radionuclides such as ^{79}Se, ^{135}Cs and ^{242}Pu, whose travel times in the geosphere are very much longer than their half-lives, the geosphere provides complete containment (Baker *et al.* 1995). For those radionuclides that are not completely contained by the geosphere such as the ^{238}U chain radionuclides, geosphere sorption is an

From: METCALFE, R. & ROCHELLE, C. A. (eds) 1999. *Chemical Containment of Waste in the Geosphere.* Geological Society, London, Special Publications, **157**, 101–116. 1-86239-040-1/99/$15.00 © The Geological Society of London 1999.

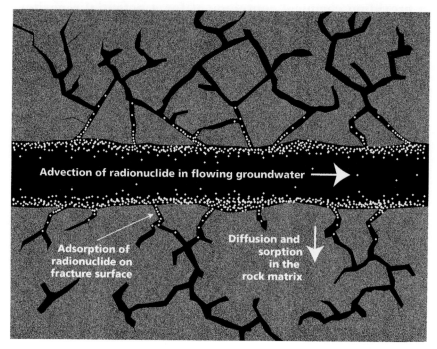

Fig. 1. Schematic diagram showing the conceptual model of the groundwater pathway in the BVG. Radionuclides are transported through the fracture network by advection with the groundwater. Two mechanisms retard migration relative to the flowing groundwater: sorption and rock–matrix diffusion.

important determinant of the radiological risk arising in the biosphere.

Hydrogeological testing (Sutton 1996) has demonstrated that the limited groundwater flow in the volcanic rocks of the Borrowdale Volcanic Group (BVG), the potential host rock at the Sellafield site, takes place predominantly through an interconnected network of fractures (Michie 1996). Little groundwater flow takes place through the blocks of rock between the fractures, because of the very low matrix permeability. Consequently, the conceptual model of the groundwater pathway in the BVG is that the radionuclides will be transported by advection with the groundwater flowing through the fracture network (Fig. 1). Two mechanisms will act to retard the migration of the radionuclides relative to the flowing groundwater: sorption and rock-matrix diffusion. Nirex undertakes an extensive research programme (the Nirex Safety Assessment Research Programme, NSARP) that addresses both of these aspects. Some geological aspects of this programme are reported in Jefferies *et al.* 1993.

The research on geosphere sorption comprises three components:

- laboratory experimental studies, to provide sorption data for performance assessment modelling and to build understanding of sorption processes;
- geochemical modelling, to build understanding of the sorption processes and to enable extrapolation of experimental sorption data to other geochemical environments;
- observation of the behaviour of natural radionuclides in groundwater systems, to identify and characterize additional radionuclide retardation processes that operate on timescales longer than can be observed in the laboratory.

The laboratory experiments are undertaken on representative samples of rock from possible flowpaths between repository and biosphere. In the case of radionuclide sorption in the BVG, the subject of this paper, two types of rock samples have been studied.

- Fracture-infill material from flowing zones (Milodowski *et al.* 1995). These samples represent the surfaces in direct contact with the groundwater flowing through the fracture network. Although volumetrically less significant than the rock matrix, it is important to have an understanding of sorption to these types of samples. Fracture-infill and flow-zone mineralogies are typically dominated by carbonate and/or clay with associated iron oxide (predominantly haematite).

Samples of rock matrix from between fractures, comprised mainly of tuffs and ignimbrites (Michie 1996). These are the most volumetrically important rock types within the BVG. Access of radionuclides to the rock matrix is achieved by diffusion (Lever & Bradbury 1985). Sorption of the radionuclides on pore surfaces in the rock matrix provides substantial retardation for the radionuclides (Nirex 1995). These rocks have a silicate-dominated mineralogy: quartz, chlorite, feldspar with minor ilmenite and secondary haematite and carbonate minerals.

The paper includes a brief description of the experimental techniques in use at AEA Technology. A comparison of crushed and intact-rock techniques is given and some of the difficulties inherent in working with intact-rock samples of low porosity and permeability are described. Results are presented that illustrate the sorption behaviour of each of the elements studied within the BVG. The principal minerals responsible for radionuclide sorption are identified. In the final section, the methods used to study the impact on radionuclide sorption of degradation of organic materials from the radioactive waste and of alkali disturbance from the cementitious backfill are described and illustrative results are presented.

Treatment of retardation in the performance assessment

In current performance assessments, radionuclide retardation in the geosphere is represented as a sorption process. Data acquired within the NSARP are used to support the representation of sorption within assessment models of the long-term performance of the repository. Sorption is generally quantified using a sorption coefficient, which is the ratio of the concentration of radionuclide sorbed on the solid (per unit mass) to the concentration in solution (per unit volume). It is assumed that sorption is linear (i.e. the sorption coefficient is independent of radionuclide concentration) and reversible. The assumption of linearity is considered reasonable for the expected range of concentrations predicted along possible flowpaths from the repository to the

biosphere. For example, recent experiments examining sorption onto haematite showed that the sorption coefficient was broadly independent of radionuclide concentration (Fig. 2). Evidence from a study of the distribution of natural uranium series elements in groundwater and rocks at Sellafield suggests that there may be some irreversible incorporation of natural radionuclides in the structure of secondary iron oxide phases (Longworth et al. 1997). The current performance assessment approach does not take credit for such additional retardation mechanisms. Thus, if the linear sorption model is appropriately parameterized, it is expected to result in overestimation of nuclide concentrations in groundwater. It is considered that this is likely to lead to an overestimate of calculated risk.

In Nirex performance assessment calculations, sorption coefficients are represented as probability density functions (PDFs) (Baker et al. 1995). The PDFs are elicited by groups of experts using laboratory data obtained under appropriate conditions. The methodology has been described elsewhere (Nirex 1994). In order to support a PDF, there must be an understanding of how sorption coefficients vary as a function of parameters such as geosphere mineralogy and geochemical conditions. Such an understanding is achieved within the research programme by undertaking experimental studies on the influence of key parameters of interest. Additionally, insight obtained from geochemical modelling can allow extrapolation of sorption coefficients to conditions where experimental measurements are unavailable. Thermodynamic expressions can be used to calculate the aqueous speciation of the radionuclide of interest, and to represent interactions of those species with mineral surfaces. Geochemical modelling codes such as HARPHRQ (Haworth et al. 1995) can thus be used to model sorption as a function of key parameters such as solution chemistry.

As will be discussed in the following sections, sorption is studied using a variety of different experimental techniques. The rock samples used are either machined to the required dimensions or crushed to a fine powder. In crushed-rock samples, a large surface area will have been generated that might not be available in samples that are left intact. Sorption is a surface phenomenon, and measurements of the sorption coefficient will thus be influenced by the experimental technique employed. PDFs are intended to represent sorption in situ, i.e. sorption onto intact rock. Many data used to generate PDFs are acquired using crushed-rock samples. It is thus necessary to develop a methodology for derivation of sorption coefficients for intact

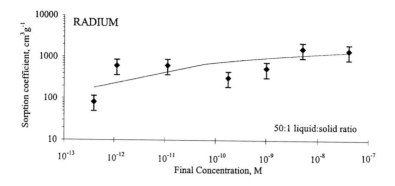

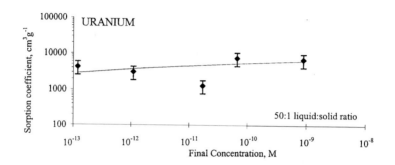

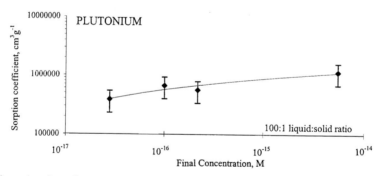

Fig. 2. Isotherm data for radium, uranium(VI) and plutonium sorption onto haematite (using the batch-sorption technique, liquid : solid ratio indicated on diagrams). The general form of the data is consistent with a linear sorption mechanism. For information, error bars corresponding to 40% of the measured value are shown (see text for further discussion of experimental variability).

samples from those measured using crushed rock. This methodology requires quantification of the amount of 'available' surface area in intact-rock and crushed-rock samples. Consideration of the mineralogy of the sorbing surface is also required when interpreting differences between intact-rock and crushed-rock samples (e.g. new surfaces may have been exposed in crushed-rock samples).

Such studies are not straightforward and are the focus of continuing effort within the NSARP. This effort concentrates in the following areas:

- a detailed sample characterization programme including
  - measurement of surface area as a function of grain size and sample dimensions
  - mineralogical analysis of different grain-size fractions
- sorption experiments carried out using crushed-rock and intact-rock techniques.

In recent assessments (Baker *et al.* 1995), it was assumed that the surface area available for sorption in intact-rock samples is less than the surface area available in experiments using crushed rock. The PDFs elicited on the basis of batch-sorption experiments are therefore modified by an appropriate factor. The factor is estimated by considering data collected so far within the two areas mentioned above (the sample characterization programme and sorption experiments using different techniques) and taking into account simple geometric models of porosity structure.

Experimental techniques in use for the measurement of sorption coefficients

Radionuclide sorption coefficients can be measured using a variety of different experimental

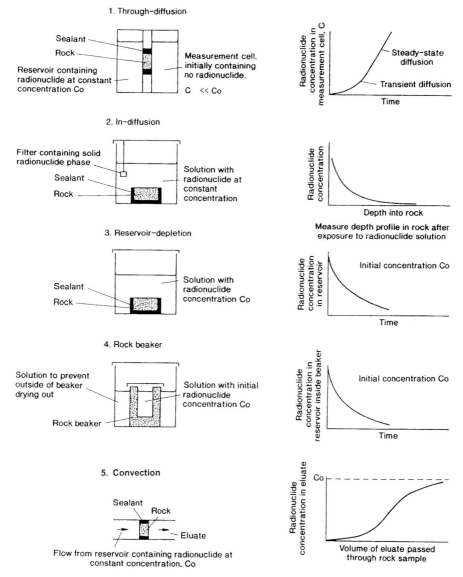

Fig. 3. Techniques for measuring radionuclide sorption onto intact samples of rock.

techniques (Lever 1986; Berry 1992). The most common method is the batch-sorption technique, in which samples of crushed rock are contacted with a solution containing a known concentration of radionuclide. As sorption proceeds, the radionuclide concentration in solution falls to a final steady value, which is measured and the sorption coefficient calculated. Issues such as sorption onto vessel walls and the statistics of radiochemical analysis are taken into account.

The principles of techniques involving the use of intact samples of rock, such as through-diffusion, in-diffusion, reservoir-depletion, rock-beaker and convection have been described in detail elsewhere (Lever 1986; Berry 1992; Triay *et al.* 1993), and are summarized schematically in Fig. 3.

Unfortunately, no single technique is ideal for the measurement of the sorption of a strongly sorbing radionuclide onto rocks. Batch sorption has the advantage of being a relatively rapid experimental technique. However, crushed-rock samples have unrealistically high surface areas and freshly exposed mineral surfaces (Bradbury *et al.* 1986). Also, batch sorption is carried out at relatively high liquid : solid ratios (Cremers *et al.* 1984; Higgo & Rees 1986) and requires careful phase separation prior to analysis of solution radionuclide concentrations (Higgo 1986).

Intact-rock techniques involve more realistic surface areas and liquid:solid ratios, but the experimental equipment is generally more complex and timescales are longer. It is recognized that machining of the samples to the required size and shape may result in some alteration to the rock surface. This effect could be important in the case of strongly sorbing radionuclides where the depth of penetration will be very small over typical laboratory timescales (e.g. 1–2 years).

For the study of the change in sorption coefficients as a function of a wide variety of perturbing parameters, it is generally accepted that the only practicable experimental technique is some form of batch methodology (Chapman & McKinley 1987). The approach adopted in the Nirex programme has been, therefore, to conduct a large number of batch experiments to study the effect of varying parameters such as radionuclide concentration, ionic strength, pH, redox potential and concentration of organic degradation products. These data provide a strong foundation on which elicitation of PDFs for repository performance assessment can be based. In order to assess surface availability issues, these batch experiments are complemented by a smaller number of experiments with intact-rock samples.

Techniques in use for the study of mineralogical controls on sorption

To supplement the measurement of sorption coefficients, and to provide information on the importance of mineralogy on sorption, surfaces of BVG samples have been studied after exposure to radionuclide solutions (principally actinides). Following exposure to actinide-bearing solutions, polished rock samples have been studied using a variety of surface analytical techniques (Berry *et al.* 1991; 1993). Radiographic methods, principally α-autoradiography, and ion beam techniques such as nuclear microprobe analysis and secondary-ion mass spectrometry (SIMS) have been used to measure qualitatively and quantitatively the distribution of actinides on the sample surfaces. Nuclear microprobe analysis involved firing a 5 MeV beam of ^{4}He ions at the surface and measuring the energy of the recoil beam (Rutherford back-scattering, RBS), and also analysing the resulting X-rays (Particle-induced X-ray emission, PIXE). RBS was used to quantify surface loadings, both surface and sub-surface concentrations being measured. PIXE has been used to confirm the identity of mineral phases when using the nuclear microprobe. SIMS analysis involved sputtering the surfaces with either a gallium or an ^{18}O beam and analysing the ions generated using a mass spectrometer.

In addition to the identification of the important sorbing mineral phases, defocused ion-beam techniques can be used to provide depth profiles for strongly sorbing radionuclides that have short penetration depths. These profiles can be used to derive diffusion data and sorption coefficients, subject to assumptions regarding access to porespace within the sample.

Control of geochemical conditions

Experimental conditions must be controlled carefully to ensure that the data acquired are relevant to the geochemical system of interest. It is particularly important to control the concentration of carbonate in experimental solutions because the aqueous speciation of some key radioelements, such as uranium(VI) is extremely sensitive to this parameter and this can have a strong impact on sorption behaviour (Hsi & Langmuir 1985).

All work has therefore been carried out under controlled conditions to simulate as closely as possible those anticipated around a potential repository. Solution compositions are based on groundwater analyses from relevant depths within Sellafield boreholes. These groundwater analyses are back-corrected using geochemical modelling techniques in order to allow for perturbation to the chemistry during the sampling process (Bond & Tweed 1995). In this way it can be ensured that the synthetic solutions used are as closely representative as possible of the groundwater chemistry *in situ*. The experiments are carried out in gloveboxes under a mixed nitrogen/carbon dioxide atmosphere. The amount of carbon dioxide in the glovebox atmosphere is adjusted to minimize loss or gain of carbonate by the experimental solutions.

Measurement of sorption coefficients for BVG samples

Tuff samples from the BVG have very low diffusivities (typically less than 10^{-13} m^2 s^{-1}) and porosities (typically less than 0.05) (Jefferies *et al.* 1993). Consequently, through-diffusion and convection experiments using strongly sorbing radionuclides would require prohibitively long times, up to 100 years. It is therefore necessary to use transient diffusion techniques (in-diffusion, reservoir depletion, rock beakers, see Fig. 3) in which a single parameter incorporating sorption and diffusive transport is calculated. In order to derive sorption coefficients from such experiments it is necessary to measure the diffusion coefficient separately using non-sorbing tracers such as tritiated water.

The interpretation of results from reservoir-depletion experiments involving actinides has been complicated by radionuclide sorption onto both the rock and the sealant used to coat the

sample, and vessel walls. This prompted development of the rock-beaker technique, which was investigated by workers at Los Alamos National Laboratory, USA, to study sorption onto porous tuff from Yucca Mountain (Triay *et al.* 1993). For work involving the BVG, the rock-beaker technique has been found to be the most appropriate intact-rock sorption technique.

Using the batch-sorption technique, a large amount of data has been acquired describing the sorption behaviour of radionuclides of interest. These data are described in the following sections. As mentioned earlier, another key area of interest is the comparison of sorption coefficients measured using crushed and intact-rock samples. A comparison of sorption coefficients measured using different experimental methods is shown in Table 1. It can be seen that sorption coefficients measured using the rock-beaker technique tend to be lower than those measured using the batch-sorption technique. This is consistent with the hypothesis that available surface area will be reduced in an intact-rock sample. However, at present, it is not possible to quantify this effect robustly because the range of measured values is wide, particularly in the case of batch-sorption experiments.

Two aspects contribute to the observed range of measured sorption values, and a rigorous comparison of sorption coefficients measured using the different techniques requires that these sources of variability are quantified for each case.

Sample variability. A large number of samples have been studied using the batch technique, and the wide range shown by the data partly reflects the different mineralogies of individual samples.

Table 1. *A comparison of BVG sorption results from experiments using crushed-rock and intact-rock techniques*

Radioelement	Sorption coefficient (cm^3 g^{-1})	
	Batch-sorption technique (50:1 liquid:solid ratio)	Rock-beaker method
Uranium (VI)	<10–110	no detectable sorption
Plutonium	1100–~10^5	2×10^4–7×10^4
Caesium	260–380	7.3–180

Values shown cover a number of experiments, samples include both tuff matrix and fracture infill. The experimental conditions were designed to represent, as closely as possible, the conditions *in situ* (near-neutral pH and solution compositions designed to represent *in situ* groundwater chemistry). Experiments were carried out under a controlled N_2/CO_2 atmosphere to minimize loss or gain of CO_2 from the solution (see text for further explanation). In selected experiments, *Eh* measurements were made and gave values of approximately +150 mV (v. Standard Hydrogen Electrode).

A more limited number of samples have been studied using intact-rock techniques.

Experimental variability. Variation in replicate measurements on the same sample can be large, especially in the case of strongly sorbing radionuclides. This variability can only be estimated in the case of batch-sorption experiments where replicate experiments have been carried out. Statistical analyses suggest that (at the 95% confidence interval) sorption coefficients generally lie within 15% of the mean. However, for strongly sorbing nuclides this value increases to 40%, due to the very low final radionuclide solution concentrations which are close to detection limits and therefore subject to larger errors.

Summary of far-field sorption data

Far-field sorption data are those acquired under conditions anticipated to match those expected *in situ* within the rock of interest. Additional effects imposed by the presence of the repository are considered separately. Batch-sorption data have been obtained for sorption onto a large number of BVG samples. As already mentioned,

two different sample groups have been studied: fracture-infill material and the rock matrix between fractures.

Figure 4 shows some typical results for sorption onto BVG tuffs. For the elements studied, highest sorption is invariably measured in the case of uranium(IV) and plutonium. Sorption coefficients are often in excess of $10^4 \, cm^3 \, g^{-1}$. Uranium(VI), radium and thorium are moderately sorbing, giving sorption coefficients around $100 \, cm^3 \, g^{-1}$. Figure 5 compares sorption coefficients measured for samples of tuff and fracture-infill material. Thorium sorption was very similar onto both types of sample. Uranium(VI) and plutonium sorption was significantly reduced onto the fracture-infill material. This could be a reflection of the carbonate-rich mineralogy of the fracture-infill material; carbonates do not have a high sorptive capacity (Berry *et al.* 1991).

Mineralogical controls on sorption

The sorptive capacity of individual minerals has been investigated by carrying out selected batch-sorption studies on single minerals of interest.

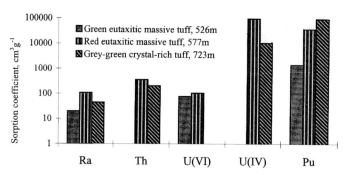

Fig. 4. Typical results for sorption onto tuff samples from the Borrowdale Volcanic Group. The data shown are for samples taken from Borehole 2, Sellefield, the depths given in the legend are metres below rotary table.

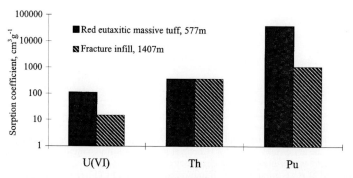

Fig. 5. A comparison of sorption coefficients measured for tuff and fracture infill assemblages. The data shown are for samples taken from Borehole 2, Sellafield, the depths given in the legend are metres below rotary table.

Table 2. *RBS data of uranium(IV) and plutonium surface loadings on tuff samples from the BVG*

Sample depth (m)*, borehole	Description of mineralogy	Measured surface loading (ng cm^{-2})
Uranium(IV)†		
622, Borehole 4	Silica-alkali feldspar matrix	2–6
	Altered plagioclase	20
	Ilmenite (FeTiO$_3$)	70–120
814, Borehole 4	Chlorite-haematite matrix	35–50
	Chlorite	50–60
Plutonium		
526, Borehole 2	Chloritized matrix	15–60
622, Borehole 4	Silica-alkali feldspar matrix	4–5
	Feldspar	3–7
	Anatase (TiO$_2$)	35–70
763, Borehole RCF3	Calcite (fracture infill)	10
	Chloritized matrix	30–35
	Chlorite	60–70
	Haematite-rich matrix	40
665, Borehole RCF3	Chloritized matrix	20–30

* The depths given refer to metres below the rotary table.
† Measured under strongly reducing conditions, $-350\,\text{mV}$ (v. Standard Hydrogen Electrode).

These experiments have also helped build understanding of sorption mechanisms, and the influence of key parameters, such as pH and solution carbonate concentration. For site-specific rock samples, however, investigation of mineralogical controls is undertaken primarily through surface analytical techniques, as mentioned earlier. The results show that sorption of uranium(IV), plutonium and thorium onto a variety of tuff samples from the BVG is not uniform but is dominated by specific minerals. Quantitative RBS results for uranium(IV) and plutonium are given in Table 2. RBS surface loading data were quantified by comparison with a thorium standard sample (McMillan *et al.* 1986). The studies show that higher concentrations of radio elements are associated with haematite, ilmenite and chlorite. Sorption onto other minerals (calcite and feldspar minerals) and quartz-rich matrices were generally up to an order of magnitude lower. RBS depth profiles have shown that the actinide is often on, or close to, the mineral surface (most activity is confined to distances of less than 10 nm from the sample surface).

Due to solubility limitations for thorium in solution (typically 10^{-11} M), it was not possible to sorb sufficient thorium on the mineral surfaces for subsequent quantitative analysis by RBS. However, sorbed thorium was measured qualitatively by SIMS and was seen to sorb preferentially on the same mineral phases as uranium(IV) and plutonium. In similar experiments involving uranium(VI) sorption onto the same intact tuff samples, surface loadings were below the limit of detection for this technique.

Iron oxides are considered to be key sorbing phases for many radionuclides. Many of the BVG samples contain significant haematite (Fe$_2$O$_3$), and enhanced sorption is associated with the presence of this mineral. Iron oxides can also form during the course of an experiment by the incongruent dissolution of iron-bearing calcites (carbonates are reactive phases and even the most careful control on experimental conditions cannot prevent such a reaction occurring). These 'experimental' iron oxides are also strong sorbing phases. Figure 6 shows an optical image of a vitric tuff sample and the corresponding α-autoradiograph. The autoradiograph shows the distribution of plutonium on the tuff surface. Plutonium is concentrated on individual crystals in the sample as well as the iron oxide-coated calcite veins running through the sample. Although a consequence of experimental conditions in this instance, such a process is also observed in natural systems.

Sorption onto iron oxide has also been identified as an important process with respect to the distribution of natural uranium in the BVG at Sellafield (Longworth *et al.* 1997). Most of the natural uranium is fixed within primary mineral phases, but small amounts have been mobilized during water–rock interactions and are associated with secondary haematite. Measurements using sequential extraction techniques have suggested that natural uranium is

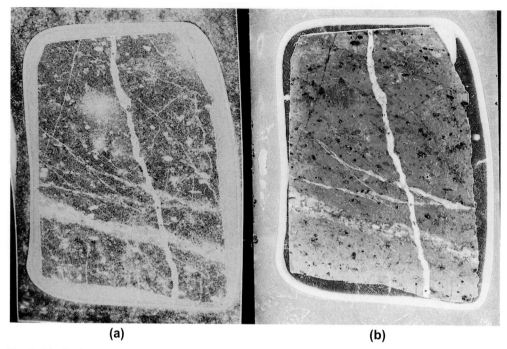

(a)   **(b)**

Fig. 6. Distribution of plutonium on a vitric tuff sample: (**a**) α-autoradiograph showing plutonium distribution; (**b**) corresponding optical micrograph.

not only sorbed on the iron oxide surfaces, but is also incorporated more deeply in the mineral lattice (Longworth *et al.* 1997). The absence of this additional mechanism in the laboratory could be due to the much shorter timescales involved, or the fact that experimental conditions are carefully controlled to minimize mineral reaction (dissolution, precipitation) during the experiments.

It has been found that sorption in batch experiments cannot be related easily to the volumetric or weight proportions of major minerals in the sample. This may be due to the influence of minor phases which are difficult (if not impossible) to identify using standard mineralogical analyses techniques, and the fact that the mineral volume or weight might not be an accurate measure of the contribution of that mineral to the surface properties of the rock in question.

Studies of other rock types using surface analytical techniques have demonstrated that the presence of iron oxide films on the surfaces of mineral grains can exert a strong influence on the distribution of sorbed radionuclides (Berry *et al.* 1991). Geochemical models based on sorption onto iron oxide surfaces have been used successfully to interpret experimental data obtained

within the NSARP (e.g. Bond & Tweed 1991). Modelling allows interpretation of data supported by knowledge of the aqueous speciation of sorbing elements, and understanding of possible sorption reactions at mineral surfaces.

The effect of repository-derived perturbations on radioelement sorption onto BVG

Effect of the presence of organic degradation products

Some waste contained within the repository will contain organic materials, such as cellulose. Under the anaerobic, alkaline conditions of the near-field, degradation of cellulose produces a range of water-soluble organic acids (Greenfield *et al.* 1994) which can form aqueous complexes with actinides (Moreton 1993). The formation of such complexes can stabilize the radionuclide in solution, causing an increase in solubility and a reduction in sorption. Organic molecules may also adsorb onto rock surfaces, possibly reducing the sorptive capacity of the rock towards radionuclides. The NSARP therefore includes studies to investigate the impact of cellulosic degradation products on radionuclide sorption in the geosphere (e.g. Baston *et al.* 1994*a*,*b*; 1995).

Batch-sorption studies of the effect of important organic materials on radionuclide sorption have been conducted:

- using authentic cellulosic degradation products (ACDP), prepared from a leachate solution formed after heating (at 80–100°C for 30 days) a mixture of wood or tissue, cement and water under anaerobic conditions;
- using the important cellulosic degradation product, iso-saccharinic acid (ISA). This

product has been shown to affect significantly radionuclide sorption behaviour within the near field of the repository (Greenfield *et al.* 1994).

The experiments have been carried out at both high and near-neutral pH values, and at various concentrations of ISA (ranging from 10^{-7} M to 10^{-3} M) and ACDP (diluted and undiluted).

These studies have included sorption of plutonium, uranium and thorium onto a wide range of BVG rock types, both matrix tuff and fracture mineral assemblages. Some typical results are shown in Figs 7 to 9. The effect of organic material on radionuclide sorption varied

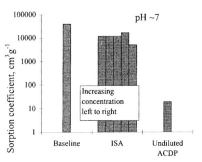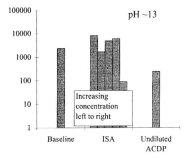

Fig. 7. Results illustrating the effect of organic degradation products on plutonium sorption onto BVG tuff. Range of ISA concentrations studied, 10^{-7} M to 10^{-3} M. The data shown are for a tuff sample taken from 577 m (below the rotary table), Borehole 2, Sellafield.

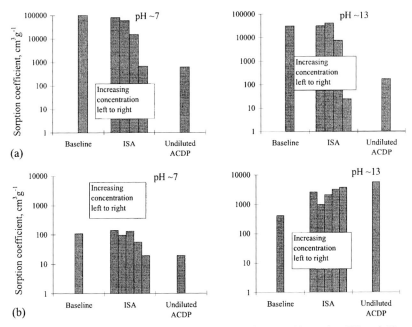

Fig. 8. Results illustrating the effect of organic degradation products on (**a**) uranium(IV) and (**b**) uranium(VI) sorption onto BVG tuff. Range of ISA concentrations studied, 10^{-7} M to 10^{-3} M. The data shown are for a tuff sample taken from 577 m (below the rotary table), Borehole 2, Sellafield.

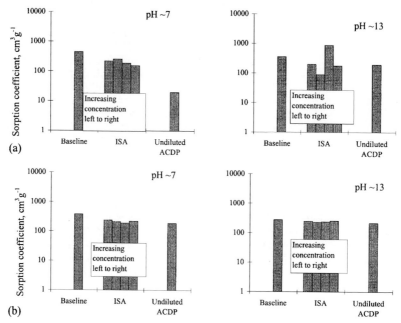

Fig. 9. Results illustrating the effect of organic degradation products on thorium sorption onto (**a**) and (**b**) fracture infill from the BVG. Range of ISA concentrations studied, 10^{-7} M to 10^{-3} M. The data shown are for tuff and fracture infill samples taken from 1407 m (below the rotary table), Borehole 2, Sellafield.

according to the concentration of organic material present, the pH and the rock type used in the experiment. Some of the main observations are as follows.

- Of the radioelements studied, plutonium and uranium(IV) are most sensitive to the presence of organic materials. Thorium is least sensitive; thorium sorption onto some rock types is completely unaffected by organic material (e.g. fracture infill material shown in Fig. 9).
- As expected, sorption is usually reduced in the presence of organic materials. Where different concentrations of organic materials were studied, results show that reductions in sorption become more significant as the organic concentration is increased.
- In the case of uranium(VI) at high pH it appears that sorption is enhanced by the presence of organic material. The reasons for this are unclear at present.
- Data for the ACDP solution and the highest ISA concentration are generally very similar, confirming that ISA is an important component of the ACDP solution. However, this is not the case for the sorption of plutonium and thorium at neutral pH, where ACDP solution appears to have a much greater

effect than 10^{-3} M ISA. It would appear that there is another component within the ACDP solution which affects these elements at neutral pH. Further work is in progress to identify this component.

Effect of alkaline conditions

Cement-equilibrated water derived from the repository will react with the surrounding rock and form an 'alkaline-disturbed zone' (ADZ) around the repository. The main reactions that are expected to occur within the ADZ are dissolution of primary silicates, and precipitation of hydrated calcium silicates (CSH phases) and possibly zeolites (Rochelle *et al.* 1992). As such water–rock interaction proceeds, the pH of the repository-derived water will be buffered toward lower values and will eventually reach the near-neutral values typical of unperturbed far-field groundwaters. It is likely that the hydrogeological and radionuclide retardation properties of the ADZ will be different from those of the unperturbed geosphere. The NSARP therefore includes work to evaluate the size of the ADZ and the extent to which its properties will be perturbed. A summary of the ADZ sorption programme is presented below.

Table 3. *Actinide sorption onto BVG samples under ADZ conditions*

Radioelement	Sorption coefficient (cm^3g^{-1})			
	Far-field	Alkaline-disturbed zone		
		No prereaction (pH 11–13*)	Pretreated (ENF) (pH 8–10)	Pretreated (YNF) (pH 10–12)
	(pH 7–8)			
Tuff				
Uranium(VI)	65–110	$18–5.4 \times 10^4$	7700	1100
Plutonium	$1500–2.5 \times 10^5$	$1800–7.7 \times 10^4$	5000	3.0×10^4
Thorium	210–450	240–5800	350–680	–
Fracture Infill				
Uranium(VI)	4	$1.0 \times 10^4–4.4 \times 10^4$	50	490–2000
Plutonium	1200	$5.5 \times 10^4–1.0 \times 10^5$	2600	$2.8 \times 10^4–5.9 \times 10^4$
Thorium	380	–	280	–

Where ranges of values are given, more than one experiment is represented.
* In these experiments the pH changed during the experiments (probably due to alkaline water–rock interaction during the experiments). The starting pH was 12.5. The range given in the table represents the range of pH values measured at the end of the experiments.

Using the batch-sorption technique, experiments have been carried out to examine sorption within the ADZ. The following types of experiment have been carried out.

(1) Sorption onto untreated rock samples in contact with simulated evolved near-field (ENF) groundwater ($Ca(OH)_2$-rich, pH 12.5 at 25°C, Bond & Tweed 1995). ENF groundwater is the water produced by interaction of far-field groundwaters with Nirex Reference Vault Backfill after the soluble alkali metal hydroxides have been leached out. These experiments, in which some degree of alkaline water–rock interaction will occur during the sorption experiment, simulate the case where radionuclide migration into the geosphere is concurrent with the development of the ADZ.
(2) Sorption onto rock samples pretreated with simulated ENF groundwater (pH 12.5 at 25°C). The solutions used in the experiments simulate the solution composition measured at the end of the pretreatment process, and have pH in the range 8–10. These experiments simulate the case where radionuclides migrate into the geosphere after development of the ADZ.
(3) Sorption onto rock samples pretreated with simulated young near-field (YNF) porewater ($NaOH/KOH/Ca(OH)_2$ solution, pH 13 at 25°C, Bond *et al.* 1995). YNF porewater is intended to represent fluids emerging from the cementitious repository at very early stages in its evolution. The solutions used in the experiments

simulate the solution composition measured at the end of the pretreatment process, and have pH in the range 10–12. These experiments also simulate the case where radionuclides migrate into the geosphere after development of the ADZ.

Pretreated samples were supplied by the British Geological Survey, Keyworth. These samples had been exposed to alkaline solutions at elevated temperatures (70°C) for up to four months.

Table 3 summarizes the results from ADZ sorption experiments. Sorption coefficients under ADZ conditions were generally higher than those measured under unperturbed, far-field conditions. Figure 10 shows results for sorption onto an individual sample (fracture infill from the BVG). The observed sorption behaviour largely follows the trend:

No pre-reaction > Pretreated (YNF)

> Pretreated (ENF) ≥ Unperturbed

The trend is most pronounced in the case of uranium(VI), and least pronounced in the case of thorium.

Enhanced sorption onto ADZ samples relative to unperturbed, far-field samples is believed to be related to the presence of newly formed, high-surface-area secondary precipitates. CSH phases, identified as the dominant reaction products in these types of experiments (Rochelle *et al.* 1992), could be important contributors to sorption. At pH values greater than 10, radioelement aqueous speciation is often dominated by neutral

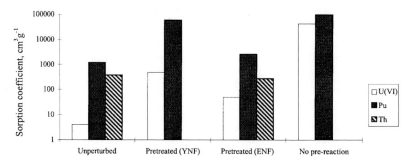

Fig. 10. Effect of alkaline conditions on the sorption of uranium(IV), plutonium and thorium onto fracture infill from the BVG. The data shown are for a fracture infill sample taken from 1407 m (below the rotary table), Borehole 2, Sellafield.

or negatively-charged hydrolysis products. Common rock-forming minerals which might contribute significantly to the sorptive capacity of a rock are iron oxides and phyllosilicates. These minerals have a point of zero charge below pH 8 (Sposito 1989). Therefore at solution pH > 10 these minerals will have a negative surface charge and tend to repel radionuclide species. CSH phases possess a high point of zero charge (Baston *et al.* 1995). The surface of a CSH phase can therefore retain a significant positive surface charge at pH 10–13 and can complex with negatively-charged radionuclides species. The experimental data for the ADZ have therefore been interpreted based on the concept of CSH phases controlling radionuclide sorption behaviour.

Experimental results suggest that thorium, uranium(VI) and plutonium sorption within the ADZ surrounding a cementitious repository will be the same as or higher than sorption in the undisturbed geosphere.

Conclusions

An extensive experimental programme, employing both batch and intact-sample techniques, is being carried out to study the sorption of radionuclides, in particular actinides, onto rock samples from the BVG. Results from experiments using single mineral phases, interpretative geochemical modelling and studies of natural radioactivity at Sellafield all contribute to an understanding that has been used to support performance assessment of the potential Sellafield repository.

The results from experimental investigations of sorption onto BVG samples have produced an extensive database that has been integral in the generation of PDFs for performance

assessment calculations. In particular, these data show that:

- actinide sorption onto BVG varies from relatively weak (uranium(VI)) to very strong (plutonium)
- plutonium sorption can be reduced significantly in the presence of high concentrations of cellulosic degradation products
- sorption onto BVG altered by cement porewater is at least as strong as onto unaltered BVG.

Probability density functions based on the programme of work summarized in the paper have been used in performance assessment of the potential Sellafield repository. The recent Nirex 95 performance assessment (Baker *et al.* 1995) demonstrated that sorption in the geosphere means that the geosphere at Sellafield would be a very effective barrier that prevents the appearance of many radionuclides in the biosphere. For some of those radionuclides that are not contained by the geosphere, sorption is important in determining the magnitude of the radiological risk arising in the biosphere. An understanding of sorption processes in the BVG and other rock units at Sellafield is therefore a key focus of the research programme being undertaken by Nirex.

This work was funded by United Kingdom Nirex Limited as part of the Nirex Safety Assessment Research Programme.

References

BAKER, A. J., JACKSON, C. P., SINCLAIR, J. E., THORNE, M. C. & WISBEY, S. J. 1995. *Nirex 95: Preliminary analysis of the groundwater pathway for a deep repository at Sellafield: Vol. 3: Calculations of risk.* Nirex Science Report **S/95/012**.

BASTON, G. M. N., BERRY, J. A., BOND, K. A., BOULT, K. A., BROWNSWORD, M. & LINKLATER, C. M. 1994. Effects of cellulosic degradation products on uranium sorption in the geosphere. *Journal of Alloys and Compounds*, **213/214**, 475–480.
——, ——, ——, ——, —— & ——1994. Effects of cellulosic degradation product concentration on actinide sorption on tuffs from the Borrowdale Volcanic Group, Sellafield, Cumbria. *Radiochimica Acta*, **66/67**, 437–442.
——, ——, BROWNSWORD, M., HEATH, T. G., TWEED, C. J. & WILLIAMS, S. J. 1995. Sorption of plutonium and americium on repository, backfill and geological materials relevant to the JNFL low-level radioactive waste repository at Rokkasho-Mura. *Scientific Basis for Nuclear Waste Management XVIII, Materials Research Society Symposium Proceedings*, **353**, 957–964.
BERRY, J. A. 1992. *A review of sorption of radionuclides under the near- and far-field conditions of an underground radioactive waste repository, Parts I–III*. UK DOE Report DOE/HMIP/RR/**92.061**.
——, BISHOP, H. E., COWPER, M. M., FOZARD, P. R., MCMILLAN, J. W., & MOUNTFORT, S. A. 1993. Measurement of the sorption of actinides on minerals using microanalytical techniques. *Analyst*, **118**, 1241–1246.
——, COWPER, M. M., GREEN, A., JEFFERIES, N. L. & LINKLATER, C. M. 1991. Sorption of radionuclides on mineral surfaces. *In: Proceedings of the 3rd International Conference On Nuclear Fuel Reprocessing and Waste Management, Sendai, Japan*. 988–993.
BOND, K. A. & TWEED, C. J. 1991. *Geochemical modelling of the sorption of tetravalent radio-elements*. Nirex Report **NSS/R227**.
—— & ——1995. *Groundwater compositions for the Borrowdale Volcanic Group, Boreholes 2 4 and RCF3, Sellafield, evaluated using thermodynamic modelling*. Nirex Report **NSS/R397**.
BRADBURY, M. H., GREEN, A., LEVER, D. A. & STEPHEN, I. G. 1986. *Diffusion and permeability based sorption measurements in sandstone, anhydrite and upper magnesian limestone samples*. UKAEA Report **AERE-R.11995**.
CHAMBERS, A. V., WILLIAMS, S. J. & WISBEY, S. J. 1995 *The near field current status in 1994*. Nirex Science Report **S/95/011**.
CHAPMAN, N. A. & MCKINLEY, I. G. 1987. *The Geological Disposal of Nuclear Waste*. Wiley, New York.
CREMERS, A., HENRION, P. & MONSECOUR, N. 1984. Radionuclide partitioning in sediments. Theory and practice. *Proceedings of the International Seminar on the Behaviour of Radionuclides in Estuaries, Renesse, The Netherlands*, 1–25.
GREENFIELD, B. F., LINKLATER, C. M, MORETON, A. D., SPINDLER, M. W. & WILLIAMS, S. J. 1994. The effects of organic degradation products on actinide disposal. *In: MISHRA, B. & AVERILL, W. A. (eds) Actinide Processing: Methods and Materials*. The Minerals, Metals and Materials Society, 289–303.

HAWORTH, A., HEATH, T. G. & TWEED, C. J. 1995. *HARPHRQ: A computer program for geochemical modelling*. Nirex Report **NSS/R380**.
HIGGO, J. J. W. 1986. *Clay as a barrier to radionuclide migration: A review*. Department of the Environment Report DOE/RW/**86.082**.
—— & REES, L. V. C. 1986. Adsorption of actinides by marine sediments. Effect of sediment/seawater ratio on measured distribution ratios. *Environmental Science and Technology*, **20(5)**, 483–490.
HSI, C-K. D. & LANGMUIR, D. 1985. Adsorption of uranyl onto ferric oxyhydroxides: Application of the surface complexation site-binding model. *Geochimica et Cosmochimica Acta*, **49**, 1931–1941.
JEFFERIES, N. L., LEVER, D. A. & WOODWARK, D. R. 1993. *NSARP Reference Document. Radionuclide transport by groundwater flow through the geosphere: January 1992*. Nirex Report **NSS/G118**.
LEVER, D. A. 1986. *Some notes on experiments measuring diffusion of sorbed nuclides through porous media*. UKAEA Report **AERE-R.12321**.
—— & BRADBURY, M. H. 1985. Rock-matrix diffusion and its implications for radionuclide migration. *Mineralogical Magazine*, **49**, 245–254.
LONGWORTH, G., LINKLATER, C. M., HASLER, S. E., MILODOWSKI, A. E. & HYSLOP, E. K. 1997. *Interpretation of uranium series measurements on Sellafield rocks and groundwaters*. Nirex Science Report **S/97/004**.
MCMILLAN, J. W., POLLARD, P. M. & PUMMERY, F. C. W. 1986. Combined RBS, PIXE and NRA examination of plutonium surface contamination on steel using the Harwell Nuclear Microprobe. *Nuclear Instruments and Methods in Physics Research*, **B15**, 394–397.
MICHIE, U. M. 1996. The geological framework of the Sellafield area and its relationship to hydrogeology. *Quarterly Journal of Engineering Geology*, **29**, S13–28
MILODOWSKI, A. E, GILLESPIE, M. R., SHAW, R. P. & BAILEY, D. E. 1995. *Flow-zone characterisation: Mineralogical and fracture orientation characteristics of the PRZ and Fleming Hall Fault Zone area boreholes, Sellafield*. Nirex Science Report No. **SA/95/001**.
MORETON, A. D. 1993. Thermodynamic modelling of the effect of hydroxycarboxylic acids on the solubility of plutonium at high pH. *Scientific Basis for Nuclear Waste Management XVI, Materials Research Society Symposium Proceedings*, **294**, 753–758.
NIREX 1994. *Post-closure performance assessment: Information Management*. Nirex Science Report No. **S/94/002**.
NIREX 1995. *Post-closure performance assessment: Modelling of groundwater flow and radionuclide transport*. Nirex Science Report No. **S/94/004**.
ROCHELLE, C. A., BATEMAN, K., MILODOWSKI, A. E., NOY, D. J., PEARCE, J. & SAVAGE, D. 1992. Reactions of cement pore fluids with rock: Implications for the migration of radionuclides. *In: KHARAKA, Y. K. & MAEST, A. S. (eds) Water-Rock Interaction*. Balkema, Rotterdam.

SPOSITO, G. 1989. *The Chemistry of Soils.* Oxford University Press, New York.

SUTTON, J. S. 1996. Hydrogeological testing in the Sellafield area. *Quarterly Journal of Engineering Geology*, **29**, S29–38

TRIAY, I. R., BIRDSELL, K. H., MITCHELL, A. J. & OTT, M. A. 1993. Diffusion of sorbing and non-sorbing radionuclides. *In: Proceedings of 4th Annual International Conference on High Level Radioactive Waste Management, Las Vegas.*

The underground sequestration of carbon dioxide: containment by chemical reactions in the deep geosphere

C. A. ROCHELLE, J. M. PEARCE & S. HOLLOWAY

British Geological Survey, Kingsley Dunham Centre, Keyworth, Nottingham, NG12 5GG, UK

Abstract: Anthropogenic emissions of carbon dioxide (CO_2) have been linked to increasing levels in the atmosphere and to potential global climate change. The capture of CO_2 from large point sources, followed by its sequestration as a supercritical fluid into the deep geosphere, is one potential method for reducing such emissions without a drastic change in our energy-producing technologies. Once emplaced underground, geochemical and hydrogeological processes will act to 'trap' the CO_2 as dissolved species and in carbonate minerals. Although dry supercritical CO_2 appears to cause little reaction with the host rocks, once dissolved in water mineral dissolution and precipitation reactions can result. From a geochemical standpoint, sandstones appear to be preferable to carbonates for sequestration operations because fluid–mineral reactions within them have a better capacity for pH buffering. However, individual host lithologies will vary in structure, mineralogy and hydrogeology, and individual sequestration operations will have to take account of local geological, fluid chemical and hydrogeological conditions. This paper summarizes some of the recent laboratory experimental, natural analogue and computer modelling approaches directed at understanding reactions involved in the chemical containment of CO_2.

Our planet is warmed by a natural greenhouse effect resulting from the presence of gases such as methane, CO_2, and water vapour in the atmosphere. Without this, global mean annual temperatures would be about $-6°C$ instead of the current $+15°C$ (Houghton 1994). The other main contributors are, in decreasing order, chlorofluorocarbons (CFCs), methane, nitrous oxide and ozone. CO_2 is not one of the most effective anthropogenic greenhouse gases, but it is emitted into the atmosphere in vast quantities. As a result it is probably responsible for about 56% of the anthropogenic greenhouse effect at present (Smith 1988). Although subject to much debate, it is now generally accepted that anthropogenic CO_2 emissions are contributing to the global rise in atmospheric CO_2 levels (e.g. Houghton 1994) and, as a consequence, to an enhanced greenhouse effect. The possibility that this will result in climate change is now causing international concern.

The thermal effect of enhanced CO_2 levels can be calculated; for example, the direct effect of a doubling of CO_2 levels in the atmosphere would lead to an average warming of just over $1°C$ at the Earth's surface (Thurlow 1990). How such heating would affect the Earth, and whether natural negative feedback mechanisms would occur, is not clear and is subject to much debate.

However, existing complex inter-relationships could alter in a warmer world, and new feedback mechanisms could be triggered (see Leggett 1990 for a summary). Measurements show that global temperatures have already risen by 0.3–$0.6°C$ this century. However, predictions of the climatic warming which could be caused by the increased concentrations of anthropogenic greenhouse gases in the atmosphere, contain a large degree of uncertainty. The scientific assessment by the Intergovernmental Panel on Climate Change (WMO/UNEP 1990) is that in a 'business as usual' scenario the global mean temperature is likely to be about $1°C$ higher in 2025 than at present, and about $3°C$ higher than at present by the end of the 21st century. In other words, they believe that anthropogenic greenhouse gases will cause a historically unprecedented global warming in the next century.

This paper will not speculate on the likely effects of an unprecedented global warming. It will focus on methodologies to enhance 'sinks' for CO_2 as ways of reducing anthropogenic emissions to the atmosphere and, in particular, geochemical considerations relevant to underground sequestration. However, it is useful to note that Woodwell (1990) highlights the single most important factor to be the rate at which climate change could occur. This could be an

From: METCALFE, R. & ROCHELLE, C. A. (eds) 1999. *Chemical Containment of Waste in the Geosphere.* Geological Society, London, Special Publications, **157**, 117–129. 1-86239-040-1/99/$15.00 © The Geological Society of London 1999.

order of magnitude faster than the most rapid climate changes of the recent geological past, and may have profound consequences on the ability of organisms and ecosystems to respond. The response of animals to adverse climate change is, initially, to migrate to more favourable climates (Huntley 1990). In the past this was possible for man. However, in the future it is quite possible that national borders, now commonly heavily defended, will be closed to 'climate refugees', possibly with disastrous implications for the affected populations.

Sources of anthropogenic CO_2 emissions and their possible control

The largest single source of anthropogenic CO_2 emissions to the atmosphere results from 'energy consumption', and is thought to be about 22 gigatonnes (Gt) annually ($= 2.2 \times 10^{13} \, \text{kg a}^{-1}$) (Turkenburg 1997). This is about 85% of the net anthropogenic CO_2 emissions. Other anthropogenic sources of CO_2 are land-use changes (about 13%), and cement manufacture (about 2%). If past trends in energy usage continue, energy-based emissions may triple over the next 50 years (Turkenburg 1997). The major contributor is fossil fuel combustion, which provides approximately 80% of the annual anthropogenic CO_2 emissions to the atmosphere (Smith 1988; Thurlow 1990). It has been estimated that a reduction of at least 60% in CO_2 emissions would be required to stabilize global atmospheric CO_2 at today's levels (WMO/UNEP 1990). It is impossible to be certain what the major sources of CO_2 emissions will be in the medium to longer term, as they depend to a large extent on political, social and technical developments. However, the investment that has been made on existing infrastructure means that the most likely scenario is that fossil fuels will continue to supply the major part of the world's energy needs well into the next century (Smith 1988). Although predictions are uncertain, long-term forecasts (Smith 1988) indicate that it may be the middle, or even the end, of the next century before CO_2 concentrations reach double the pre-industrial level.

If increased concentrations of CO_2 in the atmosphere reflect a higher rate of anthropogenic input compared to natural removal rates, and if fossil fuels continue to be the dominant source of power production, there is an advantage in enhancing natural 'sinks' of CO_2. Virtually all (more than 99.9%) of the Earth's carbon resides in the geosphere and a proportion

of this is naturally slowly cycled between the geosphere, atmosphere, biosphere and oceans as a result of volcanicity, weathering, erosion, and biological activity (the 'carbon cycle'). For example, Lee et al. (1998) estimate that on average approximately 1.5 Gt of CO_2 per year dissolves into the world's oceans from the atmosphere. If various sinks for CO_2 are to be enhanced, this sequestration must be done in such a way that the CO_2 does not reach the atmosphere for at least several hundred years (Freund & Ormerod 1997). Examples of enhancing such sinks include; reforestation (Riemer 1994; Yokoyama 1997), enhancement of oceanic algal blooms (Cooper et al. 1996; Jones & Otaegui 1997), ocean storage of CO_2 (De Baar 1992; Golomb et al. 1992; Herzog et al. 1996; Ormerod 1997; Wilson 1992) and underground storage of CO_2 (Gunter et al. 1993; Hitchon 1996; Holloway 1996a, b 1997a; Holloway et al. 1996). In many ways the first two examples seem more advantageous in that they do not involve the costs of separation, transportation and sequestration. However, the latter two examples have the advantage that they could be combined with direct extraction of CO_2 from large point sources, such as power station flue gases (for a review of separation technologies see Meisen & Shuai 1997). In other words, the CO_2 would not have to be emitted to the atmosphere in the first place. Underground sequestration also has the advantage that the carbon in the CO_2 is returned directly to the Earth's main CO_2 store – the geosphere, and once sequestered would have minimal direct impact on living organisms. As electricity generation contributes about 30% of global emissions from fossil fuel consumption (Thurlow 1990), and thus about 24% of total global anthropogenic CO_2 emissions, CO_2 capture from power-station flue gases could make a significant reduction in anthropogenic CO_2 emissions. Although such a process may have environmental benefits, the installation of new equipment for CO_2 separation and sequestration would be expensive (Holloway 1997a). However, the economics could be altered radically if separation and sequestration costs could be set against any future 'carbon tax' (e.g. Bahn et al. 1998).

The concept of CO_2 sequestration in the deep geosphere

Untreated power station flue gases generally contain between 3% and 16% CO_2 (Holloway et al. 1996) with much of the rest being nitrogen.

This mixture would require an unduly large amount of energy to compress it and, consequently, separation of CO$_2$ from the flue gases is necessary. The most practicable method appears to be 'scrubbing' using a solvent (commonly amine-based), although membrane technologies are being developed and may be available in the future. The CO$_2$ recovered would be over 97% pure, with impurities such as N$_2$, O$_2$, H$_2$, SO$_2$, H$_2$S, NO$_X$ (dependent upon the type of power station) (Summerfield et al. 1996). Once separated and dried, the CO$_2$ can be transported by pipeline for sequestration. Holloway et al. (1996) have estimated that a 500 MW coal-fired, pulverized-fuel power station (the world's most common type) with flue gas desulfurization would produce about 600 tonnes per hour ($= 167$ kg s^{-1}) of separated CO$_2$. Under near-surface conditions this equates to an enormous volume of gas (approximately 83 m^3 s^{-1}, or 1 km^3 every 4.5 months). Consideration of the phase diagram for CO$_2$ (Fig. 1) shows that relatively small increases in temperature and pressure cause it to behave as a supercritical fluid (above about 31°C and 73 bar pressure). These conditions are found naturally at depths below approximately 800 m depending upon the local geothermal gradient and reservoir pressure (Holloway & Savage 1993; Law & Bachu 1996), and are easily reached using present day drilling technologies. Supercritical CO$_2$ has an average density of about 0.7 g cm^{-3} ($= 700$ kg m^{-3}) over a wide range of potential in situ reservoir conditions (Lindeberg & van der Meer 1996) and as a consequence will occupy much less volume ($<0.3\%$) compared to its gaseous form under surface conditions. Thus, the actual storage volumes required for a CO$_2$ phase are greatly reduced, making the concept of sequestration in the deep geosphere practicable.

Underground sequestration of CO$_2$ requires injection into a porous and permeable rock (i.e. an aquifer). For example, sandstones, and to a lesser extent chalk, might be considered possible host rocks. Sequestration also requires an overlying aquiclude ('caprock') such as a clay, shale or other rock impermeable to supercritical CO$_2$ to contain the 'bubble' of CO$_2$ (which is less dense than water). Such associations occur in many parts of the world where large sedimentary basins are found. For example, lithologies being investigated for their potential to sequester CO$_2$ include: the Alberta Basin in western Canada (Gunter et al. 1993; Hitchon 1996; Law & Bachu 1996), a basin associated with the Natuna gas field beneath the South China Sea (IEA 1996a), a basin beneath the North Sea (Holloway 1996a, b; Holloway et al. 1996), and depleted oil and gas reservoirs in Texas (Bergman et al. 1997). Such locations also have the advantage of being close to major CO$_2$-producing regions, have existing infrastructure capable of undertaking large-scale CO$_2$ sequestration operations and have relatively well known regional geology. As an example of underground sequestration, one limited scale operation is already being performed beneath the North Sea, some 250 km west of Stavanger (Baklid et al. 1996; IEA 1996b, 1998). Here, the Norwegian State oil and gas company (Statoil) is exploiting the CO$_2$-rich Sleipner Vest natural gas field. The CO$_2$ has to be separated from the natural gas before it is sold, and normally this would be vented to the atmosphere. However, to avoid releases of some one million tonnes of CO$_2$ per year (approximately 3% of the annual Norwegian emissions (Baklid et al. 1996)), it is now being re-injected deep underground into a sandstone aquifer.

Although depleted oil and gas reservoirs have proven seals and could 'physically trap' CO$_2$, the volume available for storage (for data on Europe see Elewaut et al. 1996) is very small compared to regional aquifers that are capable of sustaining long-term storage operations. However, many basins are underlain by extensive, deep, sandstone aquifers which have many properties analogous to hydrocarbon reservoirs. If CO$_2$ could be injected into these, even assuming a low 'sweep efficiency' (i.e. a relatively low ability to trap CO$_2$), then they may have significant storage capacity. For example, Holloway et al. (1996) estimate that such formations in Europe may have enough capacity to sequester 700 years of current European CO$_2$ emissions.

If these aquifers are to be used, it is vital that the CO$_2$ can be contained safely within them for

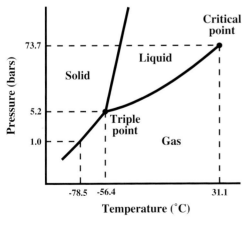

Fig. 1. Phase diagram for CO$_2$ (based upon Atkins 1983).

extended timescales. Previous workers (Gunter et al. 1993; Bachu et al. 1994) have suggested that a lateral seal is not needed in such aquifers and that two other trapping mechanisms ('hydrodynamic trapping' and 'mineral trapping') could act to contain the CO_2. In essence, 'hydrodynamic trapping' results from CO_2 migration (as either a supercritical or a dissolved phase) driven by regional-scale flow velocities (probably only a few centimetres per year). Such low flow velocities would allow sufficient time for CO_2 to dissolve into the formation water and become subject to diffusion, dispersion and convection. Bachu et al. (1994) suggested, in their studies of the Alberta Basin, Canada, that diffusion and dispersion would be the main transport mechanisms for CO_2, leading to an increased sweep efficiency and hence greater storage potential for CO_2. The second trapping mechanism, 'mineral trapping', involves geochemical reactions that sequester CO_2 in the form of carbonate minerals. The advantages of this trapping mechanism are that the minerals are stable and that the CO_2 is effectively immobilized for geologically-important timescales. Over very long timescales, mineral trapping is likely to be the dominant trapping mechanism for CO_2 (Gunter et al. 1997). The types and magnitude of reactions that will occur depend upon a variety of factors; for example, the mineralogical composition of the host rock, formation water chemistry, in situ pressure and temperature, groundwater flow rates, and the relative rates of the dominant reactions. The following sections of this paper will concentrate on some of these factors and will show how they may change over time as the sequestration process progresses and how they may influence the geochemical containment of CO_2.

Geochemical containment of CO_2

During underground sequestration operations, supercritical CO_2 will be injected into a deep aquifer and rise, as a result of buoyancy, until it reaches the overlying aquiclude where it will form a 'bubble'. The presence of supercritical CO_2 will result in chemical disequilibria and hence the initiation of reactions. It is important to understand the direction, rate and magnitude of such reactions, both in terms of their impact upon the ability of the aquifer to safely contain the injected CO_2, and in terms of the longevity of CO_2 containment. Three broad areas of reaction can be considered:

- reactions with the formation water (i.e. solution of CO_2)
- reactions with minerals in the host aquifer
- reactions with minerals in the overlying aquiclude.

Reactions with the formation water will be important because these form a relatively rapid and large sink for CO_2. Reactions with mineral phases are likely to be slower than with the formation water, but might provide a more permanent sink for CO_2 in the form of carbonate minerals. Various approaches have been used to investigate such reactions, including; laboratory experiments (e.g. Czernichowski-Lauriol et al. 1996; Gunter et al. 1997), computer modelling (e.g. Czernichowski-Lauriol et al. 1996; Gunter et al. 1997; Law & Bachu 1996; Perkins & Gunter 1995) and the study of natural analogues (Pearce et al. 1996). However, the majority of investigations have focused upon fluid–mineral reactions that might occur within the aquifer itself. It is noteworthy, however, that each of the above approaches should not be conducted in isolation. A synthesis of data and conclusions drawn between them is important in developing an overall understanding of the reactions that may occur during and after the sequestration of CO_2 in the deep geosphere.

Reactions between supercritical CO_2 and aqueous fluids

In general, reactions between liquid water and supercritical CO_2 will be faster than those involving solid phases. As such, they will be the first reactions to take place once CO_2 has been injected into the host aquifer. Typical reactions might include the following:

$$CO_2(sc) \rightleftharpoons CO_2(aq) \qquad (1)$$

$$H_2O + CO_2(aq) \rightleftharpoons H_2CO_3 \qquad (2)$$

$$H_2CO_3 \rightleftharpoons HCO_3^- + H^+ \qquad (3)$$

$$HCO_3^- \rightleftharpoons CO_3^{2-} + H^+ \qquad (4)$$

These linked reversible reactions are dependent upon the in situ temperature and pressure, and will control the solubility of CO_2. Because Eqs (3) and (4) involve the generation of H^+ they are also dependent upon the ability of the host aquifer to buffer pH. For example, addition of CO_2 to water will generate H^+ ions (i.e. it will lower the pH of the solution). However, CO_2 solubility decreases with decreasing pH. Consequently, more dissolved CO_2 can be 'trapped' in an aquifer that can maintain (buffer) the pH of the formation water compared to one

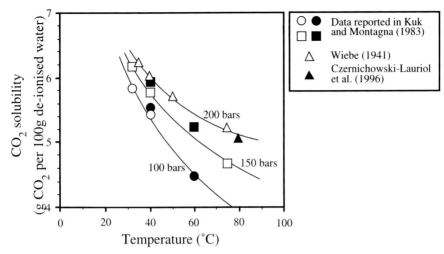

Fig. 2. CO_2 solubility in de-ionized water.

where formation water pH decreases. Previous computer simulations (Czernichowski-Lauriol *et al.* 1996; Gunter *et al.* 1993, 1997) indicate that sandstone aquifers are better at buffering formation water pH than carbonate aquifers.

Previous work with de-ionised water (van Eldik & Palmer 1982), albeit at relatively low pressures, has shown that about 99% of the total dissolved CO_2 is in the form of the dissolved gas rather than as true carbonic acid. As a consequence, reaction (1) is of key interest in understanding the dissolution of CO_2. This reaction is also likely to be slower than reactions involving ionic species, and is possibly the overall rate-limiting step in CO_2 dissolution. However, the rate of dissolution is still much faster than most fluid–mineral reactions, with equilibrium being obtained within 24 hours for a wide range of temperatures and pressures below 90 bar (Czernichowski-Lauriol *et al.* 1996; Ellis & Golding 1963; Stewart & Munjal 1970).

For a given fluid composition and likely range of typical *in situ* temperatures, CO_2 solubility increases with increasing pressure, but decreases with increasing temperature (Fig. 2). Salinity also affects CO_2 solubility, which decreases as salinity increases. For example, at 80°C and 200 bar pressure, CO_2 solubility is 1.16 mol kg^{-1} of de-ionized water and 0.98 mol kg^{-1} of 0.55 M NaCl solution (Czernichowski-Lauriol *et al.* 1996).

Once injected underground, the CO_2 is likely to migrate. However, this is likely to be very slow (only a few centimetres per year) because, once outside the influence of the injection well, dissolved and supercritical CO_2 will travel at the natural velocity of the formation water

(Gunter *et al.* 1997). Bachu *et al.* (1994) called this process 'hydrodynamic trapping', and in their studies of the Alberta Basin, Canada, suggested that diffusion and dispersion would be the main transport mechanisms for dissolved CO_2, leading to an increased sweep efficiency and hence greater storage potential for CO_2.

In summary therefore, the quantity of CO_2 that can be trapped by dissolving into the formation water will be dependent upon site-specific parameters (e.g. *in situ* pressure, temperature and formation water chemistry), and individual storage operations will have to consider these.

Reactions between supercritical CO_2 and aquifer minerals

Previous studies have used three main approaches to assess the degree of reaction between injected CO_2, formation water and minerals within the host aquifer:

- laboratory experiments
- natural analogues
- computer modelling.

Each of these has its own advantages and disadvantages. For example, detailed investigation of relatively short-term processes under simulated *in situ* conditions can be achieved by laboratory experiments, but overall reaction time is limited by the length of the study, which is seldom more than a few years. Observation of processes operating over much longer timescales can be achieved by studying rocks associated with natural accumulations of CO_2. However, samples can only be recovered by drilling and are likely to be limited in number. Geochemical computer modelling can be very useful for linking shorter-term and longer-term processes, and also has the advantage of a predictive capability. However, some codes many not cope well with CO_2 in its supercritical state, and the output from the models will always be limited by the data contained within their databases. An awareness of such advantages and limitations is important when considering the results of studies of the reactions between supercritical CO_2, formation water and aquifer minerals.

Based upon their availability and porosity–permeability characteristics, two main rock types might be considered as potential host rocks – carbonate and sandstone aquifers. However, each would respond differently to the injection of supercritical CO_2 (in terms of chemical reactions) due to their differing mineralogy. Previous studies (Czernichowski-Lauriol et al. 1996; Gunter et al. 1993; Hitchon 1996; Hutcheon et al. 1993) have shown that sandstone aquifers have a better capacity to buffer pH at higher values than carbonate aquifers. Thus, based purely upon chemical thermodynamics, sandstone aquifers have better storage capacity for CO_2 than the carbonate aquifers. Simulations by Czernichowski-Lauriol et al. (1996) illustrate this by modelling reactions that might occur within the Dogger aquifer in the Paris Basin (carbonate formation), and an oil-bearing sandstone from the North Sea. The maximum amount of CO_2 trapped in the carbonate aquifer is $1.42 \, mol \, kg^{-1}$ (equivalent to 62 g of CO_2 per kg of water) at 78°C and 200 bar pressure, but $1.89 \, mol \, kg^{-1}$ (equivalent to 83 g of CO_2 per kg of water) for the sandstone aquifer at 98°C and 200 bar pressure. It is noteworthy that even though CO_2 is less soluble at higher temperatures, more CO_2 was trapped in the sandstone aquifer. However, the types of reactive minerals within the sandstone are also important. For example, reaction of Na/K-bearing minerals will result in bicarbonate brines, whereas reaction of Fe/Ca/Mg-bearing minerals will result in carbonate mineral precipitation and a relatively unchanged formation water ionic strength (Gunter et al. 1997).

In general therefore, although both sandstone and carbonate aquifers appear suitable for CO_2 storage, the greater ability of the sandstone to sequester CO_2 makes it the favoured option. Storage in sandstone aquifers might also be more cost-effective, because fewer injection facilities would be needed to sequester a fixed amount of CO_2. This, together with the widespread occurrence of sandstones, explains why most investigations have focused on sandstone aquifers.

Laboratory experiments

Laboratory experimental investigations of the reaction of CO_2 with aquifer minerals have utilized both batch and flow-through equipment. Such experiments generally involve exposing samples of aquifer rock to CO_2-saturated synthetic formation waters for timescales of several weeks or months. The degree of reaction between CO_2, formation water and aquifer minerals is assessed by detailed mineralogical and fluid chemical analysis.

Experiments undertaken by the Alberta Research Council (Gunter et al. 1997; Hitchon 1996) simulated reactions that might occur during the injection of CO_2 into the Glauconite Sandstone aquifer within the Alberta Basin in western Canada. Samples of crushed Glauconite Sandstone were reacted with CO_2-saturated synthetic formation water in 'batch' equipment for one month. A temperature of 105°C and pressure of 90 bar were used in the experiments. This temperature is above the *in situ* temperature of the Glauconite Sandstone (54°C), but was used to enhance the rate of reaction so that any significant changes occurring could be observed over a reasonable timescale. To enhance CO_2-trapping reactions, some of the experiments were 'spiked' with selected pure minerals that were already present within the sandstone. Mineralogical analysis of the reacted solids revealed no new secondary minerals. However, the fluid chemistry of the synthetic formation water did change. Increased alkalinity indicated that fluid–mineral reactions were proceeding slowly, possibly controlled by the relatively sluggish reaction kinetics of silicate mineral dissolution at the experimental temperature.

Experiments undertaken by the British Geological Survey used a variety of rock types from the Lower Triassic Sherwood Sandstone Group (Czernichowski-Lauriol et al. 1996; Pearce et al. 1996). This is a sequence of continental red beds that forms important aquifers and petroleum reservoirs both onshore in the UK and offshore beneath the North Sea. In 'batch' experiments, small blocks of rock were partly immersed in water (either CO_2-saturated de-ionized water, or CO_2-saturated 0.55 M NaCl solution) and partly exposed to CO_2. In 'flow' experiments, short cores of sandstone was flushed with CO_2-saturated 0.55 M NaCl solution. Temperatures of 80°C and pressures of 200 bar were used in all the experiments. These pressure and temperature conditions are similar to those existing in gas fields and aquifers beneath the southern North Sea (Abbotts 1991). Although flow experiments only ran for one or two months, some batch experiments ran for up to eight months. In general, the types of reactions occurring between sandstone, CO$_2$ and water were similar in both batch and flow experiments. Dissolution of K-feldspar (in the form of etch pits) was common, especially on grains that had already undergone some authigenic dissolution. Some dissolution of dolomite cement was also observed. Small amounts of secondary precipitation were observed, with occasional aggregates of clay minerals often associated with the dissolution of K-feldspar. Small quantities of calcite also formed, being concentrated at the outlet end of the sandstone core in the flow experiments, and adjacent to the CO$_2$/water interface in the batch experiments. Crusts of halite also formed over the upper surfaces of the sandstone blocks in the batch experiments, i.e. those parts exposed to supercritical CO$_2$. These are thought to result from 'capillary-like' reaction drawing up saline aqueous fluid followed by loss of water into the supercritical CO$_2$ and precipitation of halite.

Natural analogues

The laboratory experiments above were of relatively short duration compared to the timescales that need to be considered for the long-term sequestration of CO$_2$. One way to obtain information about reactions occurring over longer timescales is to utilize natural 'experiments' that have been running for thousands, or millions, of years. These 'natural analogues' have to be chosen carefully, because such systems are much more complex and less well constrained than laboratory experiments. However, they can provide special insight into longer term processes. Natural accumulations of CO$_2$ occur in various parts of the world, but some of the best described are within the USA. Several of these are exploited commercially, with the CO$_2$ being used in enhanced oil recovery operations. One such site is the Bravo Dome CO$_2$ field in northeastern New Mexico, which is contained within the Permian Wolfcampian-Leonardian Tubb Sandstone (Broadhead 1989; Pearce *et al.* 1996). This field is thought to have originally contained about 12 trillion cubic feet (*c.* 3.4 × 10^{11} m^3) of CO$_2$ (Amoco P.C. 1990), with the ultimate recoverable CO$_2$ estimated at 5.3–9.8 trillion cubic feet (*c.* 1.5–2.8 × 10^{11} m^3) (Broadhead 1989). It was developed in the early 1980s

to supply CO$_2$ to the west Texas oilfields, and annual CO$_2$ production increased from 1 billion cubic feet (*c.* 2.8 × 10^7 m^3) in 1982, to 101 billion cubic feet (*c.* 2.8 × 10^9 m^3) in 1985. The host Tubb Sandstone is sealed by the 6 m thick Cimarron Anhydrite which, although flexed and folded, still retains good sealing characteristics across the field. The CO$_2$ is thought to have been generated from the degassing of magma, and to have entered the reservoir less than 50 000 years ago. However, other CO$_2$ fields in the USA appear to have successfully contained CO$_2$ for much longer time periods, even since the Cretaceous (Studlick *et al.* 1990).

The sandstones used in experiments reported by Czernichowski-Lauriol *et al.* (1996), and those of the Tubb Sandstone, are continental red beds with similar detrital mineralogies. The Tubb Sandstone in the Bravo Dome has been subjected to dissolution of early anhydrite, dolomite and detrital plagioclase (Nelis 1994; Pearce *et al.* 1996), which has been attributed to the early introduction of CO$_2$-rich formation waters possibly via Early Pennsylvanian and/or Tertiary faults (Nelis 1994). K-feldspar grains are corroded, and there are traces of highly corroded gypsum cement. Kaolinite, zeolites and gibbsite occur as late-stage cementing minerals (Nelis 1994; Pearce *et al.* 1996), and are closely associated with Tertiary faults, through which the CO$_2$ is thought to have migrated. Although these minerals have not been observed as reaction products in laboratory experiments it is thought that this may be due to the shorter duration of the experiments. The Bravo Dome CO$_2$ field may therefore be regarded as being at a more advanced stage of reaction than the laboratory experiments reported by Czernichowski-Lauriol *et al.* (1996), Gunter *et al.* (1997) and Hitchon (1996).

Computer modelling

In order to extend relatively short-term laboratory experiments to timescales relevant to the storage process, geochemical simulations should be undertaken. Although such an approach will not represent all the complexities of the system, the simulations are valuable because they serve to highlight important features and feedback mechanisms. Various approaches have been used to model a reactive CO$_2$-formation water-aquifer mineral system, and only a brief description is presented here.

Modelling undertaken by the Alberta Research Council (Gunter & Perkins 1993; Gunter *et al.* 1997; Hitchon 1996; Law & Bachu 1996; Perkins

& Gunter 1995) simulated reactions that might occur during the injection of CO_2 into the Cretaceous Glauconite Sandstone aquifer within the Alberta Basin in western Canada. These studies focused on CO_2 migration (as either a free phase or dissolved in formation water) as well as its reaction with aquifer minerals. Two different computer codes were used in the simulations, STARS and PATHARC.94. The multiphase model STARS was used to simulate likely hydrogeological processes occurring over 30 years of CO_2 injection (Law & Bachu 1996). In summary, the modelling predicted that a significant percentage (up to 25%) of the injected CO_2 would dissolve into the formation water, with the rest remaining as a supercritical phase. The CO_2 (as either a dissolved or a supercritical phase) was also predicted to migrate less than 5 km from the injection point. The code PATHARC.94 was used to simulate likely fluid–rock reactions that might have occurred if experiments (see above) had run for longer than one month. It was also used to investigate scenarios appropriate to large-scale injection of CO_2 (Gunter et al. 1997; Perkins & Gunter 1995). The fluid phase was assumed to be saturated with CO_2 throughout the simulation, and realistic rates were used for mineral dissolution. In summary, the modelling predicted that albite, biotite (as a proxy for glauconite) and K-feldspar would be dissolved whilst calcite, dolomite, kaolinite, muscovite, quartz and siderite would be precipitated. The major CO_2-trapping reactions were the precipitation of calcite and siderite, and the formation of aqueous bicarbonate ions. Simulations predicted that substantial CO_2-trapping reactions would have occurred in experiments at 105°C and 90 bar (see above) over a 6–40 year timescale. Although there is some uncertainty in the kinetic data used in the modelling, it appears that CO_2-trapping reactions under likely in situ conditions (54°C and 260 bar) would take hundreds of years to complete. However, as the residence time of fluids within deep aquifers in the Alberta Basin is measured in timescales that are 2–3 orders of magnitude larger than this, there is sufficient time for these mineral-trapping reactions to occur. Dissolution mechanisms appear to be the dominant processes earlier in the simulations and would be expected to occur closer to the injection well. Precipitation mechanisms that occur over longer timescales would be expected to occur a long way from the injection well and within a very large volume of rock. They would therefore be unlikely to reduce aquifer permeability to the detriment of the actual injection process (Gunter et al. 1997).

Modelling undertaken by BRGM (Czernichowski-Lauriol et al. 1996) used two computer codes; a geochemical simulator (CO2ROCK), and a one-dimensional coupled chemistry and transport code (CATCO2). These were used to help interpret experiments conducted at the British Geological Survey (see above), and to simulate reactions that might occur during the injection of CO_2 into Triassic sandstone aquifers beneath the North Sea. For the latter, a 150 m portion of typical aquifer was considered. The incoming aqueous phase was assumed to be saturated with CO_2 at the run conditions (98°C and 250 bar pressure). Minerals within the simulation (quartz, K-feldspar, albite, kaolinite, illite and calcite) were assumed to be in equilibrium with the fluid at all times (i.e. not limited by dissolution kinetics) until they were consumed. In summary, the modelling predicted the formation of 'reaction fronts' as the CO_2-rich formation water passed through the sandstone. These reaction 'fronts' moved away from the injection point, with that of albite removal ahead of that of illite removal. The region between these two fronts was a complex transition zone between a forward zone of kaolinite/K-feldspar dissolution and illite precipitation, and a back zone with progressive removal of illite and consequent precipitation of kaolinite/K-feldspar. As these fronts moved through the aquifer the remaining pH buffering provided by the aquifer minerals was far less efficient, and pH dropped. The main CO_2-trapping mechanism in these simulations was dissolution into the aqueous phase, and because of the reduced pH, the overall CO_2 storage capacity of the reservoir was barely enhanced once the two reaction zones had passed.

In summary, therefore, computer simulations show that reactive minerals may be very important for the sequestration of CO_2 in aquifers. These may act to buffer pH and enhance CO_2 dissolution in the formation water, or provide divalent metal ions to form secondary carbonate precipitates. They also indicate that, from a geochemical standpoint, a sandstone with a high proportion of these minerals (i.e. an 'immature' one) would make a better host aquifer than one poor in reactive minerals. Over shorter timescales, the amount of dissolved CO_2 may exceed that trapped by mineralogical reactions, making 'hydrogeological trapping' an important mechanism to consider alongside that of 'mineralogical trapping'. However, the complexity of these systems makes accurate modelling difficult. Useful future developments could include the addition of more solid phases into the simulations, the refinement of the treatment of reaction kinetics, consideration of the impact of porosity changes upon fluid flow, and the impact of reaction fronts. Further validation of codes against data from long-term experiments, or from large-scale demonstration injection projects could be one way to improve confidence in predictions and highlight areas where model refinement was needed.

Reactions between supercritical CO$_2$ and aquiclude minerals

Although reactions between CO$_2$ and aquifer minerals have been the focus of several studies, those between CO$_2$ and minerals within the overlying aquiclude ('caprock') have been investigated in much less detail. Such reactions will be important if they alter the ability of the overlying aquiclude to contain CO$_2$-rich fluids for extended timescales. Common aquicludes within sedimentary basins are likely to be dominated by clay-rich lithologies or evaporite deposits.

Experiments undertaken by the British Geological Survey used a variety of rock types from the late Triassic Mercia Mudstone Group (Czernichowski-Lauriol et al. 1996; Pearce et al. 1996). This is a complex succession of predominantly red siltstones and mudstones with locally extensive evaporites. The experiments focused on reacting mudstones and anhydrite with both dry supercritical CO$_2$ and CO$_2$-saturated solutions of varying salinity. As in the case of the aquifer samples (see previous section), a pressure of 200 bar, a temperature of 80°C and durations of up to eight month were used. In the experiments with dry supercritical CO$_2$ no obvious changes were observed. Such an apparent low reactivity between dry CO$_2$ and aquiclude samples serves to increase confidence in the capability of the overlying aquiclude to maintain a good seal for long timescales. Indeed, the very presence of natural accumulations of CO$_2$ shows that certain aquicludes can contain CO$_2$ over geologically-important timescales.

During and shortly after injection operations, aquicludes above the host aquifers would be in direct contact with supercritical CO$_2$ only. However, there would be places, such as at the edge of the CO$_2$ 'bubble', where they were in contact with CO$_2$-saturated water, and enhanced reaction might be expected. Czernichowski-Lauriol et al. (1996) report experiments designed to assess the degree of reaction on samples of mudstone and anhydrite under such conditions. To simulate conditions at the water–CO$_2$ boundary, short cores of rock were partly submerged in water, with the rest of the rock being exposed to supercritical CO$_2$. Most reaction of the mudstone samples occurred on the surfaces that had been submerged within the aqueous phase. Dolomite dissolution was identified, sometimes associated with clay mineral precipitation. Minor K-feldspar dissolution also occurred. Anhydrite samples showed a similar pattern of reaction, with dissolution features confined primarily to surfaces submerged in the aqueous phase. Associated with this dissolution was secondary precipitation of calcite which tended to be concentrated at the interface between the aqueous and CO$_2$ phases. A secondary CaSO$_4$ phase was also observed, presumably anhydrite, and this suggests that the fluid equilibrated relatively rapidly with the rock sample.

Observations of samples from the Bravo Dome CO$_2$ field in northeastern New Mexico (Broadhead 1989; Pearce et al. 1996) show that the Cimmarron Anhydrite forms an effective seal for the CO$_2$, despite evidence for anhydrite dissolution in the underlying Tubb Sandstone and in the experiments. Indeed, the Bravo Dome may have contained CO$_2$ for up to 50 000 years. Hence, although anhydrite can be corroded by CO$_2$-saturated formation waters, it appears unlikely that a large evaporite seal would be breached. Such long-term stability may reflect either, a relatively unreactive CO$_2$ 'bubble', CO$_2$-saturated water equilibrating with anhydrite or calcite precipitation effectively sealing off anhydrite from further reaction. Such relatively rapid precipitation of secondary anhydrite and calcite would be highly advantageous for the containment of CO$_2$ if it acted as a 'self-sealing' mechanism. However, as calcite has a lower molar volume than anhydrite, account would have to be taken of how secondary phase morphology affected the process.

In summary, therefore, although chemical reactions will occur between the injected CO$_2$, formation water and aquiclude minerals, these appear not to affect adversely the capacity of certain aquicludes to contain CO$_2$ for prolonged timescales. However, individual CO$_2$ sequestration operations may utilize aquicludes that differ from those already studied. It would be beneficial, therefore, for operators to take account of local formation water chemistries and aquiclude mineralogies in order to assess the likely degree of reaction prior to starting injection operations.

Safety aspects of the underground sequestration of CO$_2$ in the deep geosphere

The safety and stability of the underground sequestration of CO$_2$ are fundamental to its acceptance by society. Uncontrolled release of large volumes of CO_2 at the Earth's surface would be hazardous (Holloway 1997a, b) because this gas is heavier than air and, though not poisonous, is an asphyxiant. There are three main areas of safety that need to be considered:

- transport of CO_2 by pipeline to the injection site
- the injection plant and process of injection
- the stability of the CO_2 once injected into the subsurface.

The hazards associated with the transport and injection of large quantities of CO_2 (and other compressed gases) by pipeline are already well understood (Holloway 1997b; Holloway et al. 1996); such procedures are routinely used in enhanced oil recovery operations and there is a wealth of data. As a result, well managed installations are low risk. The inclusion of automatic, pressure-sensitive safety valves in pipelines and injection equipment would limit CO_2 losses if a rupture did occur. The successful capping of burning Kuwait oil and gas wells after the Gulf War shows that escapes from wells can be controlled even when all the safety controlling valves have been destroyed. In the unlikely event of a similar situation happening to a CO_2 injection well, such a procedure would, most probably, be more straightforward because CO_2 is non-combustible and non-poisonous.

Once injected underground, five key aspects relevant to the safe storage of CO_2 need to be considered (Holloway et al. 1996):

- escape of CO_2 to the atmosphere
- CO_2 pollution of formation water or other underground resources
- increased seismicity
- subsidence or absidence of the Earth's surface
- degradation of the storage reservoir.

A more detailed assessment of each of these can be found in Cox et al. (1996). However, previous experience based upon hydrocarbon exploitation and natural gas storage schemes, seems to indicate that such hazards can be minimized if the injection site is well chosen, the injection process is well managed and carefully monitored, and borehole sealing, once injection has been completed, is performed to a high standard.

If sequestration of CO_2 is most practicable within deep sandstone aquifers with no proven lateral seal, there is a need to ensure that the CO_2 will remain safely underground and not return to the atmosphere on relatively short timescales (thousands of years). The track record of purpose-designed underground storage of natural gas can provide useful information. In general, leakage underground is minimal so long as the injection process does not damage the overlying seal (Griffith 1990). Carbon dioxide is more chemically reactive than methane, so its effect on the overlying aquicludes needs to be addressed. Although short-term experiments do show reaction of potential aquiclude rocks with CO_2-saturated water, they show little reaction with supercritical CO_2. The latter should more closely represent conditions during and after injection operations when a 'bubble' of supercritical CO_2 has formed below the aquiclude. Indeed, natural accumulations of CO_2 show that reactions over longer timescales do not lead to wholesale dissolution and breakdown of the overlying aquicludes (i.e. if the seal was not effective the CO_2 would have escaped long ago).

Given that only otherwise unused deep aquifers containing saline formation waters in large basins can provide the storage volumes necessary for large scale underground sequestration (Holloway 1996b), it will be necessary to show that the CO_2 will not rapidly migrate updip and escape to the atmosphere. Computer modelling (Czernichowski-Lauriol et al. 1996; Gunter et al. 1997; Law & Bachu 1996; Perkins & Gunter 1995), together with previous studies (Gunter et al. 1993) show that sequestration of CO_2 into siliciclastic rock types (e.g. sandstones) is preferable to that in carbonate rock types (e.g. chalk) because of the greater potential for pH buffering, solution of CO_2 and net carbonate mineral precipitation. Over shorter timescales, the amount of dissolved CO_2 may exceed that trapped by mineralogical reactions, though hydrogeological simulations (Law & Bachu 1996) indicate that CO_2 migration (as either a dissolved or a supercritical phase) is only likely to be in the order of a few kilometres over several tens of years. Slower processes, such as carbonate mineral precipitation, will immobilize CO_2 over longer timescales, but must occur at a rate fast enough to prevent CO_2 migration and escape. Although computer simulations predict precipitation over tens to hundreds of years, the observation of small quantities of secondary carbonates on a laboratory timescale (Czernichowski-Lauriol et al. 1996) indicates that this process can be relatively rapid if conditions are suitable. However, the geochemical reactions to contain the CO_2 would be mineral-specific, and individual storage operations would have to take account of aquifer structure and hydrodynamics, mineralogy and fluid chemistry. Detailed study of a demonstration, or moderately sized storage operation (such as that recently started by the Norwegian oil company Statoil at the Sleipner Vest gas field in the North Sea) would also help in understanding how best to achieve the safe operation of much larger-scale disposal operations.

Summary

Anthropogenic emissions of CO$_2$ have been linked to increasing CO$_2$ levels in the atmosphere and to potential global climate change. Capture and sequestration of CO$_2$ from large point sources is one potential method for reducing such emissions without a drastic change in our energy-producing technologies. For some areas, sequestration of this waste CO$_2$ could be into deep aquifers containing saline formation waters within large sedimentary basins. Although the structure of the host aquifer may act to trap the CO$_2$ initially, over longer periods of time hydrodynamic and mineral trapping will become increasingly significant.

A variety of techniques can be used to investigate how geochemical reactions aid in the containment of CO$_2$ in the deep geosphere. Detailed investigation of relatively short-term processes can be achieved by laboratory experiments, where representative pressures, temperatures and flow rates can be simulated and monitored closely. Unfortunately, overall reaction time is limited by the length of the study, which is seldom more than a few years. However, significant reaction does occur over such timescales, and can be followed by studying changes in fluid chemistry and mineralogical abundances. Typical observations include partial dissolution of feldspars, dolomite and anhydrite, followed by the precipitation of clays and calcite.

The timescales over which reactions can be studied are greatly extended by the investigation of natural accumulations of CO$_2$. Although the use of such natural analogues has been limited to the Bravo Dome CO$_2$ field in New Mexico, studies show that timescales of reaction can be extended to several tens of thousands of years. Detailed observations revealed similar reaction sequences to the laboratory experiments, but with late-stage precipitation of kaolinite, zeolites and gibbsite. Such phases were not observed in the experiments, possibly as a result of their short duration. The Bravo Dome CO$_2$ field is also overlain by an anhydrite seal, and the very existence of the CO$_2$ field shows how effective this has been over the 50 000 year lifetime of this CO$_2$ field (even though laboratory experiments indicated reaction of anhydrite with CO$_2$-saturated water).

Linking of shorter-term and longer-term processes can be achieved by geochemical computer modelling, which also has the advantage of a predictive capability. In general, there is reasonable agreement between experiments, the analogue studies and the modelling, though the inclusion of a wider set of phases within the models could have improved the agreement. Reaction kinetics appear to have a major effect when trying to simulate shorter timescale processes, and the general lack of the necessary data may hinder their application in future studies. However, modelling did demonstrate that sandstone aquifers were preferable to carbonate aquifers for sequestration because of their better pH buffering capacity. It also demonstrated that over shorter timescales the amount of dissolved CO$_2$ may exceed that trapped by mineralogical reactions, making 'hydrogeological trapping' an important mechanism to consider alongside that of 'mineralogical trapping'. Over much longer timescales, however, trapping by mineralogical reactions will dominate. Further comparison of theoretical simulations with observations from experiments and natural analogues (or preferably test injections into deep sandstone aquifers) would aid in the improvement of future modelling studies.

Individual aquifers will vary in structure, mineralogy and hydrogeology, and each storage operation will have to take account of local geological, fluid chemical and hydrogeological conditions. However, results from the approaches outlined above show that fluid–mineral reactions can be beneficial when considering CO$_2$ solution into formation water, and the potential for CO$_2$ to be precipitated as carbonate minerals. As a consequence, chemical reactions with formation water and aquifer minerals can aid in the containment of the CO$_2$, and have the potential to trap it for geologically-important timescales.

The authors gratefully acknowledge useful review comments by B. Hitchon and B. Gunter, together with other useful suggestions by D. Savage, which have helped to improve this paper. This paper is published with the permission of the Director of the British Geological Survey (NERC).

References

ABBOTTS, I. L. 1991. *United Kingdom Oil and Gas Fields, 25 Years Commemorative Volume.* Geological Society of London Memoir 14.

AMOCO, P. C. 1990. *Bravo Dome Update – Carbon Dioxide Production in New Mexico.* Bravo Dome Operations Centre, Amoco Production Company.

ATKINS, P. W. 1983. *Physical Chemistry* (2nd edn). Oxford University Press, Oxford.

BACHU, S., GUNTER, W. D. & PERKINS, E. H. 1994. Aquifer disposal of CO_2: hydrodynamic and mineral trapping. *Energy Conversion and Management*, **35**, 269–279.

BAHN, O., FRAGNÈRE, E. & KYPREOS, S. 1998. Swiss energy taxation options to curb CO_2 emissions. *European Environment*, **8**, 94–101.

BAKLID, A., KORBØL, R. & OWREN, G. 1996. Sleipner Vest CO_2 disposal, CO_2 injection into a shallow underground aquifer. *Society of Petroleum Engineers*, **36600**, 269–277.

BERGMAN, P. D., WINTER, E. M. & CHEN, Z.-Y. 1997. Disposal of power plant CO_2 in depleted oil and gas reservoirs in texas. *Energy Conversion and Management*, **38** (supplementary volume), S211–S216.

BROADHEAD, R. F. 1989. *Bravo Dome CO_2 Field – U.S.A. New Mexico*. Treatise of Petroleum Geology Atlas of Oil and gas Fields, The American Association of Petroleum Geologists, A-011.

COOPER, D. J., WATSON, A. J. & NIGHTINGALE, P. D. 1996. Large decrease in ocean-surface CO_2 fugacity in response to in situ fertilization. *Nature*, **383**, 511–513.

COX, H., HEEDERIK, J. P., VAN DER MEER, B. VAN DER STRAATEN, R., HOLLOWAY, S., METCALFE, R., FABRIOL, H. & SUMMERFIELD, I. 1996. Safety and stability of underground CO_2 storage. *In*: HOLLOWAY, S. (ed.) *The Underground Disposal of Carbon Dioxide*. Final Report of Joule Project Number **CT92-0031**, chapter 5.

CZERNICHOWSKI-LAURIOL, I., SANJUAN, B., ROCHELLE, C., BATEMAN, K., PEARCE, J. & BLACKWELL, P. 1996. Inorganic geochemistry. *In*: HOLLOWAY, S. (ed.) *The Underground Disposal of Carbon Dioxide*. Final Report of Joule II Project Number **CT92-0031**, chapter 7.

DE BAAR, H. J. W. 1992. Options for enhancing the storage of carbon dioxide in the oceans: a review. *Energy Conversion and Management*, **33**, 635–642.

ELEWAUT, E., KOELEWIJN, D., VAN DER STRAATEN, R. *et al.* 1996. Inventory of the theoretical CO_2 storage capacity of the European Union and Norway. *In*: HOLLOWAY, S. (ed.) *The Underground Storage of Carbon Dioxide*. Final Report of Joule II Project Number **CT92-0031**, chapter 4.

ELLIS, A. J. & GOLDING, R. M. 1963. The solubility of carbon dioxide above 100°C in water and in sodium chloride solutions. *American Journal of Science*, **261**, 47–60.

FREUND, P. & ORMEROD, W. G. 1997. Progress toward storage of carbon dioxide. *Energy Conversion and Management*, **38** (supplementary volume), S199–S204.

GOLOMB, D. S., ZEMBA, S. G., DACEY, J. H. W. & MICHAELS, A. F. 1992. The fate of CO_2 sequestered in the deep ocean. *Energy Conversion and Management*, **33**, 675–683.

GRIFFITH, H. G. 1990. Storage in depleted fields. *In*: *Underground Storage of natural Gas and LPG*. Economic Commission for Europe Energy Series 3, United Nations, New York, 193–198.

GUNTER, W. D., PERKINS, E. H. & McCANN, T. J. 1993. Aquifer disposal of CO_2-rich gases: reaction design for added capacity. *Energy Conversion and Management*, **34**, 941–948.

——, WIWCHAR, B. & PERKINS, E. H. 1997. Aquifer disposal of CO_2-rich greenhouse gases: extension of the time scale of experiment for CO_2-sequestering reactions by geochemical modelling. *Mineralogy and Petrology*, **59**, 121–140.

HERZOG, H. J., ADAMS, E. E., AUERBACH, D. & CAULFIELD, J. 1996. Environmental impacts of ocean displosal of CO_2. *Energy Conversion and Management*, **37**, 999–1005.

HITCHON, B. (ed.) 1996. *Aquifer Disposal of Carbon Dioxide, Hydrodynamic and Mineral Trapping – Proof of Concept*. Geoscience Publishing, Alberta.

HOLLOWAY, S. 1996a. An overview of the Joule II project 'The Underground Disposal of Carbon Dioxide'. *Energy Conversion and Management*, **37**, 1149–1154.

—— (ed.) 1996b. *The Underground Disposal of Carbon Dioxide*, Final Report of Joule II Project Number **CT92-0031**.

——1997a. An overview of the underground disposal of carbon dioxide. *Energy Conversion and Management*, **38** (supplementary volume), S193–S198.

——1997b. Safety of the underground disposal of carbon dioxide. *Energy Conversion and Management*, **38** (supplementary volume), S241–S245.

—— & SAVAGE, D. 1993. The potential for aquifer disposal of carbon dioxide in the UK. *Energy Conversion and Management*, **34**, 925–932,

——, HEEDERIK, J. P., VAN DER MEER, L. G. H. *et al.* 1996. *The underground disposal of carbon dioxide*, Summary Report of Joule II Project Number **JOU2-CT92-0031**.

HOUGHTON, J. T. 1994. *Global Warming: The Complete Briefing*. Lion, Oxford.

HUNTLEY, B. 1990. Lessons from climates of the past. *In*: LEGGETT, J. K. (ed.) '*Global Warming – The Greenpeace Report*'. Oxford University Press, Oxford, 133–148.

HUTCHEON, I., SHEVALIER, M. & ABERCROMBIE, H. J. 1993. pH buffering by metastable mineral–fluid equilibria and evolution of carbon dioxide fugacity during burial diagenesis. *Geochimica et Cosmochimica Acta*, **57**, 1017–1027.

IEA 1996a. CO_2 capture and storage in the Natuna LNG project. *Greenhouse Issues*, **22**, 1.

——1996b. Pioneering CO_2 reduction. *Greenhouse Issues*, **27**, 1.

——1998. Sleipner aquifer storage of CO_2. *Greenhouse Issues*, **34**, 4.

JONES, I. S. F. & OTAEGUI, D. 1997. Photosynthesis greenhouse gas mitigation by ocean nourishment. *Energy Conversion and Management*, **38** (supplementary volume), S367–S372.

KUK, M. S. & MONTAGNA, J. C. 1983. Solubility of oxygenated hydrocarbons in supercritical carbon dioxide. *In*: *Chemical Engineering at Supercritical Fluid Conditions*. Ann Arbour Science, 101–111.

LAW, D. H.-S. & BACHU, S. 1996. Hydrogeological and numerical analysis of CO$_2$ disposal in deep aquifiers in the alberta sedimentary basin. *Energy Conversion and Management*, **37**, 1167–1174.

LEE, K., WANNINKHOF, R., TAKAHASHI, T., DONEY, S. C. & FEELY, R. A. 1998. Low interannual variability in recent oceanic uptake of atmospheric carbon dioxide. *Nature*, **396**, 155–159.

LEGGETT, J. K. 1990. The nature of the greenhouse threat. *In*: LEGGETT, J. K. (ed.) *Global Warming – The Greenpeace Report*. Oxford University Press, Oxford, 14–43.

LINDEBERG, E. & VAN DER MEER, L. G. H. 1996. Reservoir modelling and enhanced oil recovery. *In*: HOLLOWAY, S. (ed.) *The Underground Disposal of Cabon Dioxide*. Final report of Joule II Project Number **CT92-0031**, Chapter 6.

MEISEN, A. & SHAUI, X. 1997. Research and development issues in CO$_2$ capture. *Energy Conversion and Management*, **38** (supplementary volume), S37–S42.

NELIS, M. K. 1994. *Bravo Dome II Petrographic Study*. Report of the Amoco Production Company.

ORMEROD, B. (ed.) 1997. *Ocean Storage of Carbon Dioxide, Workshop 4 – Practical and Experimental Approaches*. Report of the IEA Greenhouse Gas R&D Programme.

PEARCE, J. M., HOLLOWAY, S., WACKER, H., NELIS, M. K., ROCHELLE, C. A. & BATEMAN, K. 1996. Natural occurrences as analogues for the geological disposal of carbon dioxide. *Energy Conversion and Management*, **37**, 1123–1128.

PERKINS, E. H. & GUNTER, W. D. 1995. Aquifier disposal of CO$_2$-rich greenhouse gases: Modelling of water–rock reaction paths in a siliciclastic aquifier. *In*: KHARAKA, Y. K. & CHUDAEV, O. V. (eds) *Proceedings of the 8th International Symposium on Water–Rock Interaction – WRI-8*. Balkema, Rotterdam, 895–898.

RIEMER, P. 1994. *The Utilisation of Carbon Dioxide from Fossil Fuel Fired Power Stations*. Report of the IEA Greenhouse Gas R&D Programme, **IEAGHG/SR4**.

SMITH, I. M. 1988. *CO$_2$ and Climate Change*. IEA Coal Research report **IEACR/07**.

STEWART, P. B. & MUNJAL, P. K. 1970. The solubility of carbon dioxide in pure water, synthetic sea water and synthetic sea-water concentrates at -5 to 25°C and 10 to 45 atm pressure. *Journal of Chemical and Engineering Data*, **1591**, 67–71.

STUDLICK, J. R. J., SHEW, R. D. BASYE, G. L. & RAY, J. R. 1990. A giant carbon dioxide accumulation in the Norphlet Formation, Pisgah Anticline, Mississippi. *In*: BARWIS, J., MCPHERSON, J. G. & STUDLICK, J. R. J. (eds) *Casebooks in Earth Sciences: Sandstone Petroleum Reservoirs*. Springer, New York, 181–203.

SUMMERFIELD, I. PUTTER, J., KAARSTAD, O. & SCHWARTZKOPF, T. 1996. Quantities and quality of CO$_2$ which may become available from fossil fuel fired plant in the European Union and Norway. *In*: HOLLOWAY, S. (ed.) *The Underground Disposal of Carbon Dioxide*. Final report of Joule II Project Number **CT92-0031**.

THURLOW, G. (ed.) 1990. *Technological Responses to the Greenhouse Effect*. Watt Committee Report Number **23**, Elsevier, London.

TURKENBERG, W. C. 1997. Sustainable development, climate change, and carbon dioxide removal (CDR). *Energy Conversion and Management*, **38** (supplementary volume), S3–S12.

VAN ELDIK, R. & PALMER, D. A. 1982. Effects of pressure on the kinetics of the dehydration of carbonic acid and the hydrolysis of CO$_2$ in aqueous solution. *Journal of Solution Chemistry*, **11**, 339–346.

WIEBE, R. 1941. The binary system carbon dioxide–water under pressure. *Chemical Reviews*, **29**, 475–481.

WILSON, T. R. S. 1992. The deep ocean disposal of carbon dioxide. *Energy Conversion and Management*, **33**, 627–633.

WMO/UNEP 1990. *Science Assessment of Climate Change*. Working Group 1 Report of the Intergovernmental Panel on Climate Change, WMO, Geneva.

WOODWELL, G. M. 1990. The effects of global warming. *In*: LEGGETT, J. K. (ed.) '*Global Warming – The Greenpeace Report*'. Oxford University Press, Oxford, 116–132.

YOKOYAMA, S.-Y. 1997. Potential land area for reforestation and carbon dioxide mitigation effect through biomass coversion. *Energy Conversion and Management*, **38** (supplementary volume), S569–S573.

Geochemical interactions between landfill leachate and sodium bentonite

A. J. SPOONER[1] & L. GIUSTI[2]

[1] *Fossil Fuels and Environmental Geochemistry (Postgraduate Institute),
Newcastle Research Group (NRG), The Drummond Building,
University of Newcastle, Newcastle upon Tyne NE1 7RU, UK*
[2] *School of the Environment, University of Sunderland, St. George's Way,
Stockton Road, Sunderland, Tyne & Wear SR2 7BW, UK*

Abstract: The disposal of municipal, commercial and industrial waste by landfill has been, and is, widely used around the world. Within the landfill, leachate is generated with the input of atmospheric water, and by mixing with the moisture already present from the decomposition of waste materials. The potentially hazardous components of leachate have the ability to pollute surface and groundwaters unless stringent criteria are met for the location of landfills, suitable materials are used to build basal barriers and capping systems, and modern leachate collection and disposal techniques are employed. Different types of barriers have been used, including naturally occurring soil/clay liners, compacted clay liners, geosynthetic materials, and composite sealing systems composed of a mineral layer and a synthetic membrane. One of the substances most commonly used in the sealing of landfill sites, either on its own or in combination with synthetic materials, is sodium bentonite. Despite its widespread use, there is some evidence that this material may not be capable of attenuating some of the substances contained in landfill leachate, and that the clay mineral structure itself may be damaged as a result of long-term exposure with the various organic/inorganic components of municipal landfill leachate. In particular, bentonites tend to collapse and shrink once exposed to certain toxic effluents. The stability, chemical resistance and permeability of a bentonite clay liner appear to be significantly affected by landfill leachates and other inorganic/organic solutions used in experimental work. The main aims of this paper are: (i) to describe the properties of bentonite and municipal solid waste leachates, and (ii) to summarize the available information on the interactions between bentonite and leachate.

Ever since the first landfills became operative in the UK in 1912 (Hasan 1995), leachate has been an environmental problem. These landfills exploited soils and rock formations rich in clay, relying upon their natural characteristics for the containment of contaminants present in the leachates. Due to the low permeability of the clay minerals they contain (1×10^{-8} and 1×10^{-11} cm/s), formations such as the Oxford Clay in the UK and Fort Benton in the USA were the primary target of landfill engineers. It was hoped that clay minerals present *in situ* in rock formations underlying the chosen landfill sites would reduce the flow of leachate. In addition, the clay formations were relied on to attenuate the potentially harmful substances contained in the leachates by a variety of sorption processes: absorption, adsorption and cation exchange.

As suitable natural sites meeting all the necessary requirements became hard to find, and due to the introduction of more stringent legislation (e.g. USEPA 1983; USEPA 1989), it became necessary to modify these sites by means of complex containment barrier systems: artificial liners (compacted clay liners constructed from natural soil, bentonite and/or synthetic material), composite liners (composed of a mineral layer of natural 'engineered' clay or of swellable smectite clay, and a geosynthetic membrane) (Daniel *et al.* 1993; Tchobanoglous *et al.* 1993) and/or geosynthetic/geotextile liners made from fibres of synthetic polymers (Anon 1986; Bhatia *et al.* 1990; Maule *et al.* 1993; Mollard *et al.* 1996). These types of liners were introduced primarily to: (i) reduce the flow of leachate escaping into surrounding strata, (ii) attenuate the potentially polluting constituents of leachate, and (iii) develop a controlled leachate collection and treatment system. The advantages and disadvantages of bentonite and other materials in their environmental use as artificial landfill liners are summarized in Table 1.

From: METCALFE, R. & ROCHELLE, C. A. (eds) 1999. *Chemical Containment of Waste in the Geosphere.* Geological Society, London, Special Publications, **157**, 131–142. 1-86239-040-1/99/$15.00 © The Geological Society of London 1999.

Table 1. *The advantages and disadvantages of containment liners*

Compacted clay liners	Bentonite clay liner	Geosynthetic clay liners	Composite clay liners
Advantages	*Advantages*	*Advantages*	*Advantages*
• Practically impossible to puncture • Considerable leachate attenuation capacity • Flexible • Requires no jointing	• Equal containment to clay liner, but doesn't need as much space • Self-sealing to certain extent • Can swell to create hydraulic barrier	• Negligible permeability • Quick to lay • Thin and easy to make • Unlikely to burn through during welding • High elongation at break • High or Low Density Polyethylene	• Added protection against leakage • Hydraulic properties less important than single layer liners • Clay component can complement ageing geomembranes
Disadvantages	*Disadvantages*	*Disadvantages*	*Disadvantages*
• Need close control of moisture content to achieve optimum compaction and low permeability • May not be acceptable where absolute containment is required • Uncertain resistance to caustic and some organic substances	• May not be acceptable where absolute containment is required • Uncertain resistance to caustic and some organic substances • Can crack/shrink in presence of organic liquors/leachates and high ionic strength solutions • May be affected by microbial degradation	• Thicker sheets are less flexible, especially in cold temperatures • Expansion may create rippling in hot conditions • May swell from organic uptake from leachates • Contaminants may 'diffuse' through the membranes • Chlorinated solvents have deleterious effects • Can go brittle if weathered	• May suffer from internal erosion, with subsequent loss of strength • May have greater risk of failure than a single clay liner barrier • Uncertain resistance to caustic and some organic substances

Modified from North West Waste Disposal Officers Sub-Group (1988).

Sodium bentonite is one of the most common materials used in containment barriers as it exhibits all the desirable properties that are required for the safe containment of the landfill leachate. However, despite the world-wide use of clays and geosynthetic membranes in waste disposal sites, it has been acknowledged that they are all susceptible to damage from potentially harmful leachates (Yanful *et al.* 1988; Mott & Weber 1991; Christensen 1992; Seymour 1992; Wagner 1993). Many researchers have conducted experiments using leachates, as well as other inorganic/organic liquors, on mineral liners, including sodium bentonite (Griffin *et al.* 1976; Garlanger *et al.* 1987; Alther *et al.* 1988; Fang & Evans 1988; Farquhar & Parker 1989; Davies *et al.* 1996; Thomas *et al.* 1996) in order to assess their suitability and durability in the landfill environment. Interaction experiments have studied parameters such as sorption capacity, cation exchange capacity (CEC), dilution and dispersion, precipitation by change of pH, degradation and plasticity (Wangen & Jones 1984; Bath 1993;

Bright *et al.* 1996). Cation exchange capacity (CEC), absorption and adsorption are seen to be the three most commonly recognized parameters which can affect the way in which clay liners operate.

Some authors (Fiebeger & Schellhorn 1994) have expressed concern about the long-term resistance and stability of bentonitic clays (in either single or composite form) in waste disposal linings. The use of clays such as kaolinite and illite in composite liners, rather than high swelling clay minerals like sodium montmorillonite is becoming more and more common, because these clay minerals are less electrochemically active, and they have been found to be more resilient to chemical attack (Fiebeger & Schellhorn 1994). These clays, coupled with synthetic membranes and geotextiles, are now seen as the new generation of technically advanced and complex containment barriers. However, they too have been associated with their own sets of problems (e.g. leakage, 'clogging', 'blocking'), which can reduce their overall effectiveness (Table 1). Leachate collection systems play an important role in the landfill system, and must operate for the entire landfill-site existence. If leachate collection systems clog or block, due to particulate and/or biological particles, accumulating liquors will force through the liner due to the stress and strain created by the accumulating head. Liquors may also diffuse through the liner itself after a period of time (Bhatia *et al.* 1990; Daniel 1990; Landreth 1990; Mlynarek *et al.* 1990; Montero & Overmann 1990; Rohde & Gribb 1990).

Geosynthetic materials provide a myriad of applications that can be made by coupling them with natural clays. The combination of HDPE (high-density polyethylene) and clay may reduce the risk of liner failure, but synthetic materials have their own recognized disadvantages (e.g. Emcon Associates 1983; Haxo *et al.* 1985). As stated by Maule *et al.* (1993), even if a synthetic liner appears to be stronger and less permeable, the risk of total liner failure due to an improperly installed liner may be higher than that of a single clay liner. Geosynthetic/composite liners have been developed which have varying degrees of effectiveness against chemical degradation or permeability (Giroud & Bonaparte 1989; Daniel 1993; Daniel *et al.* 1993; Tchobanoglous *et al.* 1993), but the chemical resistance of geosynthetic materials, other than geomembranes, is not well documented (Cassidy *et al.* 1992). Geosynthetic liners may be susceptible to damage from biological attack. Research carried out by Griffin (1985) indicated that after a number of years of burial, membranes may be prone to an increased rate of breakdown due to bacterial degradation. It is possible that liners under strain (i.e. improperly placed or uneven liners) may be more readily attacked by microorganisms.

Leachate chemistry

Chemical changes to leachate quality have been monitored since the mid 1960s. It has been noted that leachate chemistry is highly variable within a landfill, and that variations have to be expected even within a single tipping cell (Robinson 1995). Previously, information on leachate chemistry was obtained from the analysis of single samples or from sporadic periods of monitoring, or from relatively dilute unrepresentative samples. It was demonstrated that the major polluting components of landfill leachate are chemical oxygen demand (COD), biological oxygen demand (BOD) and ammoniacal nitrogen (ammonia), whereas heavy metals or other leachate constituents were found to be of relatively minor importance (Robinson 1995).

Decomposition of refuse is highly complex, with microbiological, physical and chemical processes acting all at once. With the percolation of water into the waste mass, adding to the already present moisture and liquids from the biological and physical breakdown of waste, the municipal solid waste leachate formed becomes a solvent for the organic, inorganic and biological contaminants that it contains. Leachates vary in composition from site to site, and within each site, and are affected by age of the landfill, variations of site operation, geological location, climate and degree of water ingress. It is thus possible to state that diverse liquors are produced and that no characteristic leachate composition exists. The chemical composition of landfill leachates from the UK and the USA are summarized in Tables 2 and 3.

Three common controlling phases of waste decomposition have been identified (Robinson 1995):

Table 2. *Composition of leachates from new and mature sites*

Determinand	New landfill (<2 a) Range (mg l^{-1})	New landfill (<2 a) Typical (mg l^{-1})	Mature landfill (>10 a) Range (mg l^{-1})
BOD$_5$	2000–30 000	10 000	100–200
TOC	1500–20 000	6 000	80–160
Ammoniacal N	10–800	200	20–40
Nitrate	5–40	25	5–10
pH	4.5–7.5	6	6.6–7.5
Alkalinity (CaCO$_3$)	1000–10 000	3 000	200–1000
Ca	200–3000	1 000	100–400
Mg	50–1500	250	50–200
K	200–1000	300	50–400
Na	200–2500	500	100–200
Cl	200–3000	500	100–400

After Tchobanoglous *et al.* (1993).

Table 3. *Chemical characteristics of landfill leachate*

Element	Ranges for American and European landfill sites[1] (mg l^{-1})	Ranges for other English landfill sites[2] (mg l^{-1})
Na	50–4000	9.43–670
Mg	50–1150	34.2–250
K	10–2500	11.5–279
Ca	10–2500	80–934
Cr	0.03–1.6	0.01–0.1
Mn	0.4–50	0.01–63.5
Fe	0.4–2200	0.48–153
Ni	0.02–2.05	0.05–0.1
Cu	0.004–1.4	0.01–0.1
Zn	0.05–170	0.01–0.8
Cd	0.0005–0.14	0.005–0.01
Pb	0.008–1.02	0.04–0.2
Al	ND	ND
Hg (μg l^{-1})	0.2–50	0–0.9
NH$_4$	1–1500	0.4–270
NO$_3$	0.1–50	0.2–4.19
SO$_4$	10–1200	<5–584
Alkalinity (CaCO$_3$)	300–11 500	250–3400
Cl	30–4000	15–810
pH	5.3–8.5	6.4–8.5

ND Not detected.
[1] Modified from Andreottola *et al.* (1994).
[2] Adapted from Robinson (1995). Sites were located in Tyne and Wear, County Durham, Northumberland, North Yorkshire and Leicestershire.

Phase 1 – mineralization

Aerobic breakdown of wastes over short period of time in shallow fill cells. Here, oxygen present in waste materials being disposed of is depleted, while significant amounts of CO_2, H_2 and water are produced.

Phase 2 – hydrolysis, acidogenesis and acetogenesis

This phase consists of three steps.

Hydrolysis. Anaerobic and associated micro-organisms break down the primary components (e.g. proteins, cellulose) of the waste so that insoluble polymers are turned into soluble organic compounds.

Acidogenesis. Alcohols and fatty acids form from the products broken down by hydrolysis, and are made available to acetogenic bacteria.

Acetogenesis. Carboxylic acids (e.g. acetic acid), CO_2 and H_2 form from the products described in the previous step.

Leachates produced during phase 2 are high in organic content (carbohydrates, proteins and lipids) and are characterized by high BOD (>10 000 mg l^{-1}), BOD:COD > 0.7, acidic pH, pungent odour, and very high ammonia (typically up to 1000 ppm NH$_3$). High levels of metals (e.g. Fe, Zn, Ca and Na), chloride, sulphate and nitrate are present. Temperatures within the landfill also rise from 4°C to 21°C, although biologically active sites can attain temperatures as high as 50°C).

Phase 3 – methanogenesis

At this stage, leachate is termed 'methanogenic' and is anaerobic. Bacteria (methanogens) become a major part of the environment and play a large role in removing soluble organic species such as fatty acids. A balance between the amount of gas (CO_2 and CH_4) produced from the organic compounds, and the various bacteria from phases two and three is attained, whilst the breakdown of waste continues. As the landfill ages, hydrolysable fractions of the waste are consumed until volatile fatty acid production reduces. No more fatty acids will be produced until organic matter is once again hydrolysed. This may occur over time, but not to the same extent as when the landfill was young. Compounds such as lignin, which are highly resistant to decomposition, are often found intact when analysed after many tens of years. Finally, waste that is degradable will break down leaving solid materials behind, the production of gases will cease, and *Eh* levels will change as air and water enter the system again. Here, reactivation of the once immobilized materials within the anaerobic confines of the landfill site may take place.

Whilst the recent national study into the variation of some 4000 landfill leachates (Robinson 1995) goes a long way in helping us understand more about overall leachate formation and compositions, it can be concluded that there is no such thing as a 'typical' leachate with a specific age and composition, merely a large number of different liquors reflecting the complex systems that are generated within areas of each individual landfill up to the time of sampling. For this reason, detailed and accurate characterizations of leachates are very difficult to obtain, and experiments devised to test the suitability and long-term behaviour of liners in the presence of leachates are equally difficult to control.

The liner

Clay minerals are hydrous silicates, largely composed of magnesium, iron, aluminium, silicon and potassium, and to a lesser extent sodium and calcium. They exhibit platy morphology and perfect [001] cleavage. Plasticity is also a characteristic of clay minerals, and is largely due to the negative charges on the clay surface and consequent attraction for polar fluids such as water. Due to their large surface area, clay minerals can modify the properties of a soil. When montmorillonite is added to a soil to form a bentonite-enhanced soil (BES) liner, even a small percentage (1–3 wt%) of this mineral will allow for swelling and dispersive properties to be exhibited (Goldman et al. 1990; Mollins et al. 1996).

Sodium bentonite is a natural product formed from the alteration of volcanic rhyolitic ash, after deposition in shallow marine environments (Grim & Güven 1978; Jepson 1984), or as a result of hydrothermal alteration of igneous rocks (Grim & Güven 1978). It is widely mined at sites near Fort Benton, Wyoming (from where its name originates), South Dakota and Montana (USA), although bentonite is not only found in these regions. Clays exhibiting similar structures have also been mined in many other parts of the world (e.g. Germany, Japan, China). It is composed largely of smectites, with the clay mineral montmorillonite accounting for about 60–90% of its composition (Grim 1953). The remainder is usually composed of other minerals such as quartz and illite. However, this can be seen as a generalization as the characteristics of bentonites depend upon the nature of the original volcanic ash from which they derive, as well as upon the salinity of the water in which they were deposited (Cowland & Leung 1991). The alteration of volcanic ash deposited in freshwater environments tends to form calcium bentonite.

Unlike other clay minerals such as illite ($K_{1-1.5}$ $Al_6Si_6O_{20}(OH)_4$), kaolinite ($Al_2Si_2O_5(OH)_4$) and calcium montmorillonite [$(1/2Ca)_{0.7}Al_{3.3}Mg_{0.7}$ $Si_8O_{20}(OH)_4.nH_2O$], sodium montmorillonite ($Na_{0.7}Al_{3.3}Mg_{0.7}Si_8O_{20}(OH)_4.nH_2O$), can swell by up to 13.8 times its dry volume in the presence of water (Goldman et al. 1990). Hoeks et al. (1987) reported the swelling capacity of various bentonites to range from 2 to 12 ml of water per gram of dry bentonite. This ability to swell is based upon the layered structure of the mineral itself, whereby it is stacked to form platelets. In the case of kaolinite and illite, these plates are held strongly together by van der Waals forces and strong hydrogen and potassium interlayer bonds respectively, and are able to preclude inter-layer water uptake, thus exhibiting very little inter-layer swelling (Fig. 1). The interlayer spacing (d) of pure sodium smectite can vary from about 2.0 nm to many thousands of nanometers, whereas calcium smectite has a spacing range of 1.0–2.0 nm. In its environmental use as a waste disposal liner, the degree of swelling of bentonite depends on the compressive stress of the overburden material. The free swell of typical Wyoming sodium and calcium bentonites is 1400–1600% and 125%, respectively (Grim 1962). Chemicals such as soda ash or polymers are often added to natural bentonite in order to improve its swelling capa-city and water absorbency. Due to their high water absorbing capacity, sodium smectites normally have a higher water content, larger shrink–swell potential and lower hydraulic conductivity. Other properties of sodium smectites include higher liquid limits and plasticity indices than those of calcium smectites

As sodium montmorillonite mineral particles are <2 μm in size, have a high surface area of 50–120 $m^2 g^{-1}$ (increasing to 700–840 $m^2 g^{-1}$ as the clay swells), and a cation exchange capacity of 80–150 meq/100 g (Goldman et al. 1990), sodium bentonite has the potential to sorb very high concentrations of metals onto available surface sites (Mitchell 1976).

In terms of its use as liner material, kaolinite has a lower affinity for water, has lower dispersive properties and is more permeable. However, due to its low CEC (3–15 meq/100 g) and surface area (10–20 $m^2 g^{-1}$) it is less electrochemically active and also exhibits the least plasticity. The higher permeability of kaolinite has led to fears of shortened breakthrough time. However, as this clay exhibits less attenuation, its behaviour may be less affected by the interaction with organic/inorganic chemicals, and by moisture variations.

Illite shows intermediate properties of montmorillonite and kaolinite as far as chemical activity, swelling, surface area (65–100 $m^2 g^{-1}$) cation exchange (10–40 meq/100 g). Since illite has strong inter-layer potassium bonding, this prevents the sorption of water (as in the case of kaolinite), organic liquids and cations (Deer et al. 1966). On the contrary, as the van der Waals bonding forces between the 2:1 layers of sodium montmorillonite are weak, water and other polar fluids are able to enter much more easily between the weakly cohesive unit cell layers, thus causing the interlayer spaces to expand and the clay to swell (Fig. 1). Sodium montmorillonite is therefore favoured over its counterpart, calcium montmorillonite, as swell in the latter is limited due to the presence of

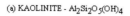

(a) KAOLINITE - $Al_2Si_2O_5(OH)_4$

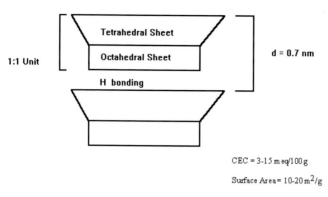

d = 0.7 nm

CEC = 3-15 m eq/100 g

Surface Area = 10-20 m^2/g

(b) ILLITE - $K_{1-1.5}Al_6Si_6O_{20}(OH)_4$

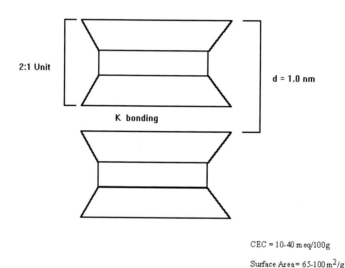

d = 1.0 nm

CEC = 10-40 m eq/100g

Surface Area = 65-100 m^2/g

Fig. 1. Simplified structure of clay minerals (**a**) kaolinite, (**b**) illite and (**c**) smectite (from Grim 1962; Goldman *et al.* 1990).

divalent cations (e.g. Ca^{2+}) (Yong and Warkentin 1975; Landreth 1980).

The low permeability of sodium bentonite clay is, in large part, due to this ability to swell and thus fill any void spaces created when mixed with other components of the liner, i.e. sand or soils (Goldman *et al.* 1990). Low permeability coupled with high surface area and cation exchange capacity mean that sodium bentonite has the ability to attenuate metals, making it an attractive barrier component. However, there is some concern that bentonite may undergo adverse changes to its chemical and physical properties, resulting in increasing permeability and potential breakthrough. These fears arise because of the variable, unpredictable, and possibly aggressive nature of some landfill leachates. In some countries (e.g. Germany) there has been an increasing demand for non-smectitic clays such as kaolinite and illite because these have low permeability (though not as low as those of smectites) and less pronounced long-term stability problems (Fiebeger & Schellhorn 1994).

Two types of changes may bring about increases to liner permeability, and thus affect the liner's integrity: firstly, shrinkage and cracking of the clay, and, secondly, decomposition of the liner material itself. Shrinkage will occur not only when water is lost from the unit cell layers, i.e. when temperature increases and dehydration

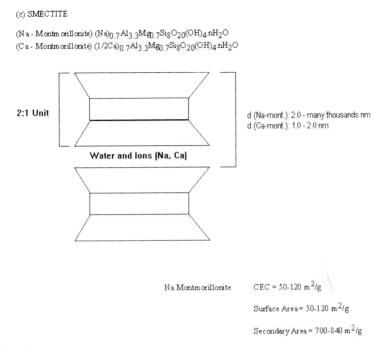

(c) SMECTITE

(Na - Montmorillonite) $(Na)_{0.7}Al_{3.3}Mg_{0.7}Si_8O_{20}(OH)_4 \cdot nH_2O$

(Ca - Montmorillonite) $(1/2Ca)_{0.7}Al_{3.3}Mg_{0.7}Si_8O_{20}(OH)_4 \cdot nH_2O$

2:1 Unit

Water and Ions (Na, Ca)

d (Na-mont.): 2.0 - many thousands nm
d (Ca-mont.): 1.0 - 2.0 nm

Na Montmorillonite $CEC = 50{-}120 \ m^2/g$

Surface Area $= 50{-}120 \ m^2/g$

Secondary Area $= 700{-}840 \ m^2/g$

Fig. 1. (*continued*).

takes place, but also if the inter-layer water is exchanged for organic solvents, such as benzene. Shrinkage can also occur if cations bridge the inter-layers. Resultant synresis cracks form due to the loss of inter-layer waters, and the build up of tension between the clay particles (Tucker 1994). Cation exchange is seen to affect the double layer thickness of the clay mineral, thus significantly affecting the permeability of clay. This may happen if the total concentration of cations in leachate is very high, or if divalent or trivalent cations such as calcium and magnesium replace the inter-layer cation (i.e. monovalent sodium in sodium montmorillonite) in exchangeable sites of the clay (Batchelder 1995). Typically though, high exchange capacity clay minerals such as montmorillonite are more greatly affected than those with lower exchange capacities (e.g. illite), and thus cation exchange can be seen as a function of increased surface area and charge (Goldman *et al.* 1990).

Studies on interactions between clay liner materials and leachate or other inorganic/organic solutions

A search of the literature reveals that very few long-term field studies of landfill leachate/liner

interactions have been published. Laboratory-based experiments mimicking field situations have been the preferred option as they can be completed over a relatively short period of time.

Farquhar & Parker (1989) reported examples of 4–15 year long field monitoring studies carried out in the USA and Canada. In these cases, municipal solid waste leachate did not affect appreciably the permeability of clay liners composed of mixtures of minerals such as illite, chlorite, quartz, feldspar, carbonates and smectite. Sodium and chloride proved to be the most mobile ions. All heavy metals were quite efficiently removed by the top 15–20 cm of the liner.

Grantham & Robinson (1988) measured a very low infiltration rate of leachate through a bentonite landfill liner, whereas Hoeks *et al.* (1987) found an increase in permeability of similar material from $10^{-9}\,\mathrm{m\,s^{-1}}$ to $10^{-7}\,\mathrm{m\,s^{-1}}$.

No significant increases in permeability of a silty clayey liner permeated with municipal landfill leachate were found by Fang & Evans (1988) over a period of three to six months. In fact, a slight decrease in permeability with time was observed and was attributed to precipitation of dissolved solids present in the leachate, to the plugging of voids by suspended solids in the leachate, and to secondary compression. However, significant changes in the physical properties of the liner were observed as the bentonite (2.6% Na, 1.3% Ca) fraction mixed with soil material was increased.

The formation of fractures and joints has been observed in clay material exposed to municipal solid waste leachate (Wellner et al. 1985) and the most common explanation given for their occurrence is the possible substitution of Na present in swelling clays. Also, organic acids (e.g. acetic acid) are known to be able to attack the clay matrix of liners and cause increased permeability. In general, dilute (<0.1 wt%) concentrations of many organic compounds do not normally increase (and in some cases even reduce) the permeability of clay barriers, whereas concentrated solutions (≥ 0.5 wt%) of acetic acid, acetone, aniline, methanol, benzene and xylene can cause a pronounced increase in permeability as a result of cracking and pore enlargement in both swelling and non-swelling clays (Brown & Anderson 1983; Brown & Thomas 1984; Evans et al. 1985; Alther et al. 1988; Fang & Evans 1988).

The addition of organobentonites to conventional sand/bentonite liners can help prevent liner failures caused by nonpolar organic liquids (Li et al. 1996). Organobentonites are produced by exchanging organic, quaternary ammonium cations for the inorganic ions present on the mineral surface of bentonite and in the interlamellar spaces. The clay mineral surfaces, normally hydrophilic, become organophilic thus increasing their affinity for nonionic organic solutes.

On the basis of laboratory experiments, many workers (e.g. Griffin et al. 1976; Griffin 1983; Fuller & Korte 1976; Farquhar & Constable 1978; Weinberg et al. 1985; Yong et al. 1986) concluded that heavy metals are normally removed quite efficiently in clay material by precipitation as carbonates and/or hydroxides and by sorption processes. The removal efficiency increases with pH, positive Eh, higher temperature, higher CEC of the barrier material and with increasing contact time.

Examples of contaminants which are poorly attenuated by clay barriers include chloride and sodium (Griffin et al. 1976), some organic acids (Weinburg et al. 1985), and neutral, non-polar organic substances (Brown & Anderson 1983).

Griffin et al. (1977) showed the strong pH dependency of the sorption by kaolinite and montmorillonite of cationic (Pb, Cu, Hg, Cr^{3+}) and anionic (As, Se, Cr^{6+}) species of heavy metals present in different landfill leachates and inorganic solutions. Sorption of the former increased as pH increased, whereas the removal of the latter decreased as the pH was raised. The greatest anion adsorption occurred in acid solutions. It was inferred that a high mobility of As^{5+} and Se^{4+} should be expected from land disposal of wastes under alkaline conditions. Montmorillonitic clays or soils were found to be able to remove substantially more cationic heavy metals than kaolinitic clays or soils.

Ten-month long laboratory simulations (Griffin et al. 1976) using landfill leachate and different proportions of kaolinite, Ca-montmorillonite, illite and pure quartz sand indicated that Pb, Cd, Zn and Hg can be strongly attenuated by small amounts (2%) of clays; poor attenuation was found for Cl, Na, and dissolved organic compounds, and moderate attenuation in the case of K, Mg, Si and Fe. Finally, Ca, B and Mn were mobilized from the clay minerals during the experiments.

It has been found that clays and soils may be damaged by ions which are present in high concentrations in leachates (Landreth 1980). Leachates with pH < 3 and pH > 11 are usually the most aggressive. When a concentrated acidic solution or leachate passes through a clayey soil, hydraulic conductivity sometimes declines but then increases later (Nasiatka et al. 1981; Peterson & Gee 1986; Bowders & Daniel 1987; Daniel 1993). Sandy soils mixed with small percentages of bentonite are very susceptible to attack by acids as these soon dissolve the bentonite grains.

The experiment carried out by Spooner (1995) with municipal solid waste leachate on rigid-walled permeameters containing Wyoming sodium bentonite showed a reduction of hydraulic conductivity from 10^{-8} m s^{-1} to 10^{-7} m s^{-1} over a period of 100 days, and a parallel mobilization of relatively high concentrations of Na, Mg, Ca, K Fe and Al from the solid phase into the leachate pointing to a possible breakdown of the clay material. During the experiment, the leachates became strongly anaerobic as indicated by a decrease in Eh values from +18.5 mV to −315 mV; conductivity increased from 9.79 to 11.14 mS cm^{-1} and pH increased from 7.6 to 8.5.

One aspect, rarely mentioned in the literature, is the role which micro-organisms can play in the breakdown of clay minerals, in different environmental conditions. Leaching experiments with micro-organisms have shown that they can cause congruent clay mineral dissolution (McFarlane & Bowden 1992; McFarlane et al. 1993). Also Kostka et al. (1996) obtained evidence of microbial reduction of structural Fe^{3+} in smectite.

In the natural environment, chemical alteration of montmorillonite to secondary minerals is considered to be quite slow, but it cannot be excluded that the decomposition of clay minerals may be accelerated within the confines of a landfill site due to the aggressive nature of the leachate present and to intense microbial activity.

Ultimately, if leachate breakthrough does occur from within the landfill, either due to pollutant migration through the liner or by liner failure, it is possible that chemical and biological constituents may be removed, by the process of filtration, or sorptive mechanisms, by the underlying strata (Tchobanoglous et al. 1993). In a survey of landfill failures undertaken by Roche (1996) it was indicated that there is a very significant frequency of surface and groundwater pollution incidences in the UK. About one third of the 4000 sites included in this study were considered to be containment sites, with many of these having a clay liner as a containment barrier. However, it has been found that if the rock formations underlying a landfill are enriched in clay minerals and the breakthrough liquor is low in trace metals, then there should be minimal threat to the underlying strata (Warith 1987).

Summary

Those features that identify montmorillonite as a low permeability barrier, namely the high surface area and inter-layer chemistry, also make it susceptible to damage. Shrinkage to the liner by the use of high-ionic strength solutions can be reasonably predicted in the laboratory, but actual leachates may be too difficult to model accurately and extrapolate to long-term liner stability.

The results of the experiments with bentonite and leachate reported in the literature have helped our understanding of the properties of bentonite, and its attenuating potential of some pollutants. However, it is very difficult to reproduce environmental conditions in the laboratory so that some of the results obtained cannot be readily extrapolated to real situations. It is difficult to compare data obtained with different experimental settings and to take into account the possible artefacts that may be caused by leachate handling, storage conditions and sample preparation techniques before analysis. Given the interferences and leachate matrix effects commonly observed during leachate analysis, the reliability of the data obtained must also be carefully considered (DeWalle et al. 1981; Daly & Farooq 1982; Lund et al. 1992).

As reactions between leachates of low ionic strength and mineral phases are less significant than those occurring in the presence of more concentrated liquors (e.g. solutions of pure organic compounds such as benzene), it is quite likely that low ionic strength solutions used in some laboratory studies may not adequately model the aggressive nature of landfill leachates.

Landfill leachate has the potential to cause significant chemical alteration of bentonite. The observed changes in hydraulic conductivity may be attributable to the leaching/solubilization of the mineral phase, swelling/shrinking of the liner due to cation exchange, to the effects of microbial decomposition and to redox mediated solubilization of the clay minerals.

Studies on the permeability of sodium bentonite indicated that pure organic compounds can be detrimental to the liner's performance. Cracking was found to occur, when compounds such as alcohols (e.g. methanol) and ketones (e.g. acetone) were introduced in high concentrations. Strong acids (e.g. hydrofluoric and phosphoric acids) and bases (e.g. NaOH) can also be quite aggressive on soil minerals and cause variations of hydraulic conductivity. However, the problem posed by organic solvents is probably exaggerated in the case of domestic landfills, as these substances normally represent a minimal proportion of the components of leachate in domestic waste landfill sites.

A low percentage (2–3%) of high-swelling clays is capable of reducing the permeability of clay liners as a result of the filling of voids between larger particles by the expanded bentonite. However, given the high degree of swelling that may be induced by leachates, significant increases in permeability can be expected, especially when relatively higher percentages of bentonite are mixed with other material in the liner.

The chemical suitability of any liner should be consistent with the chemicals that are going to come into contact with it (Jepson 1984), and testing should be carried out by the supplier of the liner to ensure that chemical compatibility is correct for each site (North West Waste Disposal Officers Sub-Group 1988). If chemicals harmful to clay minerals and geomembranes/geotextiles are introduced, breakthrough may occur, and rapid pollution of groundwater becomes possible.

The authors would like to thank C. Rochelle, J. Hinchliff and an anonymous reviewer; their constructive criticism was greatly appreciated.

References

ALTHER, G. R., EVANS, J. C. & ANDREWS, E. 1988. Organic fluid effects upon bentonite. *5th National Conference on Hazardous Wastes and Hazardous Materials*, Las Vegas, Hazardous Materials Conference Research Institution, **5**, 210–214.

ANDREOTTOLA, G. & CANNAS, P. 1994. Chemical and biological characteristics of landfill leachate. *In*: CHRISTENSEN, T. H., COSSU, R. & STEGMANN, R. (eds) *Landfilling of Waste: Leachate*. Elsevier, London, 65–88.

ANON. 1986. Selecting and specifying HDPE liner material. *Hazardous Waste Consultant*, 1.21–1.25.

BATCHELDER, M. 1995. *The performance of mudrocks for leachate containments*. PhD Thesis, University of London.

BATH, A. H. 1993. Clays as chemical and hydraulic barriers in waste disposal: Evidence from pore waters. *In*: MANNING, D. A. C., HUGHES, P. L. & HUGHES, C. R. (eds.) *Geochemistry of Clay–Pore Fluid Interactions*. Chapman & Hall, London, 316–330.

BHATIA, S. K., QURESHI, S. & KOGLER, R. M. 1990. Long term clogging behaviour of non-woven geotextiles with silty and gap-graded sands. *In*: KOENER, M. (ed.) *Geosynthetic testing for waste containment applications*. ASTM STP 1081, American Society of Testing Materials, Philadelphia, 285–297.

BOWDERS, J. J. & DANIEL, D. E. 1987. Hydraulic conductivity of compacted clays to dilute organic chemicals. *Journal of Geotechnical Engineering*, **113**(12), 1432–1448.

BRIGHT, M. I., THORNTON, S. F., LERNER, D. N. & TELLAM, J. H. 1996. Laboratory investigations into designated high-attenuation landfill liners. *In*: BENTLEY, S. P. (ed.) *Engineering Geology of Waste Disposal*. Geological Society, London, Engineering Geology Special Publication, **11**, 159–164.

BROWN, K. W. & ANDERSON, D. C. 1983. *Effects of organic solvents on the permeability of clay soils*. EPA-600/2–83, 016.

BROWN, K. W. & THOMAS, J. C. 1984. Conductivity of three commercially available clays to petroleum products and organic solvents. *Hazardous Waste or Hazardous Material*, **1**, 545–553.

CASSIDY, P. E., MORES, M., KERWICK, D. J., KOECK, D. J. VERSCHOOR, K. L. & WHITE, D. F. 1992. Chemical resistance of geosynthetic materials. *Geotextiles and Geomembranes*, **11**, 61–98.

CHRISTENSEN, T. H. 1992. Attenuation of leachate pollutants in groundwater. *In*: CHRISTENSEN, T. H., COSSU, R. & STEGMANN, R. (eds) *Landfilling of Waste: Leachate*. Elsevier, London, 441–483.

COWLAND, J. W. & LEUNG, B. N. 1991. A field trial of a bentonite liner. *Waste Management and Resources*, **9**(4), 277–291.

DANIEL, J. L. 1990. Is microstructure important to the performance of geosynthetics? *In*: PEGGS, I. D. (ed.) *Geosynthetics: Microstructure and Performance*. ASTM STP 1076. American Society for Testing and Materials, Philadelphia, 1–3.

DANIEL, D. E. (ed.) 1993. *Clay liners – Geotechnical Practice for Waste Disposal*. Chapman & Hall, London, 137–163.

——, SHAN, H. Y. & ANDERSON, J. D. 1993. Effects of partial wetting on the performance of the bentonite component of a geosynthetic clay liner. *In*: *Proceedings of the Geosynthetics '93 Conference*.

Vancouver, Canada, 30 Mar–1 Apr. 1993. Industrial Fabrics Association International (IFAI), **3**, 1483–1496.

DALY, E. L. JR & FAROOQ, S. 1982. Reliability of analytical methods for anaerobic municipal solid waste samples. *Journal of the Water Pollution and Control Federation*, **54**(2), 187–192.

DAVIES, M. C. R., RAILTON, L. M. R. & WILLIAMS, K. P. 1996. A model for adsorption of organic species by clays and commercial landfill barrier materials. *In*: BENTLEY, S. P. (ed.) *Engineering Geology of Waste Disposal*. Geological Society, London, Engineering Geology Special Publication, **11**, 273–278.

DEER, W. A., HOWIE, R. A. & ZUSSMAN, J. 1966. An Introduction to the Rock-Forming Minerals. *In*: GOLDMAN, L. J., GREENFIELD, L. I., DAMLE, A. S., KINGSBURY, G. L., NORTHEIM, C. M. & TRUESDALE, R. S. 1990. *Clay Liners for Waste Management Facilities: Design, Construction and Evaluation*. Pollution Technology Review (178), Noyes Data Corporation, New Jersey, 1–19.

DEWALLE, F. B., ZEISIG, T., SUNG, J. F. C. *et al.* 1981. *Analytical methods evaluation for applicability in leachate analysis*. Municipal Environmental Research Laboratory Office of Research and Development. US Environmental Protection Agency, Cincinnati, Ohio, 1–12.

EMCON ASSOCIATES. 1983. *Field Verification of Liners from Sanitary Landfills*. US EPA Report No. **EPA-600/2-83-046**.

EVANS, J. C., KUGLEMAN, I. J. & FANG, H. Y. 1985. Organic fluid effects on the strength, compressibility and permeability of soil-bentonite slurry walls. *In*: KUGLEMAN, I. J. (ed.) *Toxic and Hazardous Wastes*. Technomic Publishing, Lancaster, PA, 275–291.

FANG, H. Y. & EVANS, J. C. 1988. Long term permeability tests using leachate on a compacted clay liner material. *Ground Water Contamination: Field Methods*, ASTM STP 963, 397–404.

FARQUHAR, G. J. & CONSTABLE, T. W. 1978. *Leachate contaminant attenuation in soil*. Waterloo Research Institute, Project No. 2123, University of Waterloo, Waterloo, Canada.

—— & PARKER, W. 1989. Interactions of leachates with natural and synthetic envelopes. The landfill – reactor and final storage. *In*: BACCINI, P. (ed.) *Proceedings of the Swiss Workshop on land disposal of solid wastes*. Held Gerzensee, Switzerland, 14–17 Mar. 1988. Springer, Berlin, Lecture notes in Earth Sciences, **20**, 175–200.

FIEBEGER, W. & SCHELLHORN, M. 1994. Industrial minerals in soil sealants and waste dump linings. *Industrial Minerals*, **317**, 41, 43, 45–47.

FULLER, W. H. & KORTE, N. 1976. Attenuation mechanisms through soils. *In*: GENETELLI, E. J. & CIRELLO, J. (eds.) *Gas and Leachate from Landfills*. US EPA Report **EPA-600/9-76-004**, Cincinnati.

GARLANGER, J. E., CHEUNG, F. K. & TANNOUS, B. S. 1987. Quality control testing for a sand bentonite liner. *In*: COWLAND, J. W. & LEUNG, B. N. 1991. A field trial of a bentonite liner. *Waste Management and Resources*, **9**(4), 277–291.

GIROUD, J. P. & BONAPARTE, R. 1989. Leakage through liners constructed with geomembranes – Part II: composite liners. *Geotextiles and Geomembranes*, **8**, 71–111.

GOLDMAN, L. J., GREENFIELD, L. I., DAMLE, A. S., KINGSBURY, G. L., NORTHEIM, C. M. & TRUESDALE, R. S. 1990. *Clay Liners for Waste Management Facilities: Design, Construction and Evaluation*. Pollution Technology Review (178), Noyes Data Corporation, New Jersey, 1–47.

GRANTHAM, G. & ROBINSON, H. 1988. Instrumentation and monitoring of a bentonite landfill liner. *Waste Management and Research*, **6**, 125–139.

GRIFFIN, G. J. L. 1985. Biodegradable plastics. *In*: SEAL, K. J. (ed.) *Biodeterioration and Biodegradation of Plastics and Polymers*. Occasional Publication (1), The Biodeterioration Society, 11–26.

GRIFFIN, R. A. 1983. *Mechanisms of natural leachate attenuation*. Notebook for Sanitary Landfill Design Seminar, Department of Engineering and Applied Science, University of Wisconsin, Madison.

——, CARTWRIGHT, K., SHIMP, N. F. *et al.* 1976. Attenuation of pollutants in municipal landfill leachate by clay minerals, Part 1 – Column leaching and field verification. *Environmental Geology Notes*, November 1976, **78**. Illinois State Geological Survey.

——, FROST, R. R., AU, A. K., ROBINSON, G. D. & SHIMP, N. F. 1977. Attenuation of pollutants in municipal landfill leachate by clay minerals, Part 2 – Heavy metal adsorption. *Environmental Geology Notes*, April 1977, **79**. Illinois State Geological Survey.

GRIM, R. E. 1953. *Clay Mineralogy*. McGraw Hill, New York.

——1962. *Clay Mineralogy*. McGraw Hill, New York.

—— & GÜVEN, N. 1978. *Bentonites: Geology, Mineralogy, Properties and Uses*. Developments in Sedimentary Series (24). Elsevier, Oxford.

HASAN, S. E. 1995. *Geology and Hazardous Waste Management*. Prentice Hall, Englewood Cliffs, New Jersey.

HAXO, H. E., JR, HAXO, R. S., NELSON, N. A., HAXO, P. D., WHITE, R. M., DAKESSIAN, S. & FONG, M. A. 1985. *Liner Materials for Hazardous and Toxic Wastes and Municipal Solid Waste Leachate*. Noyes Data Corporation, New Jersey.

HOEKS J., GLAS, H., HOFKAMP, J. & RYHINER, A. H. 1987. Bentonite liners for isolation of waste disposal sites. *Waste Management and Research*, **5**, 93–105.

JEPSON, C. P. 1984. Sodium bentonite: still a viable solution for hazardous waste containment. *Pollution Engineering*, **16**(4), 50, 52–53.

KOSTKA, J. E., STUCKI, J. W., NEALSON, K. H. & WU, J. 1996. Reduction of structural Fe(III) in smectite by a pure culture of *Shewanella Putrefaciens* strain MR-1. *Clays and Clay Minerals*, **44**(4), 522–529.

LANDRETH, R. E. 1980. *Lining of waste impoundment and disposal facilities*. United States Environmental Protection Agency, Municipal Environmental Resources Laboratory, Cincinnati, Ohio.

——1990. Service life of geosynthetics in hazardous waste management facilities. *In*: PEGGS, I. D. (ed.) *Geosynthetics: Microstructure and Performance*. ASTM STP 1076, American Society for Testing and Materials, Philadelphia, 26–33.

LI, J., SMITH, J. A. & WINQUIST, A. S. 1996. Permeability of earthen liners containing organobentonite to water and two organic liquids. *Environmental Science and Technology*, **30**(10), 3089–3093.

LUND, U., RASMUSSEN, L., SEGATO, H. & ØSTFELDT, P. 1992. Analytical methods for leachate characterisation. *In*: CHRISTENSEN, T. H., COSSU, R. & STEGMANN, R. (eds) *Landfilling of Waste: Leachate*. Elsevier, London.

MCFARLANE, M. J. & BOWDEN, D. J. 1992. Mobilisation of aluminium in the weathering profiles of the African surface in Malawi. *Earth Surface Processes and Landforms*, **17**, 789–806.

——, —— & GIUSTI, L. 1993. Some aspects of microbially-mediated aluminium mobility in weathering profiles in Malawi – The implications for groundwater quality. *In*: GUERRERO, R. & PEDRÓS-ALIÓ, C. (eds) *Trends in Microbial Ecology*, 677–680.

MAULE, J., LOWE, R. K. & MCCULLOCH, J. L. 1993. Performance evaluation of synthetically lined landfills. *Tappi Journal*, **76**(12), 172–176.

MITCHELL, J. K. 1976. *Fundamentals of Soil Behaviour*. Wiley, New York.

MLYNAREK, J., ROLLIN, A. L., LAFLEUR, J. & BOLDUC, G. 1990. Microstructural analysis of a soil/geotextile system. *In*: PEGGS, I. D. (ed.) *Geosynthetics: Microstructure and Performance*. ASTM STP 1076, American Society for Testing and Materials, Philadelphia, 137–146.

MOLLARD, S. J., JEFFORD, C. E., STAFF, M. G. & BROWNING, G. R. J. 1996. Geomembrane landfill liners in the real world. *In*: BENTLEY, S. P. (ed.) *Engineering Geology of Waste Disposal*. Geological Society, London, Engineering Geology Special Publications, **11**, 165–170.

MOLLINS, L. H., STEWART, D. I. & COUSENS, T. W. 1996. Predicting the properties of bentonite–sand mixtures. *Clay Minerals*, **31**, 243–252.

MONTERO, C. M. & OVERMANN, L. K. 1990. Geotextile filtration performance test. *In*: KOENER, M. (ed.) *Geosynthetic testing for waste containment applications*. ASTM STP 1081, American Society for Testing and Materials, Philadelphia. 273–284.

MOTT, H. V. & WEBER, W. 1991. Diffusion of organic contaminants through soil–bentonite cut-off barriers. *Journal of the Water Pollution and Control Federation*, **63**, 167–176.

NASIATKA, D. M., SHEPHERD, T. A. & NELSON, J. D. 1981. Clay liner permeability in low pH environments. *In*: *Proceedings of the Symposium on Uranium Mill Tailings Management*, Colorado State University, Fort Collins, Colorado. 627–645.

NORTH WEST WASTE DISPOSAL OFFICERS SUB-GROUP. 1988. *Guidelines on the Use of Landfill Liners*. Lancashire Waste Disposal Authority.

PETERSON, S. R. & GEE, G. W. 1986. Interactions between acidic solutions and clay liners: permeability and neutralisation. *In*: JOHNSON, A. I., FROBEL, R. K., CAVALLI, N. J. & PETTERSSON, C. B. (eds) *Hydraulic Barriers in Soil and Rock*. ASTM STP 874, American Society for Testing and Materials, Philadelphia, 229–245.

ROBINSON, H. D. 1995. *A review of the comparison of leachates from domestic wastes in landfill sites*. Department of the Environment, Report No. **CWM/072/95**.

ROCHE, D. 1996. Landfill failure survey: a technical note. *In*: BENTLEY, S. P. (ed.) *Engineering Geology of Waste Disposal*. Geological Society, London, Engineering Geology Special Publications, **11**, 379–380.

ROHDE, J. R. & GRIBB, M. M. 1990. Biological and Particulate Clogging of Geotextile/Soil Filter Systems. *In*: KOERNER, M. (ed.) *Geosynthetic Testing for Waste Containment Applications*. ASTM STP 1081, American Society for Testing and Materials, Philadelphia, 299–312.

SEYMOUR, K. 1992. Landfill lining for containment. *Journal of the Institute of Water and Environmental Management*, **6**, 389–396.

SPOONER, A. J. 1995. *Leachate: Rock Interaction Geochemistry*. BSc Hons Dissertation, University of Sunderland, Sunderland, UK.

TCHOBANOGLOUS, G., THEISEN, H. & VIGIL, S. A. 1993. *Integrated Solid Waste Management: Engineering Principles and Management Issues*. McGraw-Hill, New York, 406–479.

THOMAS, H. R., REES, S. W., KJARTANSON, B., WAN, A. W. L. & CHANDLER, N. A. 1996. Modelling in situ water uptake in a bentonite–sand barrier. *In*: BENTLEY, S. P. (ed.) *Engineering Geology of Waste Disposal*. Geological Society, London, Engineering Geology Special Publications, **11**, 215–222.

TUCKER, M. E. 1994. *Sedimentary Petrology: An Introduction to the Origin of Sedimentary Rocks*. 2nd edn. Blackwell Scientific, London.

USEPA (United States Environmental Protection Agency). 1983. Hazardous waste management systems; permitting requirements for land disposal facilities. Federal Register, July 26 1983.

USEPA (United States Environmental Protection Agency). 1989. Requirements for hazardous waste landfill design, construction and closure. **EPA/625/489/022**, US Environmental Protection Agency, Cincinnati, Ohio.

WAGNER, J. F. 1993. Conception of a double mineral base liner. *In*: ARNOULD, M., BARRES, M. & COME, B. (eds) *Geology and Confinement of Toxic Wastes*. Geoconfine '93, Balkema, Rotterdam.

WANGEN, L. E. & JONES, M. M. 1984. The attenuation of chemical elements in acidic leachates from coal minerals waste by soils. *Environmental Geology Water Science*, **6**(3), 161–170.

WARITH, M. A. 1987. *Migration of leachate solution through a clay soil*. PhD Thesis, McGill University, Canada.

WEINBERG, R., HEINZE, E. & FORSTNER, U. 1985. Experiments on specific retardation of some organic contaminants by slurry trench materials. *In*: ASSINK, J. W. & VAN DEN BRINK, W. J. (eds) *Contaminated Soil, Proceedings of the First International TNO Conference on Contaminated Soil*, 11–15 November, Utrecht, The Netherlands. Nijhoff, Dordrecht.

WELLNER, W. W., WIERMAN, D. A. & KOCH, H. A. 1985. Effects of landfill leachate on the permeability of clay soils. *Proceedings of the Eighth Annual Madison Waste Conference*, September 18–19, Wisconsin.

YANFUL, E. K., WAYNE NESBITT, H. & QUIGLEY, R. M. 1988. Heavy metal migration at a landfill site, Sarnia, Ontario, Canada-1. Thermodynamic assessment and chemical interpretations. *Applied Geochemistry*, **3**, 523–533.

YONG, R. N. & WARKENTIN, B. P. 1975. Soil properties and behaviour, geotechnical engineering 5. *In*: LANDRETH, R. 1980. *Lining of waste impoundment and disposal facilities*. United States Environmental Protection Agency, Municipal Environmental Resources Laboratory, Cincinnati, Ohio.

——, WARITH, M. A. & BOONSINSUK, P. 1986. Migration of leachate solution through clay liner and substrate. *In*: *Hazardous and Industrial Solid Waste Testing and Disposal: Sixth Volume*. ASTM Special Technical Publication, **933**, Philadelphia.

The geochemical engineering of landfill liners for active containment

S. F. THORNTON,[1] M. I. BRIGHT,[2] D. N. LERNER[1] & J. H. TELLAM[2]

[1] *Groundwater Protection & Restoration Group, Dept. of Civil & Structural Engineering, University of Sheffield, Mappin St, Sheffield S1 3JD, UK*
[2] *Hydrogeology Research Group, School of Earth Sciences, University of Birmingham, Edgbaston, Birmingham B15 2TT, UK*

Abstract: The performance of engineered containment barriers presently used to line UK landfills is uncertain over the design lives of most sites. Failure of key components of the barrier technology can be expected within a timescale under which leachate will still present a threat to groundwater resources. Additional environmental control measures are therefore necessary to mitigate the effects of these future impacts. A methodology for the geochemical engineering design of high-attenuation landfill liners using low-cost natural and industrial waste materials which are available in the UK is presented. Anaerobic laboratory column experiments are used to test the performance of various materials as liners in chemically attenuating contaminants in a methanogenic leachate spiked with a suite of heavy metals and organic micropollutants. The basic liner design recipe which maximizes the attenuation of key contaminants in leachate has been confirmed. Variability in material composition does not significantly affect liner performance provided this recipe is used. Important design issues (liner exhaustion and attenuation reversibility effects) are examined in the study. These liners may be routinely constructed from measurements of a limited suite of material properties using standard laboratory procedures. The application of these liners within the current framework of landfill design is outlined.

Landfill design in the UK has changed substantially in the last decade and nearly all new landfills are now lined with engineered low permeability barriers to minimize the escape of leachate from the site. This policy of 'concentrate and contain' now largely supersedes the former approach of 'dilute and attenuate' as the primary means of leachate management and pollution control at most modern sites. Traditionally, landfills operated on the dilute and attenuate principle were designed to allow leachate to leak from sites into the underlying strata where contaminants would be attenuated by a suite of physical, biological and chemical processes (Ross 1985; Mather 1989). Emphasis in this case remained on the exploitation of the properties of the aquifer to ameliorate the potential impact of leachate on nearby water resources. Many older landfills continue to operate safely on this basis and recorded examples of serious groundwater pollution from such sites are comparatively rare (Robinson & Lucas 1984, 1985; Harris 1988). However, the attenuation of leachate within the subsurface is difficult to predict, and in response to increasing legislative pressure landfill design and operation has now shifted towards the increasing dependence on

technological solutions for environmental protection (Bandy 1988; Barsby 1992).

Important uncertainties exist with the current engineering approach. In particular, the performance of barriers and leachate management systems is unproven over the design lives of most sites (typically $>100\,a$) and failure of key components guaranteeing site security can be expected (Mather 1992; Rodic & Goossens 1993; Suter *et al.* 1993). Many engineered containment barriers comprise compacted *in situ* or remoulded clays (e.g. bentonite) which are combined with synthetic membranes (e.g. HDPE) in elaborate double or composite liner systems (e.g. Monteleone 1990; Seymour 1992; Coulson 1993). In most cases the design criteria require that the clay or mineral liner component of such systems is 1 m thick and compacted to achieve a permeability no greater than $10^{-9}\,\mathrm{m/sec^{-1}}$ (Barsby 1992; Murray *et al.* 1992; Seymour 1992). Despite the implementation of rigorous QC and QA procedures during barrier construction (e.g. Anon 1988; Pohl 1991; Barsby 1992), these permeability specifications may be difficult to achieve and test in the mineral layer under field conditions (Pierce *et al.* 1986; Daniel & Brown 1988). Mechanical defects and hydraulic

From: METCALFE, R. & ROCHELLE, C. A. (eds) 1999. *Chemical Containment of Waste in the Geosphere.* Geological Society, London, Special Publications, **157**, 143–157. 1-86239-040-1/99/$15.00 © The Geological Society of London 1999.

imperfections often remain in composite liners following placement (Elsbury & Sraders 1989), and faults (e.g. punctures and tears etc.) introduced during waste emplacement may reduce the effectiveness of the synthetic membrane used in these barriers (Buss *et al.* 1995). Diffusion of contaminants through both synthetic and mineral liners remains the most serious long-term problem (Johnson *et al.* 1989; Mott & Weber 1991). This process can result in the early breakthrough (<30 a) of contaminants through typical (1 m thick) clay liners (Daniel & Shackelford 1989) and the preferential transport of problematic species (e.g. organic micropollutants) through synthetic membrane liners (Buss *et al.* 1995).

Given the problems identified with the present landfill liner technology, it is necessary to try and accommodate the effects of inevitable leachate leakage within the liner design. This may be achieved by designing barriers to maximize the chemical attenuation of leachate rather than relying on physical containment alone. Conceptually, this approach is based on the controlled or engineered leakage of leachate from sites using natural attenuation to reduce contaminant loadings in groundwater. The potential for this has been demonstrated in many studies of leachate polluted aquifers (e.g. Williams 1988; Christensen *et al.* 1994). The attenuation liner concept simply incorporates the principles of the dilute and attenuate approach within the engineering framework of current landfill liner design. Instead of providing only a physical barrier to flow, the liner is modified to actively attenuate major contaminants. As a technique for the *in situ* treatment of leachate, the application of such reactive barriers has received only limited attention in the UK (Robinson & Lucas 1985) and no design guidelines exist for the geochemical engineering of fieldscale systems within a UK context. The present project was initiated to address some of these issues, with the following objectives:

- to develop a methodology for the chemical design of high attenuation landfill liners
- to test the robustness and performance of the methodology
- to examine some of the issues affecting the attenuation of key leachate contaminants.

The principles underlying the development of attenuation liners have been previously described in a related paper (Thornton *et al.* 1993) and some preliminary results of the research programme have been presented (Bright *et al.* 1996). The present paper summarizes the results of the research and implications for the wider application of these liners.

Approach to study

Contaminants of interest

Although the composition of landfill leachate is highly variable (e.g. Robinson & Gronow 1993), all leachates contain four specific groups of contaminants requiring attention. These are NH_4, heavy metals, natural dissolved organic matter (DOM) fractions (e.g. volatile fatty acids) and xenobiotic organic micropollutants (XOMs) such as petroleum derived aromatic hydrocarbons, halogenated aliphatic compounds and pesticides. The hazard posed varies for each contaminant group and NH_4 in particular is of key concern due to the very high concentrations found in most leachates and the low water quality standards set for this species. Heavy metals and XOMs are normally found at lower concentrations than NH_4, but they are of concern due to their risk to human health and toxicity when present at levels much lower than those measured in most leachates.

Controls on contaminant attenuation by natural materials

Several field and laboratory studies have identified the general physico-chemical controls on leachate contaminant migration through subsurface media (Griffin & Shimp 1976; Griffin *et al.* 1976; Korte *et al.* 1976; Mather 1989; Yong *et al.* 1990; Lyngkilde & Christensen 1992; Schwarzenbach *et al.* 1993; Christensen *et al.* 1994). In the context of the development of attenuation liners, the emphasis remains on the important properties of natural materials (e.g. clays, soils, aquifer sediments) which regulate the attenuation of the key contaminants in leachate. These properties are summarized in Table 1.

In principle the material properties identified in Table 1 allow an attenuation liner to be engineered for any given landfill. The ideal liner recipe which maximizes the attenuation of the key contaminants in leachate is deduced to be: high clay and organic carbon content plus adequate alkaline pH buffering capacity.

Experimental design

Experimental materials

The properties outlined in Table 1 were used to qualitatively evaluate a wide range of compositionally diverse low-cost and/or waste materials for testing as liners (Thornton *et al.* 1993). From

Table 1. *Summary of properties of natural materials which control the attenuation of key contaminants in leachate*

Property	Contribution
Clay mineral content	A high clay mineral content is correlated with an increased particle surface area and sorption capacity for both inorganic and organic contaminants. Permeability is also reduced, increasing the residence time of leachate within the liner, and is important for contaminants (e.g. XOMs) which may be attenuated by processes (e.g. degradation) which are kinetically controlled
Cation exchange capacity (CEC)*	Provided mainly by clay minerals and is primarily responsible for the sorption of NH_4 and heavy metals. Variations tend to reflect changes in particle surface area (smectite: $800\,m^2\,g^{-1}$; kaolinite: $10\,m^2\,g^{-1}$) and extremes in clay mineral composition (e.g. high CEC smectite v. low CEC kaolinite clays)
Solid phase organic carbon content (f_{oc})	Primarily responsible for the sorption of XOMs which may be theoretically predicted from the f_{oc} and solubility of individual hydrophobic compounds
pH buffering capacity	Natural materials may possess either an acidic or an alkaline pH buffering capacity which represents the ability to maintain a constant pH under continuous inputs of alkaline or acid solutions, respectively. Maintenance of an alkaline pH is important for reducing the mobility of many heavy metals and organo-metallic complexes in landfill leachate (particularly acidic, acetogenic leachate) and promoting the biodegradation of both natural organic fractions and XOMs. The alkaline pH buffering capacity is generally correlated with the availability of $CaCO_3$ in natural materials
Solid phase oxide content	Mn and Fe oxyhydroxides provide an important source of solid phase oxidants in natural materials and their reduction by anaerobic landfill leachate has been correlated with the increased biotransformation of XOMs

* Surface area data from Rowell (1994).

this initial screening exercise seven contrasting substrates were provisionally selected for the experiments to provide an expected wide range in the properties of interest (Table 2).

These substrates include two waste products (Black Coal Measures Clay (BCMC) and slate spoil), two aquifer materials (Triassic Sandstone and Chalk), an existing landfill liner (Oxford Clay) and a commercially available quartz sand. All the materials collected were air dried and the clays and slate spoil were ground to the consistency of powder, for compositional analysis and experimental purposes, using a mill. No further preparation was undertaken.

Table 2. *Materials initially selected for testing in liner recipes*

Material	Formation	Source	Company
Oxford Clay		Brogborough landfill site, Woburn, Bedfordshire	Shanks and McEwan Ltd
Coal Measures (Fire) Clay			
Red marl	Etruria Marl Series, Upper Coal Measures	Walleys Quarry, Silverdale, Stoke-on-Trent, Staffordshire	Redland Building Products
Black marl	Black Band Series, Middle Coal Measures	Kidsgrove, Stoke-on-Trent, Staffordshire	Birchenwood Brickworks
Slate spoil	Llangollen Slate Belt, Wenlock and Ludlow Series, Silurian	Llangollen, N. Wales	–
Triassic Sandstone	–	Bromsgrove, W. Midlands	Sandy Lane Quarry
Quartz Sand (Chelford '30)	Pleistocene glacio-fluvial sand	Lower Withington, Chelford, Cheshire	Filterzand Wessen (UK), Newcastle-under-Lyme, Staffordshire
Chalk	Middle Chalk	Hillington, NW Norfolk	–

Methanogenic (M-phase) leachate was used to test the performance of the attenuation liner recipes, as this matrix is likely to be the fluid in contact with the liner for most of the lifetime of the landfill. This composition is also more stable for experimental purposes than acetogenic leachate. 1200 l of fresh M-phase leachate was collected from a representative UK domestic waste disposal site and sampled from a capped cell containing waste of approximately seven years of age. The leachate was spiked with a range of heavy metals (Cr, Cd, Ni, Zn) and XOMs (BTEX compounds, chlorinated aromatic and aliphatic hydrocarbons and pesticides). A 2200 ml stock solution of the heavy metals, comprising 2000 mg l^{-1} Zn and 1000 mg l^{-1} Cd, Cr and Ni, was added to the leachate to provide theoretical concentrations of 10 mg l^{-1} Zn and 5 mg l^{-1} Cd, Cr and Ni when diluted. The XOM spike was prepared from high purity chemicals (Aldrich chemical company). The heavy metals and XOMs added in the spike were not present in significant quanities in the leachate used, and were selected on the basis of their presence in other UK landfill leachates (Robinson & Gronow 1993) and their variable geochemical and biochemical behaviour. Freshwater (tapwater) was also used in the experiments (see below). The compositions of the M-phase leachate and freshwater used in the study is presented in Table 3. Equilibrium concentrations of the heavy metals (Cd, Cr, Ni, Zn) added to the leachate are lower than the theoretical values expected. This is attributed to precipitation of metal carbonate fractions during storage prior to flushing of the experimental columns.

Table 3. *Composition of the methanogenic leachate and freshwater used in the experiments*

Species	Leachate	Freshwater
pH	7.2	6.65
Electrical conductivity (EC) (μS^{-1}cm^{-1})	14 000	93
Eh (mV)	64	nd
Alkalinity (as CaCO$_3$)	4 600	10
NH$_4$	1 029	bdl
Ca	125	5.1
Mg	135	0.97
K	490	0.63
Na	1 340	6.06
Cl	1 965	11.4
SO$_4$	52	4.05
Mn	0.17	bdl
Fe	13.9	0.25
Si	17.8	0.41
Cd	2.65†	bdl
Cr	0.39†	bdl
Ni	4.34†	bdl
Zn	2.25†	bdl
Chemical oxygen demand (COD)	1 060	bdl
Benzene*	385†	nd
Toluene*	558†	nd
1,2-Dichlorobenzene (DCB)*	393†	nd
1,2,3-Trichlorobenzene (TCB)*	151†	nd
Hexachlorobenzene (HCB)*	47†	nd
1,1,1-Trichloroethane (TCA)*	676†	nd
Trichloroethene (TCE)*	598†	nd
1,1,2,2-Tetrachloroethane (TeCA)*	166†	nd
Tetrachloroethene (TeCE)*	279†	nd
Naphthalene*	120†	nd
Hexachlorobutadiene (HCBD)*	277†	nd
Lindane*	87†	nd
Dieldrin*	16†	nd

All concentrations in mg l^{-1} unless stated otherwise.
* Concentrations in μg l^{-1}.
† Equilibrium concentrations after spiking.
bdl below detection limits.
nd. not determined.

Experimental apparatus

Laboratory column experiments were used to investigate the interactions between selected liner materials and the spiked leachate. A schematic of the experimental set-up is presented in Fig. 1. The apparatus supplied freshwater and leachate to the columns under conditions isolated from the atmosphere. The leachate was stored anaerobically under a 50:50 CO$_2$:N$_2$ atmosphere in sealed, refrigerated tanks at 4°C to minimize compositional changes (Korte *et al.* 1976). A gas pressure of 30 kPa was used to supply leachate to the experimental columns and the apparatus is designed to transfer leachate out of the active tanks (A and B) only. The passive tanks (P) provide an equilibrium gas phase composition of volatile organic fractions which is transferred to the active tanks. This gas phase

compensates for headspace volatilization losses from the leachate which would otherwise occur as the active tanks empty during the experiment. All metal components in contact with leachate were 316 grade stainless steel to minimize contamination. The freshwater was pumped from a 200 l HDPE drum to a header tank and delivered by gravity under a constant head (3 m). A separate circuit was configured within the

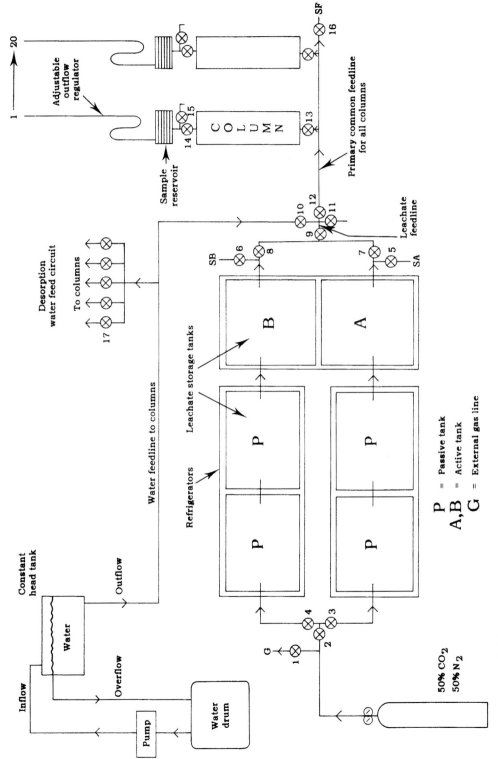

Fig. 1. Schematic of the laboratory apparatus.

main freshwater feedline to enable flushing of leachate-loaded columns for the contaminant desorption phase of the study. The freshwater and leachate compositions were monitored during the experiments at sampling ports (SA, SB and SF) in the main circuit (Fig. 1).

All columns (20) were constructed from 1 m lengths of UPVC fitted with endcaps and the set comprised 18 standard diameter (9.9 cm ID), one narrow diameter (6.5 cm ID) and one wide diameter (15 cm ID) columns. All columns were operated in upflow mode and samples were collected in a 3 m stainless steel coil (8 mm ID) attached to the adjustable outflow regulator (AOR). Samples were removed through a valve (No. 15) at the base of the coil after filling (Fig. 1). The height of the AOR was adjusted for each column to provide a flow rate of 140 ml over a 24-hour period. This provided an average residence time of 25, 8 and 45 d for the standard, narrow and wide diameter columns, respectively. The experiments were conducted in a laboratory in which the temperature was 18–21°C.

Each column was manually packed and reweighed to determine individual packing densities and approximate void volume. All columns were flushed with one pore volume of freshwater prior to leachate percolation.

Analytical procedures

Freshwater and leachate samples

All freshwater and leachate samples were filtered through a Whatman® 1.2 μm glass fibre (pre)-filter (as necessary) and Whatman® 0.45 μm membrane filter. The samples were processed and analysed by standard methods (APHA 1985) using the procedures summarized in Table 4.

Solid phase analysis

Duplicate subsamples of the seven liner materials initially selected were analysed for pH (van Reeuwijk 1987), organic carbon (OC) content (Gaudette *et al.* 1974) and carbonate content (van Reeuwijk 1987). Exchangeable bases were extracted using an unbuffered 1 N NH_4Cl solution and 1 N KCl was used to release exchangeable NH_4 (van Reeuwijk 1987). The CEC of each material was determined by summation of the exchangeable bases, with correction of Ca analyses for dissolution of native carbonate by the extract. Metal oxide fractions were extracted and partitioned into 'amorphous' and 'crystalline' phases according to the procedures developed by Canfield *et al.* (1992) and Kostka & Luther (1994). The metals in each extract were analysed by ICP-AES.

Design of column experiments

Of the seven test materials initially selected, slate spoil and the red Coal Measures Clay were rejected for use in the column experiments. This was because preliminary compositional analysis indicated that the effects of their range of properties (e.g. CEC, organic carbon content, pH buffering capacity) could be adequately investigated with the remaining substrates. Selected geochemical properties of the materials subsequently used in the column experiments are presented in Table 5. These illustrate the compositional diversity in the suite of materials tested which includes substrates possessing one (e.g. Chalk), several (e.g. BCMC, Oxford Clay) or none (e.g. quartz sand) of the key properties of interest.

These contrasting substrates were blended in different proportions to create a series of 20 liner

Table 4. *Summary of processing and analytical procedures for freshwater and leachate samples*

Parameter	Preservation and processing	Method of analysis
pH, EC	No filtration or preservation; measured immediately	Probe
Eh	No filtration or preservation; measured in a flowcell	Probe
Alkalinity	No filtration or preservation; measured immediately	Titration to pH 4.5 end-point
Metals	Filtered and acidified to pH < 2 with HNO_3	ICPS
Cl, NH_4	Filtered; no preservation	Colorimetry
NO_3, NO_2	Filtered; preserved with 1 drop 0.25% $HgCl_2$	Colorimetry
SO_4	Filtered; no preservation	Ion chromatography
COD	No filtration; acidified to pH < 2 with H_2SO_4	Titration after digestion
XOMs	No filtration or preservation but addition of deuterated benzene (Benzene D5) and 1 bromo-3 chloropropane as internal standards for analysis of volatile compounds. Extraction of samples with pentane and addition of acenaphthene as internal standard for analysis of semi- to non-volatile compounds	GC-MS
Gases	Samples collected in glass ampoules	GC-MS

Table 5. *Selected geochemical properties of the test substrates*

Substrate	pH	EC*	Organic carbon†	CaCO₃†	CEC‡	Mn†§	Fe†§
Quartz sand	6.5	68	0.0073	<1	0.16	0.00013	0.011
Triassic Sandstone	4.3	25	0.026	<1	3.24	0.0076	0.78
Black Coal Measures Clay (BCMC)	2.4	4750	2.61	<1	18.54	nd	1.61
Oxford Clay	7.4	3200	3.92	14.75	20.43	0.0045	0.76
Chalk¶	na	na	0.0545	87.8	0.44	na	na

$*(\mu S\,cm^{-1})$.
† Values in wt%.
‡ (meq $100\,g^{-1}$).
§ MnO_2 and Fe_2O_3 equivalent.
nd not detected.
na not available.
¶ Data taken from Needham (1995).

recipes for testing in the column experiments (Table 6). Individual recipes were designated 'poor', 'good' and 'improved' according to the extent to which each satisfied the ideal recipe which maximized the attenuation of the key contaminants.

The quartz sand was used to dilute the active substrates and also as a control experiment. Powdered Chalk was used as a pH buffering agent and its performance when added as a layer or admixed with the host liner substrate evaluated. This array of experiments enabled an assessment to be made of impacts of variations in one or several key substrate properties on the liner performance, and also to test the robustness of the methodology against a range of materials. The column experiments were run for up to 15 months and a selection of leachate-loaded liner recipes were flushed with oxygenated water to examine attenuation reversibility effects. Some exhausted liner recipes were also sectioned to determine the distributions of attenuated contaminants.

Interpretation of column experiments

Effluents from the column experiments were sampled periodically and analysed using the procedures previously described (Table 4). Samples (approximately 60 ml) were collected from each column at intervals corresponding to 0.1 pore volumes for the first pore volume of the leachate flush and thereafter at intervals of 0.2–0.25 pore volumes. The results are presented in terms of dimensionless concentration (C/C_0), where C and C_0 are the concentrations of species in the effluent and leachate, respectively, against the volume of leachate eluted through the liner (pore volumes). The storage system used maintained stable concentrations of analytes in the leachate over the course of the experiments. These measured concentrations were averaged to provide the input values (C_0) for normalization of the column effluent data.

The solute transport relationships in each column experiment were evaluated using an analytical solution of the advection–dispersion equation (Ogata & Banks 1961). This model was used to estimate the hydraulic parameters (kinematic velocity, dispersivity and porosity) of each column by least-squares fitting to the chloride tracer. The hydraulic parameters were then fixed for the transport simulations and evaluation of experimental solute distribution coefficients (K_d^e). The latter was estimated using the following relationship:

$$R_f^e = 1 + (\rho_b \times K_d^e/\theta) \qquad (1)$$

where ρ_b is the dry bulk density $(g\,cm^{-3})$ and θ is the effective porosity of the media. Values of ρ_b

Table 6. *Summary of liner designs tested*

Liner recipe	Substrates*
Blank	Quartz sand (>98.5% SiO₂)
Poor	Triassic Sandstone
Poor	BCMC (10, 10%) + quartz sand
Good	Oxford Clay (5, 10†, 15, 15%) + quartz sand
Improved	Oxford Clay (10, 15, 15, 20%) + Triassic sandstone
Improved	BCMC (10%) + quartz sand + CaCO₃ (admixed, layered)

* Numbers in brackets represent percentages of clay in liner recipe and designated repeat experiments.
† This recipe also used to examine leachate residence time in the liner using the narrow and wide diameter columns.

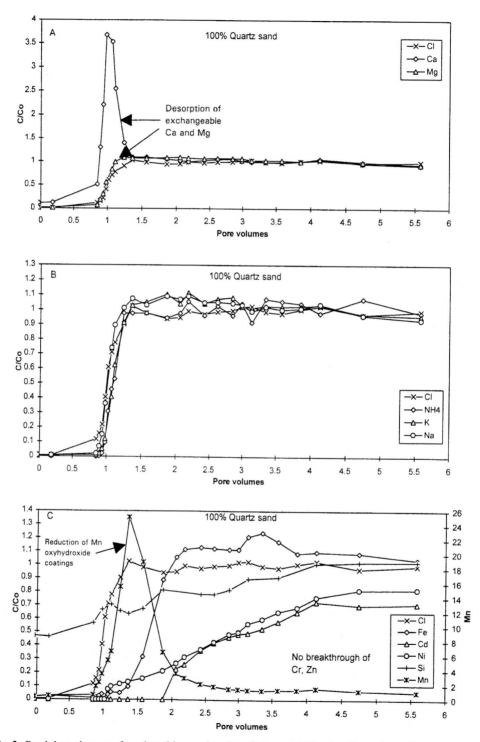

Fig. 2. Breakthrough curves for selected inorganic solutes from a control column experiment (quartz sand).

were determined for each column. Values of θ and R_f^e (experimentally determined retardation factor) were determined using the transport model (Ogata & Banks 1961) by, respectively, curve fitting of the chloride tracer profile and from the transport simulations of solutes in each experiment. This modelling approach assumes that the column properties remain constant throughout the experiment. The model was used to estimate the values of R_f^e and K_d^e from the breakthrough profiles of NH_4, natural DOM and XOMs in each experiment. Theoretical solute distribution coefficients (K_d^t) for the XOMs in each experiment were also estimated, according to the following relationship:

$$K_d^t = f_{oc} \times K_{oc} \qquad (2)$$

where f_{oc} is the fraction of organic carbon present in the liner recipe and K_{oc} is the theoretical solute partition coefficient to solid organic carbon (taken from Schwarzenbach et al. 1993).

Static metal speciation and saturation state modelling of the column effluents was also undertaken using the MINTEQA2 geochemical program (Allison et al. 1990). No allowance was made for the contribution of natural DOM fractions in the speciation modelling due to uncertainty in the composition of DOM in this leachate. This modelling was used in conjunction with solute mass balances to provide supplementary data on processes regulating contaminant attenuation.

Results and discussion

The research programme has generated a detailed, high-quality database on leachate attenuation by a variety of natural materials currently used to seal landfills (e.g. Coal Measure Clay and Oxford Clay) or which form the base of many unlined sites (e.g. Triassic Sandstone). The performance of the liner recipes in attenuating the key contaminants in this leachate can be illustrated by comparing a representative selection of contaminant breakthrough curves (BTC) from a control experiment and an ideal liner recipe. These are displayed in Figs 2 to 5 which show BTC for selected inorganic solutes and organic solutes for a 100% quartz sand column (Figs 2 and 3) and a 10% Oxford Clay–90% quartz sand column (Figs 4 and 5).

Attenuation of NH_4

Attenuation of leachate NH_4 occurred by cation-exchange reactions with the liner materials and resulted in the desorption of native exchangeable Ca (and to a lesser extent, Mg) from the host substrate (Figs 2a and 4a). The extent of NH_4 retardation increased with the proportion of the active substrate (Oxford Clay or BCMC) in the liner recipe, compared with the control experiment (Fig. 2b). Mass balance calculations also indicate that comparable amounts of NH_4 are sorbed by the BCMC and Oxford Clay recipes containing identical proportions (e.g. 10%) of these substrates. However, NH_4 is no longer attenuated once the liner's sorbed cation composition has reached equilibrium with the leachate cation composition. This occurs after three pore volumes in the column illustrated and results in NH_4 being eluted from the liner at concentrations found in the leachate (Fig. 4a).

Despite the clearly complex five-way (Ca, Mg, Na, K NH_4) ion exchange processes involved, solute transport modelling of the experimental data suggested that NH_4 transport *in these liner recipes* could be described by a simple sorption coefficient (K_d). Computed K_d values for NH_4 were found to increase with the clay content and CEC of the liner and suggested that the measurement of either property could be used to estimate NH_4 sorption by the liner, based on the K_d approach, at least over the range of clay contents used in the systems examined. The flushing of exhausted liner recipes with freshwater resulted in the desorption of previously attenuated NH_4. The data indicated that all sorbed NH_4 is released under these conditions, although concentrations are significantly lower than those in the leachate and manageable from a water quality viewpoint.

Attenuation of heavy metals

Ni and Cd in the leachate were eluted from the quartz sand control column at concentrations significantly above UK water quality standards (Table 7) (Fig. 2c). Cd was initially removed completely for almost two pore volumes in this experiment, coincident with a flush of Mn from the column and retardation of leachate Fe. Mass balance calculations suggest that the behaviour of Mn and Fe can be explained by the chemical dissolution of Mn oxyhydroxides on the quartz sand (Table 5) by leachate, according to the general reaction (Postma 1985):

$$2Fe^{2+} + MnO_2 + 4H_2O$$
$$\rightarrow 2Fe(OH)_3 + Mn^{2+} + 2H^+. \qquad (3)$$

The Cd removed over this interval is probably retained as a coprecipitate within the Fe

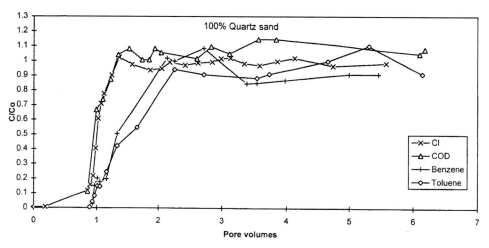

Fig. 3. Breakthrough curves for selected organic solutes from a control column experiment (quartz sand).

oxyhydroxide formed by reaction 3 (e.g. Suarez & Langmuir 1976). In contrast, the breakthrough of Ni (and Cd after the Mn flush) in this experiment was very similar to that of Si. Concentrations of these species increased slowly but steadily to near input levels over five pore volumes. This behaviour may reflect the possible association of Cd and Ni with a colloidal (Si) component in the leachate or leached from the column.

All added heavy metals (up to $10 \, \text{mg} \, \text{l}^{-1}$), with the exception of Ni, were completely attenuated by the ideal recipe (Fig. 4c). No analyses for dissolved sulphide species were undertaken but bacterial SO_4 reduction (BSR) was recorded in many experiments and provides an important sink for the heavy metal contaminants. This (BSR) was deduced from the reduction in column effluent SO_4 concentrations below input leachate values ($C/C_0 < 1.0$), column sectioning (see below) and the results of MINTEQA2 calculations which suggested that the leachate was substantially undersaturated with respect to SO_4-bearing phases modelled after breakthrough. Porewater SO_4 concentrations are initially very high due to oxidation of native pyrite within the Oxford Clay during the preliminary freshwater flushing (Fig. 4b). However, this supply is rapidly depleted by BSR ($C/C_0 < 1.0$ after four pore volumes) which results in the development of SO_4-limited (steady state) conditions after eight pore volumes. This condition limits the attenuation of additional Fe mobilized by the microbially mediated reduction of Fe oxyhydroxides on the liner substrate (Table 5) (Fig. 4b, after eight pore volumes).

Equilibrium geochemical modelling calculations and the results of column sectioning (Scott 1994) suggest that the heavy metals added to the leachate are immobilized within the liner as metal sulphides and carbonates. A small quantity of Ni (max $c. 0.9 \, \text{mg} \, \text{l}^{-1}$) was eluted from most of the liners tested. This mobile fraction may be part of a colloidal component, or reflect complexation with an organic/inorganic component which is not geochemically attenuated. The environmental significance of enhanced transport of contaminants due to association with organic and inorganic colloids in landfill leachate has recently been evaluated by Gounaris *et al.* (1993). However, porewater loadings of Ni remain low in the liners tested and the ideal liner recipe was superior to all the other substrates examined in preventing remobilization of attenuated heavy metals during freshwater flushing. Only Ni was eluted at concentrations above background values, although loadings were significantly lower than those eluted during leachate flushing.

Table 7. *UK water quality standards for selected pollutants in landfill leachate*

Parameter	Water quality standard	Units of measurement
NH_4	0.5	$\text{mg} \, \text{l}^{-1}$
Fe	200	$\mu\text{g} \, \text{l}^{-1}$
Mn	50	$\mu\text{g} \, \text{l}^{-1}$
Cd	5	$\mu\text{g} \, \text{l}^{-1}$
Cr	50	$\mu\text{g} \, \text{l}^{-1}$
Ni	50	$\mu\text{g} \, \text{l}^{-1}$
Zn	100*	$\mu\text{g} \, \text{l}^{-1}$
COD	5	$\text{mg} \, \text{l}^{-1}$

Data taken from Gray (1995).
* EC standard given.

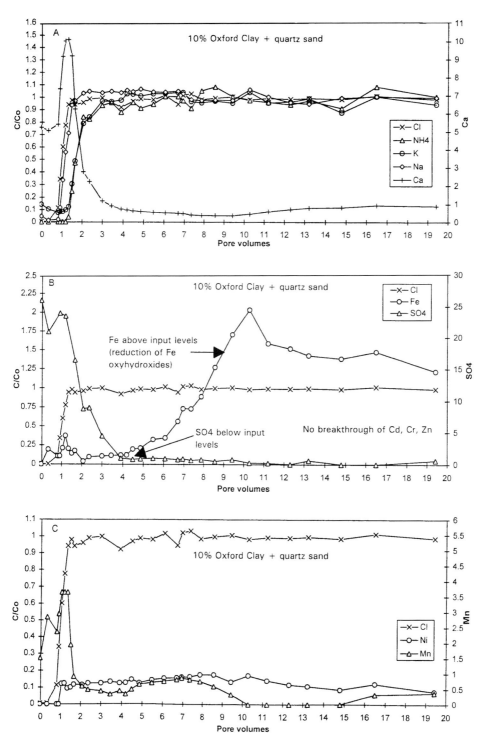

Fig. 4. Breakthrough curves for selected inorganic solutes from a 10% Oxford Clay-90% quartz sand column experiment.

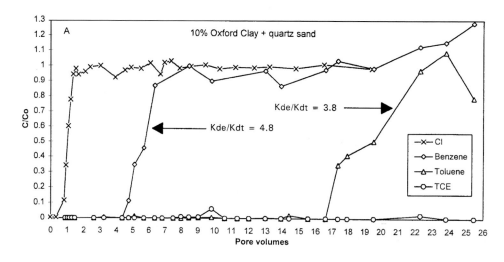

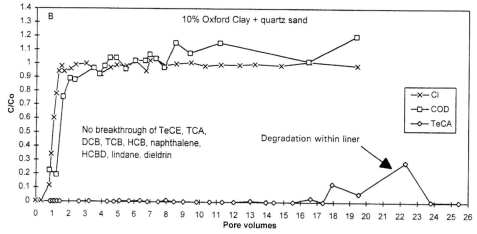

Fig. 5. Breakthrough curves for selected organic solutes from a 10% Oxford Clay-90% quartz sand column experiment.

Attenuation of dissolved organic matter and XOMs

No analyses for total organic carbon in this leachate were made and instead the natural DOM is represented by COD. This is a realistic analogue of the DOM in this type of leachate as the added XOMs do not significantly contribute to the COD measurements (Christensen *et al.* 1994). There was minimal sorption of leachate DOM (expressed as COD) by the control column (Fig. 3) and only a marginal increase in the sorption of this fraction by the ideal recipe (Fig. 5b). Degradation of DOM in this leachate was also not significant in any of the experiments. This feature probably reflects the dom-

inance of humic-like substances in the DOM pool of most M-phase leachates (Chian & De-walle 1976). These are typically more refractory and resistant to biotransformation than fatty acid components (Christensen *et al.* 1994).

In contrast, both sorption and biotransformation played a key role in attenuating the organic micropollutant load added to the test leachate. Sorption of XOMs by the ideal recipe was substantially greater than that in the control experiment (cf. Figs 3 and 5a). The experimental sorption coefficients (K_d^e) for the organic solutes were found to exceed predicted values (K_d^t) by up to a factor of five (Fig. 5a). Although benzene remained biologically recalcitrant within the timescale of the experiments many of the other added aromatic and aliphatic compounds were rapidly degraded to low concentrations (Fig. 5b). It is not possible to predict the biotransformation (e.g. half lives) of many compounds in the XOM mixture for the liner systems studied due to non-breakthrough. However, generalizations on the potential for biotransformation of these compounds based on the governing hydrochemical conditions can be made. It is therefore noteworthy that biotransformation of these contrasting XOMs occurred under a variety of anaerobic conditions (e.g. Mn-reducing, Fe-reducing and SO_4-reducing), as observed in many leachate plumes (Christensen et al. 1994). This suggests that under the conditions created within the ideal liner recipe, biotransformation of many organic chemicals of environmental concern will be quite significant and sustained by a variety of readily available oxidants.

Engineering application

This study has demonstrated that chemical attenuation has a potentially significant role to play in current landfill liner design; its contribution may be through increasing the leachate attenuation capacity of the primary containment barrier, or as a separate liner in its own right, functioning as a back-up liner beneath the former. The liner can be constructed from locally available materials, supplemented by specific additives where necessary to modify its chemical properties. Leachate quality is difficult to predict *a priori* and a generic composition can be used in the design process. Measurement of solute K_d provided an adequate estimate of NH_4 attenuation in the liners tested but this approach may not be applicable in all cases for other systems. Solid phase organic carbon content and alkaline pH buffering capacity or $CaCO_3$ content appear to provide a conservative estimate of contaminant breakthrough and liner performance for organic chemical and heavy metal attenuation. Natural DOM in leachate may not be significantly attenuated by these liners and this fraction should therefore be treated as a conservative species for liner design. The key engineering issue remains predicting where leachate leakage will occur. Uniform seepage through a containment barrier would require an attenuation liner under the entire site, whereas point source leakage would require a concentrated larger mass of attenuating material. The economic implications of each scenario differ considerably and it may be more cost effective to design sites to focus leachate flow into low points underlain by attenuating material. The conventional approach to site design is sufficiently flexible in this respect to incorporate this feature without undue loss of landfilling capacity.

Conclusion

This research study has experimentally tested a design methodology for the geochemical engineering of high attenuation landfill liners using low-cost natural and industrial waste materials. The ideal liner recipe maximizing the chemical attenuation of the key contaminants in leachate has been confirmed and these liners can be routinely constructed from available materials from measurements of a limited suite of properties using standard laboratory procedures. Material variability does not significantly affect liner performance provided the standard recipe is used and additives should be admixed within the host liner substrate. Attenuation of organic chemicals in the liners occurred by both sorption and biodegradation. Heavy metals in leachate are immobilized within the liner substrate as metal carbonates and sulphides. All the liners examined had a finite attenuation capacity for NH_4 but this is large for heavy metals, provided the system equilibrium and supply of $CaCO_3$ and SO_4 is maintained.

This work was completed with financial support from the UK EPSRC under grant GR3/F87233. H. Robinson and Aspinwall & Co. are thanked for providing site access for leachate collection. Shanks and McEwan Ltd are also thanked for providing site access for collection of the Oxford Clay samples.

References

ALLISON, J. D., BROWN, D. S. & NOVO-GRADAC, K. J. 1990. *MINTEQA2 metal speciation equilibrium model for surface and groundwater, version 3.00.* Center for Exposure Assessment Modelling, U.S. EPA, Athens, Georgia.

ANON 1988. *Guidelines on the use of landfill liners.* North West Waste Disposal Officers Landfill Liners Sub-Group Report.

APHA 1985. *Standard methods for the examination of water and wastewaters.* American Public Health Association.

BANDY, J. T. 1988. Hazardous landfill containment and reduction of hazardous waste. In: GRONOW, J. R., SCHOFIELD, A. N. & JAIN, R. K. (eds) *Land Disposal of Hazardous Waste: Engineering and Environmental Issues.* Ellis Horwood, Chichester, 297–303.

BARSBY, R. 1992. The lining of landfill sites with natural clay materials: two case histories. In: *Planning and Engineering of Landfills.* Proceedings of the Midlands Geotechnical Society Conference, 230–235.

BRIGHT, M. I., THORNTON, S. F., LERNER, D. N. & TELLAM., J. H. 1996. Laboratory investigations into designed high-attenuation landfill liners. *In*: BENTLEY, S. P. (ed.) *Engineering Geology of Waste Disposal*. Geological Society, London, Engineering Geology Special Publications, **11**, 165–170.

BUSS, S. E., BUTLER, A. P., JOHNSTON, P. M., SOLLARS, C. J. & PERRY, R. 1995. Mechanisms of leachate leakage through synthetic landfill liner materials. *Journal of the Chartered Institute of Water and Environmental Management*, **9**, 353–359.

CANFIELD, D. E., RAISWELL, R. & BOTTRELL, S. H. 1992. The reactivity of sedimentary iron minerals towards sulfide. *American Journal of Science*, **292**, 659–683.

CHIAN, E. S. K. & DEWALLE, F. B. 1976. Sanitary landfill leachates and their treatment. *Journal of the Environmental Engineering Division*, ASCE, **102**, 411–431.

CHRISTENSEN, T. H., KJELDSEN, P., ALBRECHTSEN, H. J., HERON, G., NIELSEN, P. H., BJERG, P. L. & HOLM, P. E. 1994. Attenuation of landfill leachate pollutants in aquifers. *Critical Reviews in Environmental Science and Technology*, **24**, 119–202.

COULSON, J. 1993. Geomembrane liners prevent leachate leakage. *Pollution Prevention*, **3**, 71–72.

DANIEL, D. E. & BROWN, K. W. 1988. Landfill liners: How well do they work and what is there future. *In*: GRONOW, J. R., SCHOFIELD, A. N. & JAIN, R. K. (eds) *Land Disposal of Hazardous Waste: Engineering and Environmental Issues*. Ellis Horwood, Chichester, 235–244.

—— & SHACKELFORD, C. D. 1989. Containment of landfill leachate with clay liners. *In*: CHRISTENSEN, T. H, COSSU, R. & STEGMANN, R (eds) *Sanitary Landfilling: Process, Technology and Environmental Impact*. Academic, London, 323–341.

ELSBURY, B. R. & SRADERS, G. A. 1989. Building a better landfill liner. *Civil Engineer*, **59**, 57–59.

GAUDETTE, H. E., WILSON, F. R., TONER, L. & FOLGER, D. W. 1974. An inexpensive titration method for the determination of organic carbon in Recent sediments. *Journal of Sedimentary Petrology*, **44**, 249–253.

GOUNARIS, V., ANDERSON, P. R. & HOLSEN, T. M. 1993. Characteristics and environmental significance of colloids in landfill leachate. *Environmental Science and Technology*, **27**, 1381–1387.

GRAY, N. F. 1995. *Drinking Water Quality: Problems and Solutions*. Wiley, Chichester.

GRIFFIN, R. A. & SHIMP, N. F. 1976. Effects of pH on exchange-adsorption or precipitation of lead from landfill leachates by clay minerals. *Environmental Science and Technology*, **10**, 1256–1261.

——, ——, STEEL, J. D., RODNEY, R. R., WHITE, W. A. & HUGHES, G. M. 1976. Attenuation of pollutants in municipal landfill leachate by passage through clay. *Environmental Science and Technology*, **10**, 1262–1268.

HARRIS, R. C. 1988. Leachate migration and attenuation in the unsaturated zone of the Triassic Sandstones. *In*: GRONOW, J. R., SCHOFIELD, A. N. & JAIN, R. K. (eds) *Land Disposal of Hazardous Waste: Engineering and Environmental Issues*. Ellis Horwood, Chichester, 175–197.

JOHNSON, R. L, CHERRY, J. A. & PANKOW, J. F. 1989. Diffusive contaminant transport in natural clay: a field example and implications for clay-lined waste disposal sites. *Environmental Science and Technology*, **23**, 340–349.

KORTE, N. E., SKOPP, J., FULLER, W. H., NIEBLA, E. E. & ALESII, B. A. 1976. Trace element movement in soils: influence of soil physical and chemical properties. *Soil Science*, **122**, 350–359.

KOSTKA, J. E. & LUTHER, G. W. 1994. Partition and speciation of solid phase iron in saltmarsh sediments. *Geochimica et Cosmochimica Acta*, **58**, 1701–1710.

LYNGKILDE, J. & CHRISTENSEN, T. H. 1992. Fate of organic contaminants in the redox zones of a landfill leachate pollution plume (Vejen, Denmark). *Journal of Contaminant Hydrology*, **10**, 291–307.

MATHER, J. D. 1989. The attenuation of the organic component of landfill leachate in the unsaturated zone: a review. *Quarterly Journal of Engineering Geology*, **22**, 241–246.

——1992. Current landfill design – a short term engineering solution with a long term environmental cost. *In*: *Planning and Engineering of Landfills*. Proceedings of the Midlands Geotechnical Society Conference, 168–174.

MONTELEONE, M. 1990. Landfill liner quality. *Civil Engineer*, **60**, 47–49.

MOTT, H. V. & WEBER, W. J. 1991. Diffusion of organic contaminants through soil–bentonite cut-off barriers. *Journal of the Water Pollution Control Federation*, **63**, 166–176.

MURRAY, E. J., RIX, D. W. & HUMPHREY, R. D. 1992. Clay linings to landfill sites. *Quarterly Journal of Engineering Geology*, **25**, 311–376.

NEEDHAM, S. N. 1995. *The behaviour of atrazine and simazine within the Chalk aquifer*. PhD Thesis, University of Birmingham.

OGATA, A. & BANKS, R. B. 1961. *A solution of the differential equation of longitudinal dispersion in porous media*. US Geological Survey Professional Paper, **411-A**, 7.

PIERCE, J. J., SALFOURS, G. & PETERSON, E. 1986. Clay liner construction and quality control. *Journal of the Environmental Engineering Division, ASCE*, **112**, 13–24.

POHL, D. H. 1991. Construction quality assurance for double and composite lined landfills. *In*: *Proceedings of the Annual Madison Waste Conference*, **14**, 141–151.

POSTMA, D. 1985. Concentration of Mn and separation from Fe in sediments-I. Kinetics and stoichiometry of the reaction between birnessite and dissolved FeII at 10°C. *Geochimica et Cosmochimica Acta*, **49**, 1023–1033.

ROBINSON, H. D. & GRONOW, J. R. 1993. A review of landfill leachate composition in the UK. *In*: *Sardinia 93, Proceedings of the Fourth International Landfill Symposium*. CISA, Environmental Sanitary Engineering Centre, Cagliari, Italy, 821–832.

—— & LUCAS, J. L. 1984. Leachate attenuation in the unsaturated zone beneath landfills: instrumentation and monitoring of a site in southern England. *Water Science and Technology*, **17**, 477–492.

—— & ——1985. Attenuation of leachate in a designed, engineered and instrumented unsaturated zone beneath a domestic waste landfill. *Water Pollution Research Journal of Canada*, **20**, 76–91.

RODIC, L. & GOOSSENS, L. H. T. 1993. Long term performance of landfill technology, expert judgement. *In: Sardinia 93, Proceedings of the Fourth International Landfill Symposium*. CISA, Environmental Sanitary Engineering Centre, Cagliari, Italy, 1193–1201.

ROSS, C. A. M. 1985. The unsaturated zone as a barrier to groundwater pollution by hazardous wastes. *In: Memoirs of the 18th Congress of the International Association of Hydrogeologists*. 127–141.

ROWELL, D. L. 1994. *Soil Science: Methods and Applications*. Longman.

SEYMOUR, K. J. 1992. Landfill lining for containment. *Journal of the Chartered Institute of Water and Environmental Management*, **6**, 389–396.

SCHWARZENBACH, R. P., GSCHWEND, P. M. & IMBODEN, D. M. 1993. *Environmental Organic Chemistry*. Wiley, New York, 681.

SCOTT, P. K. 1994. *An investigation into the retention of heavy metals within experimental landfill liner material*. MSc Thesis, University of Leeds.

SUAREZ, D. L. & LANGMUIR, D. 1976. Heavy metal relationships in a Pennsylvania soil. *Geochimica et Cosmochimica Acta*, **40**, 589–598.

SUTER, G. W., LUXMORE, R. J. & SMITH, E. D. 1993. Compacted soil barriers at abandoned landfills are likely to fail in the long term. *Journal of Environmental Quality*, **22**, 217–226.

THORNTON, S. F., BRIGHT, M. I., LERNER, D. N. & TELLAM, J. H. 1993. The role of attenuation in landfill liners. *In: Sardinia 93, Proceedings of the Fourth International Landfill Symposium*. CISA, Environmental Sanitary Engineering Centre, Cagliari, Italy, 407–416.

VAN REEUWIJK, L. P. 1987. *Procedures for soil analysis*. International Soil Reference and Information Centre, Technical Paper **9**. Wageningen, The Netherlands.

WILLIAMS, G. M. 1988. Integrated studies into groundwater pollution by hazardous waste. *In:* GRONOW, J. R., SCHOFIELD, A. N. & JAIN, R. K. (eds) *Land Disposal of Hazardous Waste: Engineering and Environmental Issues*. Ellis Horwood, Chichester, 37–48.

YONG, R. N., WARKENTIN, B. P., PHADUNGCHEWITT, Y. & GALVEZ, R. 1990. Buffer capacity and lead retention in some clay materials. *Water, Air & Soil Pollution*, **53**, 53–67.

Metal retention from landfill leachate by glacial clays: laboratory experiments

E. L. HURST[1] & S. P. HOLMES

University of Sunderland, School of the Environment, Benedict Building,
St George's Way, Sunderland, SR2 7BW, UK
[1] *Present address: Environment Agency, Waterside House, Waterside North,*
Lincoln LN2 5HA, UK

Abstract: Two sets of laboratory experiments were carried out to investigate the role of clay liners on the retention of metals present in landfill leachate. In the first experiment leachate was passed through clay filled columns and comparisons made of the metal content in treatment columns (leachate) with that in the control columns (de-ionized water). The metals Na, K, Mg, Ca, Al and Fe show elevated concentrations when compared with the control columns. In addition, a black sludge was observed on the upper surface of the treatment columns. On analysis this sludge was found to be Fe-rich when compared with the surrounding leachate/clay matrix. In the second experiment a simple tank system was used to investigate sludge formation at the leachate : clay liner interface. The observed sludge was found to have elevated metal concentrations, with a corresponding decrease in metal levels in leachate samples, compared with analysis before the experiment. From these observations it was concluded that the sludge may be acting as a sink for key metal contaminants from leachate.

The attenuation of landfill leachate by mineral liners is a concept in which the landfill industry has placed considerable confidence. This is reinforced by the small number of pollution incidents in the UK arising from the disposal of municipal solid waste (MSW), by virtue of appropriate geological and design considerations (Robinson & Lucas 1984; Ross 1985; Robinson & Gronow 1992). The composition of landfill leachate reflects the relatively low levels of heavy metal input in MSW sites; levels of metals such as Pb, Cd, Cu and Zn are usually found at less than 1 ppm (Ehrig 1989; Thornton *et al.* 1993; Robinson & Gronow 1995). Contamination of groundwater with metals from such a source is therefore considered unlikely. In addition, the high pH of a mature (methanogenic) landfill environment encourages precipitation and ready immobilization of heavy metals (Christensen *et al.* 1994). Although the concentration and mobility of heavy metals within the containing matrix are minimal, even these concentrations have been shown to cause organism toxicity through bioaccumulation (Alloway 1990; Vernet 1991). The discharge of metals such as these is also controlled under European and UK legislation (EC 1980; NRA 1991). In the light of these factors, the attenuation of heavy metals still merits consideration.

Preliminary studies of the interaction of leachate with natural clay mineral liners, has revealed the development of a black sludge material. The formation of a sulphide sludge at the leachate : clay interface is well documented and has been acknowledged to be a significant factor in the blinding of pores and the consequent reduction in the permeability of clay mineral liners (Bisdom *et al.* 1983; Brune *et al.* 1991). It is postulated that this material may act as a sink for metals and other contaminants in leachate, by microbially controlled precipitation. Specific micro-organisms exist that are able to reduce sulphates and initiate the precipitation of metal sulphides, in this way (Brune *et al.* 1991; Watson *et al.* 1995).

The concentration of some metals in a semi-stable sludge at the base of landfill sites may present a serious potential pollution problem. Long-term changes in environmental conditions, caused by the infiltration of rainfall or the rise of oxygenated groundwater into the unsaturated zone, may lead to the dissolution and remobilization of these metals as the matrix retaining them is degraded.

Two approaches have been employed to study the nature of heavy metal behaviour at the leachate : mineral liner interface. Firstly, changes in the heavy metal content of the clay are examined

From: METCALFE, R. & ROCHELLE, C. A. (eds) 1999. *Chemical Containment of Waste in the Geosphere.* Geological Society, London, Special Publications, **157**, 159–166. 1-86239-040-1/99/$15.00 © The Geological Society of London 1999.

as leachate is passed through large scale columns. Secondly, tank experiments are used to model the leachate : clay interface in greater detail. Although this study is primarily concerned with the interactions of the heavy metals, some discussion is also made of other metals in the leachate as they have an important role in exchange and ion competition processes.

Materials and methods

The mineral liner material for this study was obtained from a landfill site in Northumberland, UK, where there is glacial till to a depth of approximately 10 m. The columns, (polypropylene lined steel pipes, 200×250 mm), were pushed by a JCB, into the base of a cell that was being prepared for landfilling, and carefully dug out. The samples were therefore as undisturbed as possible and at field compaction and moisture levels.

Leachate was collected from a leachate sump located within a 3- to 5-year-old waste mass at a working landfill in North Yorkshire, UK. A methanogenic sample was selected for this test as it was considered to reflect prevailing conditions at a landfill in the long term, an acetogenic leachate would probably have contained higher levels of heavy metals, but the acetogenic phase is relatively short-lived (DoE 1986). A small pump powered by a 12 volt battery was used to draw the leachate from the well directly into polypropylene drums continually purged with nitrogen to maintain an anaerobic atmosphere. Leachate composition was monitored throughout the test (Table 1). An experimental rig was devised (Fig. 1) and leachate was delivered to the

Table 1. *Summary of leachate and control eluent composition including standard error*

Element	Leachate (mg l^{-1})	Control (mg l^{-1})
Na	290.3 ± 49	1.23 ± 0.6
K	450.9 ± 11	0.76 ± 0.3
Mg	209.1 ± 37	1.32 ± 0.5
Ca	66.4 ± 4	909 ± 0.7
Fe	3.67 ± 0.1	0.1 ± 0.3
Cu	0.04 ± 0.01	bd
Zn	0.81 ± 0.08	bd
Pb	0.14 ± 0.04	bd
Cd	0.02 ± 0.01	bd
Al	0.23 ± 0.05	bd
Cl	3628 ± 136	nd

bd: below detection.
nd: no data.

surface of the clay in the column, under gravity, via a plastic pipe network. The system was fed by the header tank into which the leachate was recirculated from the chilled reservoir, maintained at 3–5°C, to minimize biological degradation of the leachate. Anaerobic conditions were maintained throughout the five month experiment by flushing continually with nitrogen gas.

On completion, the seven columns were sampled at five depths through the clay core at 0, 50, 100, 150 and 200 mm. The samples were oven dried (105°C for 12 h), and ground to a fine powder, using a pestle and mortar and automated grinder. The ground samples were then digested in concentrated nitric and perchloric acid and analysed for a suite of metals by Inductively Coupled Plasma Atomic Emission Spectrometry (ICPAES) (Thompson & Walsh 1983). An identical system of control columns

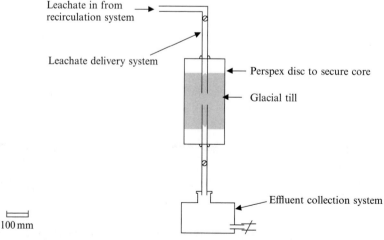

Leachate in from recirculation system

Leachate delivery system

Perspex disc to secure core

Glacial till

Effluent collection system

100 mm

Fig. 1. Column experiment layout.

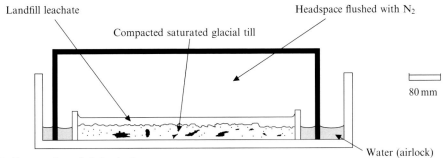

Landfill leachate

Headspace flushed with N_2

Compacted saturated glacial till

80 mm

Water (airlock)

Fig. 2. Cross section of sludge tank.

were run concurrently with the treatment columns, using deionized water as the eluent.

Samples were also analysed for chloride to determine if migration of the more mobile leachate constituents, such as Cl^-, might be detected. Samples for Cl^- analysis were prepared as above and Cl^- was extracted by a water leaching processes (Rump & Krist 1992). The resulting solutions were analysed by I7on Chromatography.

The interface of the leachate and clay liner material was modelled using a simple tank system. A perspex box ($350 \times 350 \times 50$ mm) was filled with a layer (depth 20 mm) of pre-saturated clay. The clay was collected at the same time and from the same location as the samples used in the column experiments and was pre-saturated with tap water, to minimize wetting and capillary effects. The tank was filled with leachate to a depth of 40 mm and the head space flushed with nitrogen. An oversized lid and water filled trench

around the tank provided an airlock (Fig. 2). This system maximized the surface area contact between the leachate and the clay liner, and therefore optimized sludge production and allowed the penetration of contaminants from the leachate into the liner material.

Samples were prepared for analysis in the same way as for the columns, with the exception that they were digested in concentrated nitric and hydrochloric acid and analysed by Atomic Absorption Spectrometry (AAS) and Inductively Coupled Plasma Mass Spectrometry (ICPMS).

Data analyses were conducted using analysis of variance (ANOVA), and analysis of covarianace (ANCOVA) with mathematical transformation of the data where appropriate.

Results

Initial observations on dismantling the columns revealed a black sludge at the point of entry

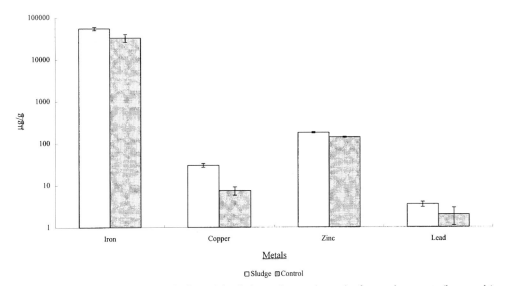

μg/g

100000

10000

1000

100

10

1

Iron   Copper   Zinc   Lead

Metals

□ Sludge ▨ Control

Fig. 3. Metal concentration (± standard error) in sludge and control samples from column tests (log_{10} scale).

of the tube into the clay material. Statistical analysis of the data obtained from the chemical analysis of the sludge when compared with the background matrix revealed that the sludge was enriched with Fe, Cu, Zn and Pb (Fig. 3). Examination of how the concentrations changed with depth showed an initial surface enrichment followed by a gradual decrease to background levels with depth (Figs 4(a) and 4(b)). However, when these changes were statistically

analysed using ANCOVA no consistent trend was determinable.

Analysis of the Mg and Ca concentrations showed a consistent increase in their concentration with depth (Fig. 5). In contrast, Na and K concentrations increased in the upper levels of the column (Fig. 6).

When all the mean element concentrations in the treatment columns and control columns (irrespective of depth) were compared using

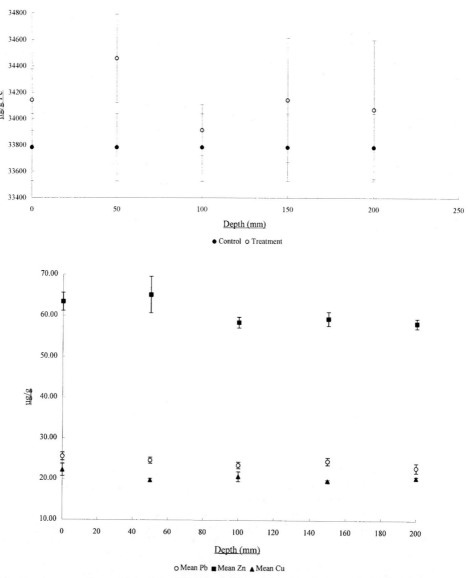

Fig. 4. Mean Fe concentrations (**a**) and mean heavy metal concentrations (**b**) in glacial till samples from treatment and control columns plotted v. depth.

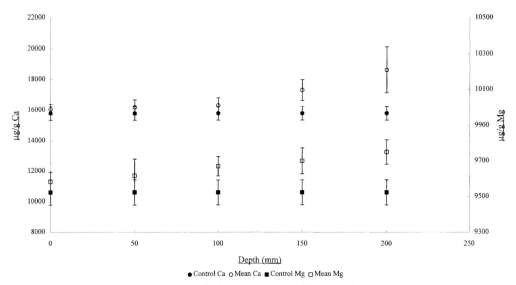

Fig. 5. Ca and Mg concentrations (± standard error) in glacial till samples for treatment and control columns plotted v. depth.

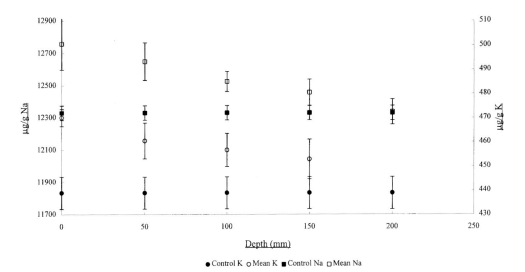

Fig. 6. Mean Na and K concentrations (± standard error) in glacial till for control and treatment columns v. depth.

individual ANOVA tests, all elements except Fe and Cd were found to be at higher concentration in the treatment matrix than in the control (Table 1). In effect, retention of selected metals from leachate is occurring as a result of interaction with the geomatrix.

Examination of changes in the chloride concentration both between the treatment and the control and for depth changes determined that the treatment column was Cl⁻-enriched in comparison to the control column and that the concentration of Cl⁻ ions was elevated with increasing depth.

Two types of material were observed on completion of the sludge tank experiment. An orange gelatinous precipitate was found around the edges of the tank, and toward the centre a filamentous grey green material, with black staining underneath. Analyses of both these materials showed an increase in metals Fe,

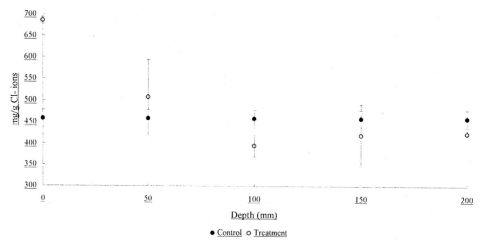

● Control ○ Treatment

Fig. 7. Mean chloride concentration (± standard error) in glacial till samples for treatment and control columns plotted v. depth.

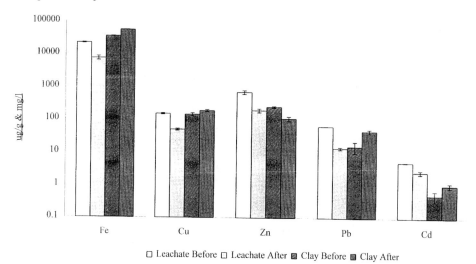

□ Leachate Before □ Leachate After ▨ Clay Before ▥ Clay After

Fig. 8. Metal concentration (± standard error) in clay and leachate before and after the sludge tank test.

Cu, Zn, Cd and Pb, and a reciprocal decrease in the concentration of metals in the leachate (Fig. 8).

Discussion

Heavy metals

These are the preliminary results of an on-going study investigating the leachate:clay liner interface. Although it has been widely acknowledged that heavy metals are readily attenuated in the first few centimetres of clay mineral liners, detail about the mechanisms and quantity of metals

that can be retained in this way are limited. These results suggest the formation of a sludge may be a key element in metal retention. There is evidence from these column experiments that attenuation processes are occurring. The black sludge at the leachate:clay interface appears to retain the majority of metals, as there is little evidence of further transport of metals at greater depth in the columns. The attenuation of metals over a relatively short distance in clay liners has been well recognized (Griffin & Shimp 1978; Christensen et al. 1994). It has been considered that this was mainly by the processes of ion exchange and precipitation. The occurrence of the sludge suggests that some or all of these metals may be only temporarily retained in this zone. Possible sequestration of metals into an organic matrix may lend them greater mobility by chelation and complexation processes, particularly if that matrix later becomes degraded. These findings are reinforced by those of the tank system, where the development and metal enrichment of sludge are apparent. The possible origins of this sludge material are three-fold. The reduction of metals and their precipitation as sulphides is a common process in anaerobic zones of the geosphere. It can take place as a purely chemical process or may be mediated by micro-organisms who are able to derive energy from such biochemical cycles (Schlesinger 1991). The source of the leachate : clay sludge may be a combination of these two reactions. However, the extensive community of micro-organisms essential to the degradation of the waste mass itself is likely to have some influence on the zone immediately below the site. The sludge could also be produced by the settling out of particulates and aggregated colloidal materials. Metals in leachate are likely to be adsorbed on the surfaces of such materials, some of which may be organic, and as these settle and accumulate they could be responsible for the generation of a sludge layer (J. Dearlove pers. comm., 1992).

Other species

Further evidence of transport in the column experiments comes from the exchange system of Ca/Mg for Na/K. The slight increase at depth of Ca/Mg, with the comparable Na/K increase in the surface layers, suggests that Na/K from the leachate is being exchanged for Ca/Mg on the geomatrix. This mimics the 'hardness halo' effect found in groundwater around some landfill sites (Griffin & Shimp 1978). The statistics do not bear out these observations, but these are clearly incipient trends. The chloride data suggests that some migration has occurred. The conservative nature of chloride means that its passage through the geomatrix is relatively unimpeded by interaction with the geomatrix. However, other components in the leachate may be able to form complexes with Cl species and this can affect the mobility of both constituents (Yong & Sheremata 1991).

This work suggests that the convention and accepted wisdom concerning metal dynamics in the leachate : clay system do hold true in large scale laboratory tests. However, there is a possibility that the mechanisms involved are perhaps more complex than was at first thought. The possible role of micro-organisms in leachate attenuation is still poorly understood although

recognized to be of significance. Further work is to be undertaken to develop an understanding of the quantitative balance in the leachate : clay system and to determine whether metals can be readily remobilized from the matrix. Field samples are also needed to evaluate whether this material is truly analogous to that reportedly found in our landfill sites.

The authors would like to acknowledge the valued assistance of A. Mistry and J. Holdsworth at Sunderland University, C. Ottley at Durham University, and B. Coles and A. Doyle at Imperial College Analytical Unit for Geology. This study constitutes part of the work towards a PhD at the University of Sunderland under the supervision of J. P. L. Dearlove, Westlakes Research Institute.

References

ALLOWAY, B. J. 1990. *Heavy Metals in Soils.* Blackie, Glasgow.

BISDOM, E. B. A., BOEKSTEIN, A., CURMI, P. *et al.* 1983. Submicroscopy and chemistry of heavy metal precipitates from column experiments simulating conditions in a soil beneath a landfill. *Geoderma*, **30**, 1–20.

BRUNE, M., RAMKE, H. G., COLLINS, H. J. & HANNERT, H. H. 1991. Incrustation processes in the drainage systems of sanitary landfills. *Proceedings: Sardinia '91 Third International Landfill Symposium. 14–18th October, 1991.* CISA, Environmental and Sanitary Engineering Centre, Cagliari, Italy.

CHRISTENSEN, T. H., KJELDSEN, P., ALBRECHTSTEIN, H. J., HERON, G., NIELSEN, P. H., BJERG, P. L. & HOLM, P. E. 1994. Attenuation of landfill leachate in aquifers. *Critical reviews in Environmental Science and Technology*, **24**(2), 119–202.

EC COUNCIL DIRECTIVE. 1980. *On the protection of groundwater against pollution caused by certain dangerous substances.* 80/68/EEC.

EHRIG, H. J. 1989. Leachate quality. *In:* CHRISTENSEN, T. H., COSSU, R. & STEGMAN, R. (eds) *Sanitary Landfilling: Process, Technology and Environmental Impact.* Elsevier, Amsterdam, 213–229.

GRIFFIN, R. A. & SHIMP, N. F. 1978. *Attenuation of pollutants in municipal landfill leachate.* EPA 600/2-78/157. Illinois State Geological Survey, Urbana.

NRA. 1991. *Policy and Practice for the Protection of Groundwater.* National Rivers Authority, HMSO, London.

ROBINSON, H. & LUCAS, J. L. 1984. Leachate attenuation in the unsaturated zone beneath landfills: instrumentation and monitoring of a site in Southern England. *Water Science and Technology*, **17**, 477–492.

—— & GRONOW, J. 1992. Groundwater protection in the UK: Assessment of the landfill leachate source term. *Journal of the Institute of Water and Environmental Management.* **16**(2), 229–236.

ROSS, C. A. M. 1985. The unsaturated zone as a barrier to groundwater pollution hy hazardous waste.

Hydrogeology in the Service of Man: Proceedings of 18th Congress of International Association of Hydrogeologists, Cambridge.

RUMP, H. H. & KRIST, H. 1992. *Laboratory Manual for the Examination of Water, Waste-water and Soil*. 2nd edn, VCH, New York.

SCHLESINGER, W. H. 1991. *Biogeochemistry: An Analysis of Global Change*. Academic, New York.

THOMPSON, M. & WALSH, J. N. 1983. *A Handbook of Inductively Coupled Plasma Spectrometry*. Blackie, Glasgow.

THORNTON, S. F., LERNER, D. N., BRIGHT, M. I. & TELLAM, J. H. 1993. The role of attenuation in landfill liners. *Proceedings Sardinia. 93. Fourth International Landfill Symposium*.

VERNET, J. P. 1991. *Heavy Metals in the Environment*. Elsevier, Amsterdam.

YONG, R. N. & SHEREMATA, T. W. 1991. The effect of chloride ions on adsorption of cadmium from a landfill leachate. *Canadian Geotechnical Journal*, **28**, 378–387.

WATSON, J. H. P., ELLWOOD, D. C., DENG, Q., MIKHALOVSKY, S., HAYTER, C. E. & EVANS, J. 1995. Heavy metal adsorption on bacterially produced FeS. *Minerals Engineering*, **8**(10), 1097–1108.

Radioactive waste containment in indurated shales: comparison between the chemical containment properties of matrix and fractures

L. DE WINDT, J. CABRERA & J. Y. BOISSON

*Institut de Protection et de Sûreté Nucléaire, DPRE-SERGD,
92265 Fontenay-aux-Roses Cedex, France*

Abstract: The chemical containment properties of poorly consolidated, plastic clays have received considerable attention, particularly during investigations related to radioactive waste disposal. However, the chemical containment properties of indurated shales have received rather less consideration. This paper summarizes those properties of consolidated shales which make them suitable for the chemical containment of radionuclides, using the example of investigations at the Tournemire research site in the south of France. The problems associated with investigating such low-permeability media are highlighted, and solutions to these problems are presented. At this research site, tectonic processes have induced centimetre-scale fractures and decimetre-scale faults in the shales, whereas excavation of a tunnel has generated metre-scale fracturing. Some low water flows through the fracture networks have been detected. However, the Tournemire shales are characterized by very low matrix pore-water contents, and any limited fluid mobility through the matrix is caused mainly by diffusion. Slowly flowing waters in fractures are chemically analogous to these matrix pore-waters. The hydrochemistry is of $Na–Cl–(HCO_3^-)$ type, and the waters are slightly alkaline, strongly reducing and of low ionic strength. The resemblance between the chemistries of pore- and fracture-waters arises from the buffering of the pore-water chemistry by matrix mineral/water interactions, and possible subsequent diffusion of the pore-water into the fractures. The chemical buffering mechanisms controlling the water chemistries (especially pH, alkalinity and *Eh*), are discussed and modelled. The chemical conditions will minimize the mobility of many radionuclides in both shale matrix and in fractures. Chemical evidence from the fracture-filling calcite is consistent with this hypothesis and demonstrates the past immobility of U and Cs. Batch experiments carried out in gloveboxes under a N_2 atmosphere and with synthetic fracture-water show weaker retention of Cs by the calcite fillings than by the shale matrix. Due to its very low porosity, the shale matrix should behave like a membrane filter for colloids, further reducing the mobility of radionuclides. In fractures, interface structures between matrix walls and calcite veins should also hinder colloid migration (except perhaps in occasional centimetre-scale apertures).

The safe disposal of radioactive wastes requires that they are isolated from the biosphere for many thousands, or even millions, of years. Many different rock types have been investigated in different countries to evaluate their suitability to host repositories of radioactive wastes (e.g. Savage 1994). In several countries, deep argillaceous formations have been investigated. Potentially, such formations are able to contain radionuclides for extremely long periods of time (of the order of 10^6 years), due to their low intrinsic permeabilities, high element sorption capacities, microparticle filtration properties, and reducing and slightly alkaline hydrochemistries, which minimize the solubilities and mobilities of many radionuclides (Gera *et al.* 1992; Higgo 1987; Savage 1994). It is

also generally assumed when developing a waste disposal concept, that mudrocks have the extra advantages of high plasticity and swelling properties, leading to fracture self-healing (Higgo 1987).

For the past twenty years, scientific and technical publications concerned with radioactive waste disposal in argillaceous media have dealt mostly with plastic clays. The most advanced investigations were undertaken in the HADES underground laboratory in the Boom clays at Mol in Belgium (Bonne 1992), although other countries also developed generic research programmes in this field (Bath *et al.* 1988; Cramer & Smellie 1994; Gautschi 1992; Gera *et al.* 1992). However, more recently, indurated argillaceous formations (such as marls, claystones or shales)

From: METCALFE, R. & ROCHELLE, C. A. (eds) 1999. *Chemical Containment of Waste in the Geosphere.* Geological Society, London, Special Publications, **157**, 167–181. 1-86239-040-1/99/$15.00 © The Geological Society of London 1999.

have also been investigated (Thury 1997; Lalieux *et al.* 1996; Kickmaier & McKinley 1997), especially in France where such media are considered as potential host rocks for disposal of high level radioactive waste. The characteristics of these formations differ, in many respects, from those of plastic clays. For example, water within indurated shales is generally located within both fractures and pore space, whereas in plastic clays, fracturing is much less important. Therefore, when developing concepts for radioactive waste repositories in indurated clay media, it is important to understand the chemical containment properties not only of the rock matrix, but also of fractures and veins. Another important difference between indurated and plastic clays is that any organic matter present within the former tends to be more mature than organic matter within less compacted, plastic clays. This distinction is important, since organic matter is potentially an important complexant which may bind with radionuclides.

The contrasts in textural, mineralogical, and organic maturity between indurated and plastic clays will also cause the two types of material to evolve differently in the future. When comparing the suitability of plastic or indurated clays as possible hosts for radioactive waste repositories, it is important to understand and predict these different behaviours. The present bias in favour of our understanding of plastic clays needs to be redressed by further studies of indurated argillaceous lithologies.

In the long term, radionuclide migration in the *far-field* of an underground repository (that is, beyond the direct influence of any engineered barrier system), will be controlled by radionuclide speciation and by mineral (co)precipitation and sorption phenomena along the preferential water flow pathways in the geological media. Therefore, any investigation of indurated clays should give a high priority to understanding the chemical processes operating in both natural and artificially induced fracture networks, which will form the main conduits for water flow. A knowledge of the water–rock interactions and, more specifically, of the chemical containment processes within the matrix and fractures, should be acquired in parallel with the characterization of hydrogeological parameters. It is the purpose of the present paper to evaluate these chemical containment processes, using information acquired at the Tournemire research site in southern France (Barbreau & Boisson 1993; Boisson *et al.* 1996). This site is the subject of on-going investigations by the French Institute for Nuclear Safety and

Protection (IPSN) aimed at developing appropriate and robust approaches for assessing potential nuclear waste storage sites within indurated argillaceous rocks.

General geological setting

Stratigraphy and structure

The Tournemire research site is situated in the Larzac Causse in southern Aveyron (Fig. 1). This site is located within sub-horizontal Jurassic sedimentary rocks which may be described by a 'three-layer model' (Cabrera 1991; Barbreau & Boisson 1993; Boisson *et al.* 1996; Fig. 1): a lower layer, 400 m thick, consisting of limestone and dolomite (Hettangian, Sinemurian and Carixian); an intermediate layer, 250 m thick, composed of shales and marls (Domerian, Toarcian and lower Aalenian); and an upper layer, 300 m thick, characterized by limestone and dolomite formations (upper Aalenian, Bajocian and Bathonian). A railway tunnel, constructed about 100 years ago, gives access to the argillaceous formation, and allows generic scientific and geotechnical investigations under conditions that are in many ways similar to those of a deep repository (Fig. 1). Galleries have been driven from the tunnel, and boreholes have been drilled to obtain samples of rock and water and to make hydraulic measurements.

The Larzac area is cut by regional E–W and NE–SW fault systems that affect all the rock formations of the Permo-Jurassic basin. This fault system shows evidence for a polyphase tectonic evolution. The major faulting event is attributed to N–S trending Pyrenean compressional tectonics (50 to 60 million years ago). Oligocene E–W extension and E–W compression have also affected this region. The Tournemire research area is delimited by two major strands of the approximately E–W fault system of which, to the north, the Cernon fault is the most important (Fig. 1) (Cabrera 1991, 1995). The detailed structure of this E–W Tournemire block is characterized by NNW–SSE trending fractures and faults. These fractures are related to Pyrenean compressional tectonics (Cabrera 1995). Within the Tournemire clayey formation, the fracturing is essentially sub-vertical and includes microfissures (millimetre-scale) and fractures (centimetre-scale) sealed with calcite, and faults (decimetre- to metre-scale), including fault breccia (Cabrera 1995). Engineering works related to tunnel construction, have generated additional fractures (millimetre- to centimetre-scale) within Excavated Disturbed Zones (EDZ) two metres thick.

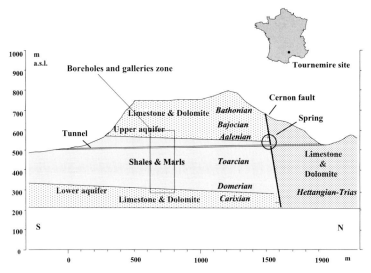

Fig. 1. Geological cross-section of the Tournemire tunnel site with inset map showing location of Tournemire.

Rock laboratory and sampling zone

The zone where *in situ* investigations and rock sampling have been undertaken is illustrated in Fig. 1. A radial set of boreholes was initially drilled perpendicular to the tunnel axis, to facilitate acquisition of data in two dimensions from within the argillaceous formation. Following borehole drilling, two lateral galleries, 30 metres long, were excavated in the shales to perform geotechnical tests and *in situ* experiments. This excavation has revealed the presence of two structurally distinct compartments separated by a major fault within the shales (Fig. 2): a fractured compartment on the west side; and an almost unfractured compartment on the east side. In contrast to the high fracture density of the western compartment, a single fracture plane was encountered in the eastern gallery. This

plane produced a weak, but long-term water flow, which permitted sampling for hydrochemical investigations. Apart from analyses of these water samples, all the other results presented in this paper are derived from pore-waters, extracted from undisturbed and fractured shale samples collected during either borehole drilling or the gallery excavations.

Hydrogeology

The Aalenian limestones represent a local aquifer, whereas the Carixian limestones are a regional aquifer. The argillaceous formation represents an aquitard characterized by very low hydraulic conductivities ($K \approx 10^{-13}\,\mathrm{m\,s^{-1}}$) and by very small apparent diffusion coefficients ($D_e < 10^{-11}\,\mathrm{m^2\,s^{-1}}$) (Boisson 1996). Within the

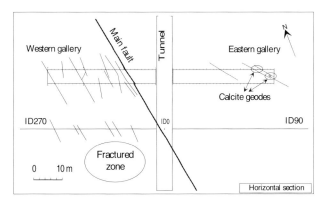

Fig. 2. Location of the natural and man-induced fracturing around the galleries.

clayey matrix of the shale, porewater contents are very low (3–5% by weight) and any water movements, though of very minor importance, seem to be driven substantially by diffusion. The few other indurated clay-rich sites which are currently being investigated in other countries, also show such extreme hydrogeological parameters (Gautschi 1996; Scholtis et al. 1996; Thury 1997), natural fracturing (Chiantore & Gera 1986; D'Allessandro & Gera 1986; Gautschi 1996; Thury 1997) and/or relatively large EDZ (Thury 1997). For instance, tests performed on Opalinus Clay yielded hydraulic conductivities in the range of 2×10^{-14} to $6 \times 10^{-13}\,\mathrm{m\,s^{-1}}$ (Gautschi 1996).

In addition, damp zones following the fracture networks have been observed in some boreholes and during the excavation of the galleries. Some of these water inflows are permanent, even though flow rates are low. These flowing zones display the most evidence for a tectonic origin. The proximity of the Aalenian aquifer associated with the drilling of the boreholes themselves could explain the fluid production in some fractures, but not all. In general, engineering works within the argillaceous formation have enhanced water circulation with permanent inflows in some cases.

Petrophysics and mineralogy

Undisturbed shales and marls

Weight loss by water evaporation (at 105°C), Hg injection and N_2-BET isotherm experiments were performed to investigate the water content and the pore space of the shales and marls. In addition to their low water content, the shales and marls exhibit no macroporosity (pore size > 50 nm), a porosity of 9% centred on a 2.5 nm pore size and specific surface areas greater than $20\,\mathrm{m^2\,g^{-1}}$. A porosity interconnectivity of 'parallel plane' type (pores of long and flat struc-

tures) gives the best fit to the BET isotherms. This result is consistent with the clay particles being orientated parallel to the stratigraphy, under the influence of directed lithostatic pressure. These observations imply water–rock interactions must have been characterized by high rock/water ratios within the shale matrix.

The mineralogical analysis has been performed by X-Ray Diffraction (XRD) on rock powder and oriented sections to improve the identification of clay minerals. The mineralogy of the Toarcian shales and marls is summarized in Table 1. Shales and marls are similar when observed with Scanning Electron Microscopy (SEM). A typical photomicrograph of a shale is presented in Fig. 3a. Clays are the most abundant minerals and are composed mostly of kaolinite and 'illite' (pure illite and interstratified illite/smectite with more than 70% of illite). Although small, the smectite component plays a role in the cationic exchange capacities (CEC) of the Tournemire shales. Calcite is the predominant carbonate and has rather high Mg, Sr, Fe and Mn contents (see Table 2). Other carbonates such as dolomite (especially in marls), ankerite and siderite are identified in smaller proportions. Detrital muscovite, biotite, albite and K-feldspar have been identified by SEM. Framboidal pyrite and organic matter are widespread in the shale's matrix. The total content of organic matter is close to 1% by weight. The organic matter was investigated using Rock-Eval pyrolysis and Gas Chromatography Mass Spectroscopy (GC/MS). Rock-Eval pyrolysis gave a T_{max} of 440°C and the organic matter was found to include $\alpha\beta$-hopanes and $\beta\alpha$-diasteranes, but no fulvic or humic acids. Furthermore, the n-alkanes present were found to have a Gaussian distribution: $nC_{16}-nC_{33}$ range of molecular weights with nC_{20} as the most abundant alkane. Thus, the organic matter is mainly mature kerogen, and has probably a very low content of functional groups suitable for complexing

Table 1. *Mineralogy of the Tournemire Toarcian shales and marls*

Material	Mica + Illite	I/S	Kaol.	Chlor.	Quartz	Calcite	Dolom. / Anker.	Sider.	Feldsp.	Pyrite	Hemat.	Organ. Carb.
Shales												
bulk	25–15	30–15	15–10	5–2	20–15	15–11	1–0	3–0	3–0	3–0	1–0	1–0
<2 µm	25–15	60–45	23–15	7–2	5–2	5–2	1–0	–	1–0	2–0	1–0	–
Marls												
bulk	20–15	25–15	15–10	5–2	10–5	28–23	7–3	5–0	3–0	1–0	1–0	1–0
<2 µm	22–10	52–35	20–10	7–2	5–2	15–10	2–0	–	1–0	1–0	1–0	–

The proportions are given in wt%.
I/S stands for interstratified illite/smectite clays.

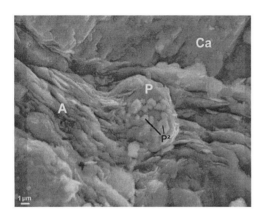

Fig. 3. (a) SEM photograph of typical shale (A: clays, Ca: calcite, P: pyrite), (b) open fracture showing leaps completely covered by macroscopic growth of calcite.

with aqueous species (such as phenol or carboxylic acids).

The Cationic Exchange Capacity (CEC) and the ion occupancies have been determined by the NH_4Ac method applying a correction factor which takes into account the carbonate dissolution. The measure of the consumption of the NH_4^+ ions yields the CEC value. Typical data are reported in Table 3. Marls have lower CEC values due to their higher carbonate contents. Ca is the predominant exchangeable species, although rather high Na occupancies are also measured.

Fracturing structure and discrete water-conducting features

Standard thin-sections petrography, cathodoluminescence petrography and scanning electron microscopy (SEM) have been used to characterize the mineral parageneses of the fracture fillings. Four types of mineral parageneses have been identified in veins (Mathieu *et al.* 1997): calcite alone; calcite and cubic pyrite; calcite and framboidal pyrite; and calcite and baryte. Calcite is the dominant mineral, and the first two parageneses are by far the most frequent ones. No relationship has been established between the mineral parageneses of the veins, their spatial orientation and the tectonic episodes. However, different conditions of formation and several episodes of calcite deposition have been identified. Clayey matrix fragments are sometimes trapped within the calcite veins.

Field observations indicate that the flowing water mostly follows preferential pathways located in decimetre-scale dilation zones where calcite geodes occur (see Fig. 3b), and to a lesser extent, in some of the subvertical fracture planes which are not fully sealed by calcite and show millimetre-scale apertures. The vein walls are brecciated on a millimetre-scale. These observations are corroborated by microsopic studies of apparently sealed fractures. Preliminary field observations made in the galleries indicate that

Table 2. *Average values of some element contents in shales and calcites*

Material	Element (ppM)							
	Ba	Ce	Cs	Fe	Mg	Mn	Sr	U
Shales (ppm)	250	80	6	36 500	13 500	320	265	2
Calcites (ppm)	1.8	3.0	0.02	9 100	1 900	1600	330	0.01

Table 3. *Typical CEC and major cation occupancy data for the Tournemire shales and marls*

Material	CEC bulk	CEC < 2 µm	Na	K	Mg	Ca
Shales	11.4	23.4	2.6	1.6	3.0	4.5
Marls	8.7	11.6	2.1	1.1	2.4	3.4

Values in meq/100 g.

these two types of preferential pathway might be connected to each other.

EDZ fracturing

The EDZ fracturing extends up to a distance of approximately two metres away from the tunnel walls. Within the EDZ, the disruption of the shale matrix is not complete, but rather propagates along macro- and micro-fissures (centimetre- to millimetre-scale) with no calcite sealing. Macroscopic crystals of gypsum occur on the fractured surfaces of the shales. But neither pyrite alteration nor neoformations of minerals such as iron oxyhydroxides or jarosite have been observed by optical and electron microscopes.

Water–rock interactions

Water–rock interactions are important controls on the chemical containment of radionuclides in the geosphere (Bruno *et al.* 1995; Gera *et al.* 1992; Higgo 1987). Radioactive elements exhibit a very wide variety of chemical behaviours, and there are large possible variations in element mobility as the chemistry of a natural water changes in response to water–rock interactions. For example, U solubility is about 10 000 times greater in oxidizing waters than in reducing ones (Casas *et al.* 1995). However, the characterization of present hydrochemistry for a given natural lithology is not sufficient to allow adequate safety assessments for any potential radioactive waste repository. Due to the rather large period of time to be considered (more than one million years) and the unavoidable chemical perturbations induced by repository construction, it is essential to also investigate the chemical processes which buffer the measured hydrogeochemical parameters, and the variation in these parameters over time. Special care is needed to determine and quantify those water–rock interactions that buffer pH, *Eh* and

alkalinity, since these parameters are particularly important controls on actinide solubility and complexation.

Water–rock interactions within the matrix of the Tournemire shales

Due to low water contents, high specific surface areas and pore structures (see the previous section), the water–rock interactions within the Tournemire shales must be characterized by strong *short range* (nanometre-scale) water (or solute) molecules–mineral interactions. Therefore, the physico-chemical characteristics of the water and its solutes will be different from that of free water which is conventionally considered to take part in water–rock interactions (Horseman *et al.* 1996). The difference arises from factors such as the very low mobility of water in thin films, the high suction potentials developed owing to water–mineral surface electrostatic interactions and the membrane filtration of anions. The 9% porosity given above must be considered as a maximum value since waters *bound chemically* to mineral surfaces are included in the estimates; in reality free waters are of most importance to the present study.

Thus, there are difficulties in the acquisition of porewater hydrochemical data that can be interpreted meaningfully in terms of solute transport. Direct porewater sampling or pH or redox potential (*Eh*) measurement cannot be carried out *in situ* in the Tournemire shales. The total moisture content is too low for a direct application of the squeezing technique (Entwisle & Reeder 1993). Instead porewater data is acquired using the leaching/cation exchange method (Baeyens & Bradbury 1994). The method has been employed using anoxic laboratory conditions to avoid pyrite oxidation which would otherwise disturb the porewater's chemical composition.

The principle of the leaching/cation exchange method is to derive an experimental data set describing the solid phases present (their mineralogy, CECs, *in situ* cation occupancies and cation exchange selectivity coefficients) along with the aqueous chloride content, which is assumed to be conservative. The porewater composition is then modelled using the thermodynamic computer code PHREEQC (Parkhurst 1995). The pH and the p_{CO_2} are both free dependent variables and one of them has to be fixed in the calculations. Taking into account data derived from *in situ* water sampling (see the following sub-section) and literature review (Baeyens & Bradbury 1994; Griffault *et al.* 1996; Pearson & Scholtis 1995), it seems reasonable to fix p_{CO_2} at a value of 10^{-2} atm. The ionic composition of the porewater was constrained using measured specific ion exchange properties, and an assumption of saturation with respect to the minerals identified in the shale. Electrical neutrality was maintained by adjusting Cl$^-$ concentrations as appropriate. Only minerals likely to exhibit fast water–rock interaction kinetics were considered (carbonates, oxides etc). In this context, the clay minerals act on the porewater chemistry solely through exchanging cations with the porewaters.

Results of models for the Tournemire shales are reported in Table 4. These models were based upon the mineralogical and CEC data presented in Tables 1 and 2 the calculated cation exchange selectivity coefficients and the concentrations of leachable Cl$^-$ present in the rocks. Preliminary selectivity coefficients (K_c(Mg/Na) and K_c(Ca/Na) close to 3 and 4 respectively) have been derived from aqueous extraction experiments. They are in rather good agreement with literature data (Baeyens & Bradbury 1994), and are mainly related to the illite and interstratified illite/smectite contents. The leachable Cl$^-$ has been extracted with pure water at a liquid/solid ratio of 10 ml g^{-1}. It is clear that the Tournemire porewater cannot be definitively described with the data available. The porewater is of Na–Cl–(HCO$_3^-$) type, with a low salinity, equivalent to a total dissolved

solid (TDS) content of 1500 mg l^{-1}, corresponding to a relatively low ionic strength of about 2.5×10^{-2} molal. The porosity is calculated from leachable [Cl$^-$] (around 17 ppm by weight of rock) and computed [Cl$^-$] concentration in the water. The computed porosity of 4.5% compares well with the petrophysical total porosity data of 9% (the total porosity must be considered here as a maximum threshold), when due account is taken of the uncertainties associated with the two methods. Due to the presence of organic matter, siderite and pyrite, and the existence of low water mobility, reducing conditions probably prevail, but were not considered in this preliminary modelling (*in situ* redox conditions are discussed below for the fracture water). Sensitivity analyses emphasize the importance of the p_{CO_2} parameter in controlling the composition of the porewaters (Baeyens & Bradbury 1994; Pearson & Scholtis 1995). From the preliminary calculations, it is clear that the Tournemire porewater cannot be described definitively with the data available. Beside better constraints on the *in situ* p_{CO_2}, the derivation of selectivity coefficients based on measured data from the Tournemire shales are necessary to obtain more accurate porewater compositions.

This modelling approach meant that carbonates, mostly calcite, played an important role in the buffering of pH. In the real mudrock system, which is isolated from the atmosphere, siderite and pyrite probably act to buffer redox conditions at reducing values (Mazurek *et al.* 1996). Clay minerals probably also play an important role in regulating the porewater chemistry, through major cation exchanges and, to some degree, due to buffering the pH through proton exchanges. Additionally, during the diagenetic history of the argillaceous formation, the evolution of organic matter is interpreted to have contributed to the attainment and maintenance of reducing conditions. The maturation of the organic matter would have caused a

Table 4. *Model describing the pore-water chemistry of the eastern gallery shales*

Temp. (°C)	pH	p_{CO_2} (atm)	TDS (mg l^{-1})	[Na]$_T$ (mmol l^{-1})	[Mg]$_T$ (mmol l^{-1})	[Ca]$_T$ (mmol l^{-1})	[Si]$_T$ (mmol l^{-1})	[F$^-$]$_T$ (mmol l^{-1})	[Cl$^-$]$_T$ (mmol l^{-1})	[HCO$_3^-$]$_T$ (mmol l^{-1})	Porosity (%)
25*	7.4†	$10^{-2.0}$†	1500	18.5	1.1	1.5	0.28	0.23	19.1	4.9	4.5

Model constraints: equilibrium between pure water and calcite (Ca control), dolomite (Mg control, dissolution only), fluorite (F control) and chalcedony (Si control). Electroneutrality is ensured by charge-balancing using Cl$^-$ $\times K^{app}$ (M^{2+}/Na) $= 3.2$ (where K^{app} is the cation selectivity constant and M^{2+} is a divalent metal cation in milli-equivalents Gaines and Thomas model). K$^+$ is not considered because of too high uncertainty on its K^{app}. Porosity calculated from leachable [Cl$^-$] (around 17 ppm by weight of rock) and computed [Cl$^-$] concentration in the water.
* Constrained parameters.
† pH kept fixed after equilibration with calcite and dolomite at p_{CO_2} of $10^{-2.0}$ atm.

progressive increase in the CO_2 partial pressure. However, the mature character of the organic matter probably means that it does not control natural porewater composition at the present. Furthermore, it appears unlikely that the organic matter will exert a great influence on the water chemistry.

Water–rock interactions in fractures

Boreholes were drilled, using dry drilling techniques, to intercept fractures showing water flows in the eastern gallery (see Fig. 2). A mechanical packer, which had a *Teflon* covering to avoid chemical contamination, was installed in each borehole. A permanent circulation of pure Ar was maintained during the time for which water accumulated within the packed-off interval (from one day to three days). Water samples were subjected to ultrafiltration to obtain samples for Al and U analysis. Aqueous concentrations of cation and anion have been determined respectively by Inductively Coupled Plasma-Mass Spectrometry (ICP-MS) and Ion Chromatography (IC). The pH in the packer chamber was typically about 8.2 and the temperature was 13°C but measurements made directly at the water inflows with *a one drop* ISFET pH meter indicated a pH of 8.0 and a temperature of 13°C. A minimum value for *in situ* p_{CO2} of $10^{-2.6}$ atm may be estimated from the above two pH data and from the total aqueous concentrations of carbonate measured in boreholes.

Analytical results describing the hydrochemistry of the waters within selected fractures are reported in Table 5. The two surrounding aquifers and the Cernon fault yielded oxidizing Ca–HCO_3^- waters with Cl^- contents of about $0.2 \, \text{mmol} \, \text{l}^{-1}$ and pH values around 7.5. By comparison, the waters sampled in the fractures within the shale have a very different hydrochemistry which indicate a strong geochemical signature derived from the shale matrix. These waters are of Na–Cl–(HCO_3^-) type, with Cl^- contents around $10 \, \text{mmol} \, \text{l}^{-1}$ and pH close to 8.0. These waters are either 'old' water (possibly connate) or inflowing waters which have been chemically modified through water–rock interactions involving the shales in the fracture walls, and/or mixing with porewaters originating in the shale matrix. In the second hypothesis, porewater diffusion can explain the relatively high concentrations of conservative chloride found in the waters sampled from the fractures.

The mineral saturation indices computed with MINTEQ (Allison *et al.* 1991) are given in Table 6. The MINTEQ database is particularly

suitable for modelling low temperature mineral–water interactions (Engi 1992; Griffault *et al.* 1996). The assumptions made *a priori* for the porewater modelling of the previous section are consistent with the chemical data for waters sampled from the fractures: calcite, dolomite, chalcedony and fluorite are calculated to be in, or near, equilibrium with the sampled waters.

The calculations imply that the waters are close to saturation with chalcedony, rather than quartz. Such a finding is typical of many natural waters in argillaceous rocks at low temperature (Griffault *et al.* 1996). It may reflect the low temperature degradation of aluminosilicates, which liberate silica to solution, coupled with the slow precipitation kinetics of quartz (Ribstein *et al.* 1985).

Fluorite has not been identified by mineralogical analysis, but it is frequently quoted as the mineral controlling the fluorine content in other marl or shale formations (Baeyens & Bradbury 1994; Pearson & Scholtis 1995).

Siderite, strontianite and magnesite are also calculated to be close to saturation (Table 6). However, these results reflect the fact that the various metal carbonates were simulated as discrete phases, rather than as calcite-dominated solid solutions. The rather high contents of Fe, Mg and Sr found in calcite from fracture fillings supports the hypothesis that the siderite, strontianite and magnesite actually occur in solid solution with calcite (Table 2).

As discussed in the previous sub-section, the existence of strongly reducing conditions within the matrix of the shales is to be be expected in view of the mineralogical composition of the shale. However, such conditions are also demonstrated by the rather high Fe contents of the waters samples from the fractures (Table 5), and the fact that a calculated *Eh* of $-240 \, \text{mV}$ is consistent with siderite and goethite saturation (Table 6). This *Eh* value is probably a maximum since even more reducing conditions are probably required to explain the presence of pyrite (Fig. 4; Griffault *et al.* 1996; Cramer & Nesbitt 1994). The HS^- contents were not measured in this first study, but aqueous concentrations of U lower than 10^{-9} M also suggest reducing conditions.

Low albite and microcline, which are identified within the matrix of the shales, are also calculated to be close to equilibrium with the water samples from the fractures. Thus, in addition to the *fast* reactions involving principally cation exchange and reactions with carbonate minerals, *slow* dissolution/precipitation reactions involving feldspars, can also be hypothesized to interpret the chemistry of water occurring throughout the shales.

Table 5. *Hydrochemistry of the eastern gallery fracture: in situ measurements and modelling results*

	Temp. (°C)	pH	p_{CO_2} (atm)	TDS (mg l⁻¹)	[Na]$_T$ (mmol l⁻¹)	[K]$_T$ (mmol l⁻¹)	[Mg]$_T$ (mmol l⁻¹)	[Ca]$_T$ (mmol l⁻¹)	[Sr]$_T$ (μmol l⁻¹)	[Fe]$_T$ (μmol l⁻¹)	[Si]$_T$ (mmol l⁻¹)	[Al]$_T$ (μmol l⁻¹)	[F]$_T$ (mmol l⁻¹)	[Cl]$_T$ (mmol l⁻¹)	[HCO$_3$]$_T$ (mmol l⁻¹)	[SO$_4^{2-}$]$_T$ (mmol l⁻¹)	[U]$_T$ (nmol l⁻¹)
Measure (packer)	13	8.20*	–	930	12.6	0.14	0.29	0.37	5.7	2.7	0.11	0.14	0.23	9.3	4.4	0.21	≤1
Modelling (*in situ*)	13	8.05*	$10^{-2.6}$	1060	13.0	0.10	0.39	0.31	–	–	0.13	0.09	0.24	9.3	6.1	–	–

Model constraints: equilibrium between pure water and calcite (Ca control), dolomite (Mg control), low albite (Na control, dissolution only), microcline (K control), chalcedony (Si control) and halloysite (Al control).

* Constrained parameters.

† The pH in the packer chamber is 8.2, but measurements directly realized at the water inflow with *a one drop* ISFET pH meter indicates a pH of 8.0 and a temperature of 13°C.

Table 6. *Hydrochemistry of the eastern gallery fracture: saturation indices (SI)*

Carbonates	SI	Oxides	SI	Alumino-silicates	SI	Others	SI
Calcite	0.0	Quartz	0.3	Gibbsite	−0.5	Fluorite	0.0
Dolomite	−0.1	Chalcedony	−0.1	Halloysite	−0.1	Gypsum	−2.7
Magnesite	−0.6	Goethite*	−0.1	Low albite	0.1		
Strontianite	−0.9			Microcline	0.2		
Siderite*	0.0			Illite	1.4		

SI computed with pH = 8.20 for the carbonates (fast equilibrium ⇒ pH of the packer chamber) and with pH = 8.05 for the other minerals (slow mechanism ⇒ pH *in situ*).
* *Eh* = −240 mV.

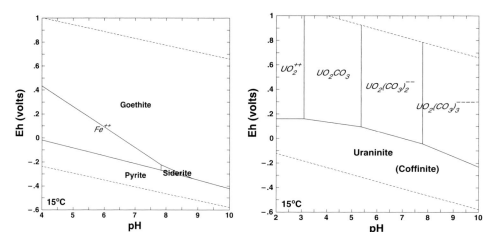

Fig. 4. Fe and U Pourbaix diagrams for the case of a simplified fracture hydrochemistry ($[Fe]_T \approx 10^{-5.5}$ molar, $[U]_T \approx 10^{-8}$ molar).

Implications for the chemical containment of radionuclides

Past containment properties

Mineralogical and geochemical studies have provided some insights into the past conditions within the Tournemire shales. Particular attention has been paid to investigations of the scale of water movement, and the chemical and structural evolution of the system relative to geological events, such as orogenic episodes (Mathieu *et al.* 1997). Important matrix buffering of the ancient fluids by the host shales and local water circulation are postulated as a cause for the origin of most veins. Direct hydraulic connections to the aquifers were not involved. The fractures were mainly filled by calcite precipitating from solutions which moved laterally from the enclosing shales. In other words, the shale behaved as a closed hydrogeological system in most cases. On the other hand, a very few exceptions to the establishment of local and buffered circulation of ancient fluids seem to exist. On-

going studies indicate that some of the fracture planes with calcite geodes may be related to circulations of mineralizing fluids with a partial origin from the limestone aquifer and with important variations in their chemistry.

The distribution of U within fracture-filling minerals and the matrix gives important insights into the past mobility of this element. A photomicrograph of induced uranium fission tracks is given in Fig. 5. A microfissure is selected for graphical convenience, but similar U fission track patterns are observed in centimetre-scale fractures. The relatively high levels of U in the shales contrast with the very small quantities of U that have evidently been incorporated chemically into the fracture-filling calcites during precipitation (Fig. 5; Table 2). There is no evidence for U having been leached from the shales or mobilized by the palaeofluids that migrated through the fractures and precipitated calcite.

There is also no evidence for the past oxidation of any of those chemical features of the veins thought to be indicative of reducing redox states at the present time: relatively high Fe (and to a

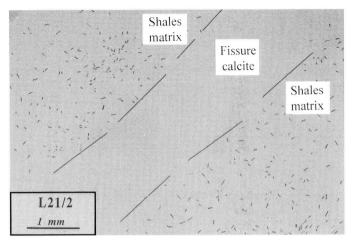

Fig. 5. Photomicrograph of induced uranium fission tracks: homogeneous distribution in shales contrasting with the very few tracks in the calcite vein.

lesser extent Mn) contents shown by the calcite fillings; the relative immobility of U; and the presence of pyrite crystals. Thus, all the palaeo-fluids that flowed through the fractures during and since the formation of the calcite veins, must have been reducing in character.

Clearly, the fluid must have been carbonate-bearing to explain the predominance of calcite as the principal fracture-filling mineral. A decrease of p_{CO_2} in the open fractures, compared with the CO_2 levels in the matrix of the shales, could explain the precipitation of the calcite.

In summary, it appears that the chemistry of the fluids that passed through the fractures, was regulated by the water–mineral interactions occurring in contact with the matrix of the shales. The pressure/temperature conditions in the past, which are not well constrained, needs further investigation. Temperatures below 60°C are derived from a preliminary study of the scarce fluid inclusions within the calcite fillings.

The calcite veins have much lower concentrations of Cs than the shale matrix. On the other hand, the chemical contrasts are much less marked in the cases of Mg and Sr (Table 2). These findings are important because ^{137}Cs is an important radionuclide in many radioactive wastes. The evidence from the Tournemire shales implies that such a nuclide is strongly retained by the shales, even during any dewatering that may accompany compaction or, potentially, heating. The most likely explanation for the contrasting behaviours of Cs, Mg and Sr is that Cs is more strongly sorbed onto illite and interstratified I/S minerals than Mg or Sr (Brouwer *et al.* 1983).

Further evidence for the ability of chemical processes within the shale to retard important radionuclides, is provided by the distribution of Ce and Ba. These elements are chemical analogues for the radiologically important elements Th and Ra respectively, which are important constituents of many radioactive wastes. The elevated levels of Ce and Ba in the shale matrix, compared to the levels in the calcite veins (Table 2) may be evidence that these elements were relatively immobile during the formation of the calcite veins. An implication is that Th and Ra would also have been relatively immobile under these conditions (undisturbed *far-field* conditions). Such investigations require, however, better characterization of the mineral sinks for Ba, Ce and other trace elements, and determination of the calcite crystallization mechanisms.

Present containment properties

The pore-waters and the water sampled from the fracture in the eastern gallery are characterized by slightly alkaline and strongly reducing conditions (*Eh* < −200 mV) with relatively low dissolved carbonate content and low ionic strength. The mobility of many radionuclides is known to be minimized under such chemical conditions (Bruno *et al.* 1995; Gera *et al.* 1992; Higgo 1987). For example, Fig. 4 presents a simplified Pourbaix diagram illustrating aqueous U speciation at a low U total concentration of 10^{-8} M. Under reducing conditions (*Eh* < 0.1 V) and near-neutral pH, solid phases are stable. Aqueous species of U will be stabilized under such redox and pH conditions only under high carbonate concentrations ($\gg 10^{-2}$ mol l^{-1}) and/or high ionic strengths.

In addition to solubility limits, allied to chemical complexation, radionuclide mobilities are also influenced by colloid migration, and by mineral sorption and ion exchange reactions. Complexes may involve other ligands besides carbonate, notably organic species.

Many previous studies of radioactive nuclide containment by chemical means have highlighted the potential role of organic matter in the retention and/or migration of radionuclides (e.g. Bruno et al. 1995; Carlsen 1989; Put et al. 1992). Organic matter of high molecular weight, which is insoluble or fixed to the clay minerals, will favour the retention of radionuclides. Complexation with organic matter has been postulated to cause the nearly irreversible uptake of some radionuclides, such as Am^{3+} (e.g. Bruno et al. 1995). In contrast, the complexation of radionuclides with mobile, liquid, aqueous or colloidal organic matter, may tend to enhance their mobility (for Am^{3+}, for instance, see Dozol & Hagemann 1993).

The ability of organic matter to form complexes with radionuclides is strongly dependent upon its maturity. As the maturity of organic matter increases, there is a corresponding loss of functional groups, such as phenol or carboxylic acid, which may otherwise bind with radionuclides. Thus, in poorly consolidated mudrocks, such as the Tertiary Boom Clay of Belgium, complexation of radionuclides by organic matter may be an important retention mechanism (e.g. Put et al. 1992). However, the data from the Tournemire shales indicate that in more consolidated argillaceous rocks, the organic matter may be predominantly immobile

kerogen, which is not anticipated to affect either retention or mobilization of radionuclides.

Sorption of radionuclides onto relatively stable but mobile colloids has also often been highlighted as a mechanism which may potentially prevent their effective chemical containment (e.g. Bruno et al. 1995; Higgo 1987). However, in shales such as those at Tournemire, colloid migration could not occur through the rock matrix, since the micro- and meso-scale pore-size is too small. Migration of microparticles (most probably of argillaceous origin due to rock disruption after the borehole drilling) was observed in millimetre- to centimetre-scale apertures present in fractures from the eastern gallery (see above and Fig 3b). These apertures are probably connected to the most preferential water pathways which are located at the interface between matrix walls and calcite veins. However, these interfaces frequently exhibit a thin microbrecciated texture which might also act as a filter, hindering the migration of microparticles. Work is in progress to better characterize the flow porosity.

Laboratory scale experiments performed under well-controlled conditions represent a complementary approach towards the characterization of the chemical containment of radionuclides in the undisturbed and fractured shales. Sorption isotherms of Cs for the shale matrix and for calcite are provided in Fig. 6. Batch experiments were carried out in gloveboxes under N$_2$ atmosphere and with synthetic fracture water which approximate in situ and far-field conditions. The liquid/solid ratio was 5 ml g^{-1}. The results show the weaker retention of Cs by calcite fillings than by the shale matrix. Interstratified I/S have strong affinities for many radionuclides (Milton & Brown 1986), whereas calcite is known to have small laboratory K_d values for

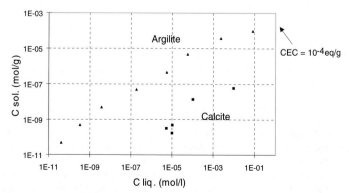

Fig. 6. Isotherms for Cs sorption on shale matrix and calcite. Batch experiments were performed under a N$_2$ atmosphere and with synthetic fracture water (liquid/solid = 5 ml g^{-1}).

many elements (Landström & Tulborg 1995; Milton & Brown 1986). It is worth mentioning that calcite has also been demonstrated to insert significant radionuclide contents within its crystalline lattice when precipitated rapidly (Landström & Tulborg 1995).

Implications for the future containment of radionuclides

The results from Tournemire illustrate that, where a mudrock has undergone progressive diagenesis, it may remain effectively closed to external sources of water, even up to the point where it is considerably lithified. Furthermore, the chemical conditions may favour the retardation of radionuclides throughout diagenesis. Provided that the mudrock remains closed to external fluids, there is no reason to suppose that chemical conditions would become less favourable to radionuclide containment than at the present, since the same mineralogical buffers that operate today, and which operated in the past, will still be present.

The data indicate that fluid movements have been extremely slow in the natural fractures, and that diffusion has been an adequate mechanism to produce similar chemical conditions in both the fractures and the rock matrix. In effect, chemical reactions within the rock matrix have buffered the chemical conditions throughout the rock. In principle, any further compaction of indurated clays in the future, following repository closure, should result in extremely low water flows owing to the low porosity. In this respect, indurated argillaceous rocks may have an advantage over many more porous, plastic lithologies, which may potentially lose water during compaction. However, the increased susceptibility to fracturing of highly consolidated clays will make them susceptible to the introduction of circulating water, originating externally. This is a particular consideration in fracture networks related to an EDZ. If water flows from an excavated repository follow *identified fracture networks* (in the EDZ or at the site scale), saturation of the sorption sites and/or chemical perturbation fronts (oxidation or high pH) could propagate within the fractures and the surrounding matrix over time (Mazurek *et al.* 1996), progressively modifying the radionuclide chemical containment along those preferential pathways. As a consequence, it is particularly important not only to identify those processes that regulate the chemical buffering of waters, and consequently radionuclide mobility, but also to evaluate the chemical buffering capacity of the argillaceous media (i.e to quantify the ability of the rock formation to resist time-integrated chemical changes). At present, most investigations of indurated argillaceous media, including those at Tournemire, are still in an early stage of acquiring such information.

Conclusions

Many types of argillaceous media are promising hosts for repositories of radioactive waste. However, to date, most emphasis has been placed on investigating plastic, poorly consolidated clay formations. Such rocks are characterized by very low porosities and permeabilities and typically exhibit an ability to 'self-heal' by plastic deformation following disturbance. In comparison, relatively little research has been undertaken into the ability of more indurated argillaceous formations to retain radionuclides in the long term. Particular problems associated with investigating such formations are their low matrix porosities and permeabilities, and the need to investigate water flow in both fractures and matrix. The on-going generic investigations at Tournemire are developing approaches for solving these problems.

The matrix waters of argillaceous rocks at Tournemire have been investigated by leaching techniques, while groundwaters have been sampled from fractures during hydraulic testing of boreholes. The chemical data acquired by these approaches have been investigated using theoretical modelling. This approach reveals that the flowing water sampled from fractures in the shale is chemically analogous to that occurring in the matrix of the mudrocks. The water is reducing in character, and capable of mobilizing many radiologically important elements that might occur in radioactive wastes.

Detailed mineralogical investigations have revealed that the hydrogeological and geochemical conditions during the diagenesis of the argillaceous formation and tectonic events have remained appropriate for the containment of several important radionuclides over a considerable period of time. However, these observations may not be readily related to radionuclide migration from an actual repository site. In this context, laboratory-scale experiments performed under well-controlled conditions represent a complementary approach towards the characterization of the chemical containment of radionuclides not naturally present in the site (or present only at trace level).

This work was carried out in the framework of the CCE-CEA Project number FI 2W CT9-0115 (1992–1996). The preliminary pore-water analysis and modelling have been performed in collaboration with Drs M. Mazurek & H. N. Waber (Rock-Water Interaction Group, University of Berne, Switzerland). The OM analysis has been realised by Professor Amblès (Labo de Chimie XII, Université de Poitiers, France). The authors finally acknowledge their colleagues Drs S. Bassot, C. Beaucaire and H. Pitsch for fruitful discussions on water–rock interactions, and the referees and editors for their valuable comments.

References

ALLISON, J. D., BROWN, D. S. & NOVO-GRADAC, K. J. 1991. MINTEQA2/PRODEF2, A Geochemical Assessment Model for Environmental Systems: Version 3.0 User's Manual. EPA/600/3–91/021, US EPA, Athens, GA.

BAEYENS, B. & BRADBURY, M. H. 1994. Physico-chemical characterisation and calculated in situ porewater chemistries for a low permeability Palfris marls sample from Wellenberg. NAGRA report NTB 94-22, Wettingen.

BARBREAU, A. & BOISSON, J. Y. 1993. Caractérisation d'une formation argileuse/synthèse des principaux résultats obtenus à partir du tunnel de Tournemire de janvier 1992 à juin 1993. Rapport d'avancement n° 1 du contrat CCE-CEA n° FI 2W CT91-0115, EUR 15736FR.

BATH, A. H., ENTWISLE, D., ROSS, C. A. M. et al. 1988. Geochemistry of porewaters in mudrock sequences: evidence for groundwater and solute movements. In: Proceedings of the International Symposium on Hydrogeology and Safety of Radioactive and Industrial Hazardous Waste Disposal (Orléans, France), BRGM document 160, 87–97.

BOISSON, J. Y. 1996. Caractérisation d'une formation argileuse- Etudes des transferts au laboratoire à partir des argilites du tunnel de Tournemire. Rapport d'avancement n° 3 du contrat CCE-CEA n° FI 2W CT91-0115, IPSN Report DPRE/SERGD 96/11.

——, CABRERA, J. & DE WINDT, L. 1996. Investigating faults and fractures in argillaceous Toarcian formation at the IPSN Tournemire research site. Joint OECD/NEA-EC Workshop on 'Fluid flow through faults and fractures in argillaceous formations', Bern, 207–222.

BONNE, A. 1992. Overview of the HADES project, Pilot test on radioactive waste disposal in underground facilities. (Proceedings of a Workshop, Braunschweig, Germany), CEC eds, EUR 13985, 223–238.

BROUWER, E, BAEYENS, B., MAES, A. & CREMERS, A. 1983. Cesium and rubidium ion equilibria in illite clay. Journal of Physical Chemistry, 87, 1213–1219.

BRUNO, J., ESCALIER DES ORES, P., KIM, J. I., MAES, A., DE MARSILY, G. & WERNICKE, R. S. 1995. Radionuclide transport through the geosphere into the biosphere. Review of the project mirage, EC report, EUR 16489.

CABRERA, J. 1991. Etude structurale dans le milieu argilleux: LEMI du tunnel de Tournemire. Rapport n°3 du contrat CEA-CABRERA.

CABRERA, J. 1995. Site de Tournemire: programme Diffusion-Isotopes, synthèse géologique des sondages ID, campagne Septembre–Décembre 1994. Rapport n°1 du contrat CEA-IPSN/ADREG.

CARLSEN, L. 1989. The role of organics on the migration of radionuclides in the geosphere. Nuclear Science and Technology Series, EUR 12024 EN.

CASAS, I., DE PABLO, J., MARTI, V., GIMENEZ, J. & TORRERO, E. 1995. UO2 leaching and radionuclide release modelling under high and low ionic strength solution and oxidation conditions. ENRESA Publicacion Tecnica Num. 02/95.

CHIANTORE, V. & GERA, F. 1986. Fracture permeability of clays: a review. Radioactive Waste Management and the Nuclear Fuel Cycle, 7(3), 253–277.

CRAMER, J. & NESBITT, W. 1994. Groundwater evolution and redox geochemistry. In: CRAMER, J. & SMELLIE, J. (eds) Final Report of the AECL/SKB Cigar Lake Analog Study, SKB 94-04, 191–207.

—— & SMELLIE, J. (eds) 1994. Final Report of the AECL/SKB Cigar Lake Analog Study. SKB 94-04.

D'ALESSANDRO, M. & GERA, F. 1986. Geological isolation of radioactive waste in clay formations: fractures and faults as possible pathways for radionuclide migration. Radioactive Waste Management and the Nuclear Fuel Cycle, 7(4), 381–406.

DOZOL, M. & HAGEMANN, R. 1993. Radionuclide migration in groundwaters: review of the behaviour of actinides, Pure and Applied Chemistry, 65, 1082–1102.

ENGI, M. 1992. Thermodynamic data for minerals: a critical assessment. In: PRICE, J. D. & ROSS, N. L. (eds) The Stability of Clay Minerals. The Mineralogical Society Series 3, Chapman & Hall, London, 267–328.

ENTWISLE, D. C. & REEDER, S. 1993. Extraction of pore fluids from mudrocks for geochemical analysis. In: MANNING, D. A. C., HALL, P. L. & HUGHES, P. R. (eds) Geochemistry of Clay-Pore Fluids Interactions. The Mineralogical Society Series 4, Chapman & Hall, London, 365–388.

GAUTSCHI, A. 1992. Characteristics of argillaceous rocks. A catalogue of the characteristics of argillaceous rocks under investigation in the radioactive waste programmes of Belgium, Canada, France, Italy, Japan, Spain, Switzerland and United Kingdom. NAGRA documentation, revision number 1.0–13/3/92.

——1996. Hydrogeology of a fractured shale (Opalinus clay) – A review. Joint OECD/NEA-EC Workshop on 'Fluid flow through faults and fractures in argillaceous formations', Bern, 199–206.

GERA, F. ANDRETTA, D., BOCOLA, W., CHIANTORE, V. & SCHNEIDER, A. 1992. State of the art report: disposal of radioactive waste in deep argillaceous formations. ENRESA Technical publication, 01/92.

GRIFFAULT, L., MERCERON, T., MOSSMAN, J. R., NEERDAEL, B., DE CANNIIERE, P., BEAUCAIRE, C., DAUMAS, S., BIANCHI, A. & CHRISTEN, R. 1996. Acquisition et régulation de la chimie des eaux en *milieu argileux: projet 'ARCHIMEDE-argiles'*, CEC eds, **EUR 17454FR**.

HIGGO, J. J. W. 1987. Clay as a barrier to radionuclide migration, *Progress in Nuclear Engineering*, **19**(2), 173–207.

HORSEMAN, S. T., HIGGO, J. J. W., ALEXANDER, J. & HARRINGTON, J. F. 1996. *Water, gas and solute movement through argillaceous media*. OECD report, **CC-96/1**.

KICKMAIER, W. & McKINLEY, I. 1997. Research in European rock laboratories. *NAGRA Bulletins*, **29**, 29–36.

LALIEUX, P., THURY, M. & HORSEMAN, S. 1996. Radioactive waste disposal in argillaceous media, *NEA Newsletter*, Spring 1996, 34–36.

LANDSTRÖM, O. & TULBORG, E. L. 1995. *Interactions of trace elements with fracture filling minerals from the Äspö Hard Rock Laboratory*. SKB technical report **95-13**.

MATHIEU, R., PAGEL, M., CLAUER, N., DE WINDT, L., CABRERA, J. & BOISSON, J. Y. Paleofluid circulation records in shales: a mineralogical and geochemical study of calcite veins from the experimental Tournemire tunnel site (France), submitted to *European Journal of Mineralogy*.

MAZUREK, M., ALEXANDER, W. R. & MACKENZIE, A. B. 1996. Contaminant retardation in fractured shales: matrix diffusion and redox front entrapment, *Journal of Contaminant Hydrology*, **21**, 71–84.

MILTON, G. M. & BROWN, R. M. 1986. Adsorption of uranium groundwater by common fracture secondary minerals, *Canadian Journal of Earth Sciences*, **24**, 1321–1328.

PARKHURST, D. L. 1995. *User's guide to PHREEQC – A Computer Program for Speciation, Reaction-path, Advective Transport, and Inverse Geochemical Calculations*. US Geological Survey Water-Resources Investigation, Report **95-4227**.

PEARSON, F. J. JR & SCHOLTIS, A. 1995. Controls on the chemistry of porewater in a marl of very low permeability. *In*: KHARAKA, Y. K. & CHUDAEV, O. V. (eds) *Water–Rock Interaction*. Balkema, Rotterdam, 35–38.

PUT, M., MONSECOUR, M. & FONTEYNE, A. 1992. Mobility of the dissolved organic material in the interstitial Boom clay water. *Radiochimica Acta*, **58/59**, 315–317.

RIBSTEIN, A., LEDOUX, L., BOURG, A., OUSTRIERE, P. & SUREAU, J. F. 1985. *Etude du colmatage des fissures en milieu granitique par précipitation de la silice*. EC Report **EUR9476FR**.

SAVAGE, D. 1994. *The Scientific and Regulatory Basis for the Geological Disposal of Radioactive Waste*. Wiley, Chichester.

SCHOLTIS, A., PEARSON,, F. J. JR, LOOSLI, H. H., EICHINGER, L., WABER, H. N. & LEHMANN, B. E. 1996. Integration of environmental isotopes, hydrochemical and mineralogical data to characterize groundwaters from a potential repository site in central Switzerland. *In*: *Proceedings of a Symposium on Isotopes in Water Resources Management AIEA (Vienne)*, **Vol. 2**, 263–280.

THURY, M. 1995. *Mt Terri project. Experiments in an existing tunnel for the hydrogeological and geochemical characterisation of an argillaceous formation (Opalinus clay)*. Project Proposal, June 1995, Swiss National Hydrological and Geological Survey, NAGRA, Geotechnical Institute Ltd.

——1997. The Mont Terri rock laboratory. *NAGRA Bulletin*, **31**, 33–44.

Experimental simulation of the alkaline disturbed zone around a cementitious radioactive waste repository: numerical modelling and column experiments

K. BATEMAN,[1] P. COOMBS,[1] D. J. NOY,[1] J. M. PEARCE,[1]
P. WETTON,[1] A. HAWORTH[2] & C. LINKLATER[2]

[1] British Geological Survey, Keyworth, Nottingham, NG12 5GG, UK
[2] AEA Technology PLC, Harwell, Didcot, Oxfordshire, OX11 0RA, UK

Abstract: One approach to describe the migration of an alkaline plume from the cementitious engineered barriers of a geological disposal facility for radioactive wastes is to employ coupled chemistry and flow computer models. Although evidence from natural systems is useful to constrain reaction mechanisms and minerals to be incorporated into such models, time-dependent information is generally lacking. A series of laboratory column experiments has been conducted in order to test the capabilities of two of the currently available, coupled models to predict product solids and output fluid compositions with time. The coupled models PRECIP and CHEQMATE were used to provide predictive calculations based upon known experimental parameters and data available from the literature. The predictions did not replicate all the variations in mineralogy observed in the experiments, primarily due to restrictions in the availability of kinetic and thermdynamic data for the range of secondary phases of interest. However, the model predictions did reproduce the general variation of secondary phases with time and distance along the columns.

Current concepts of the geological disposal of radioactive wastes envisage the use of multiple-barriers (i.e. both geological and engineered barriers) to contain the radioactive materials in the waste (Savage 1995). The barriers include the waste form itself, a container or canister, surrounded by a backfill, and finally the geosphere. Cementitious materials and concrete may be used extensively in the construction of the disposal facility itself, in the canister or as a backfill. After closure, the repository will saturate with groundwater and become part of a modified regional groundwater flow system. Groundwater equilibrated with the cement will, at some stage, migrate in to the geosphere. It is expected that the chemical composition of the cement-equilibrated porewater will evolve with time (Atkinson 1985), due to the dissolution of alkali hydroxides present in the cements. This will lead to the development of an alkaline plume (pH 13–12.5) which will migrate into the geosphere. The alkaline plume will then react with the components of the rock and this may lead to changes in physical, chemical and mineralogical characteristics of the surrounding rock. The interactions of the alkaline plume with the surrounding rock will be complex and one approach is to employ coupled chemistry and flow computer models to scope these changes.

Although evidence from natural systems is useful to constrain reaction mechanisms and minerals to be incorporated into such models, time-dependent information is generally lacking. A series of laboratory column experiments have been conducted as 'blind' test cases in order to test the capabilities of two of the currently available, coupled models to predict product solids and output fluid compositions with time. The experiments reacted single minerals of importance to the radioactive waste disposal programmes in the UK, Sweden and Switzerland, with simplified 'young' (Na–K–Ca–OH) and 'evolved' (Ca(OH)$_2$) synthetic cement porewater leachates.

The objectives of the project were as follows:

(a) To derive spatial and temporal fluid chemical and mineralogical data using laboratory experiments as a means of testing coupled chemical and flow computer models developed to model the migration of an alkaline plume from a deep geological repository.

(b) To conduct 'blind' computer modelling using physical, chemical and mineralogical data to predict the results of the experiments.

From: METCALFE, R. & ROCHELLE, C. A. (eds) 1999. *Chemical Containment of Waste in the Geosphere.* Geological Society, London, Special Publications, **157**, 183–194. 1-86239-040-1/99/$15.00 © The Geological Society of London 1999.

Description of the experiments

A schematic diagram of the equipment used is shown in Fig. 1. An experimental temperature of 70°C was chosen as a compromise between likely lower temperatures in the rock surrounding a deep cementitious repository, and the elevated temperatures necessary to achieve extensive reactions on a laboratory timescale. The equipment consisted of two reservoirs, which contained the reactant fluids under a protective nitrogen atmosphere (1 bar) to prevent ingress of CO_2, and held at room temperature. These were connected to a series of heated conditioning vessels, maintained at the experimental temperature, which were in turn connected via peristaltic pumps to the columns (internal diameter 7.5 mm, length 300 mm) immersed in two constant temperature water baths. A flowrate of approximately 1 ml hr^{-1} was maintained through the columns, and the reacted fluids were collected in sample bottles from which sub-samples were prepared for chemical analysis.

The fluids used were simplified versions of the leachates expected from a cementitious repository (Savage *et al.* 1998*a*, *b*). One represented a 'young' leachate with high pH and high levels of Na and K. The other represented a later, or 'evolved', leachate and is a saturated $Ca(OH)_2$ solution. The compositions of both fluids are given in Table 1. Both fluids were prepared from ARISTAR® grade reagents.

Mineral samples of quartz, albite, muscovite and calcite were obtained and crushed to the 250–125 μm size range. The muscovite sample was washed by hand to minimize break-up of the mica plates, whilst the remaining samples were ultrasonically cleaned.

All the experiments used single minerals apart from those using muscovite, where quartz was

Table 1. *Summary of fluid compositions*

Simple 'evolved' fluid [$Ca(OH)_2$]

pH @25°C = 12.45, [Ca^{2+}] = 2.012 × 10^{-2} mol dm^{-3}

Simple 'young' fluid [Na–K–Ca–OH]

pH @25°C = 13.22, [Na^+] = 6.52 × 10^{-2} mol dm^{-3}

[K^+] = 1.61 × 10^{-1} mol dm^{-3}

[Ca^{2+}] = 2.24 × 10^{-3} mol dm^{-3}

added in ratio 25% (by weight) quartz to 75% (by weight) muscovite. This was necessary in order to prevent the column from being blocked due to the plate structure of muscovite.

Results of the experiments

In calcite column experiments with both fluid types, dissolution was found to have occurred close to the inlet of the column. This was accompanied by the formation of small amounts of portlandite.

In the quartz column experiments using the 'young' (Na–K–Ca–OH) fluid, most dissolution occurred in the first half of the column. The Ca concentration in the output fluids dropped in the first 600 hours, at the same time the concentration of dissolved SiO_2 rose. These variations in the fluid chemistry suggest the concentration of SiO_2 in the output fluids was controlled by the availability of Ca in the reacting fluid and the subsequent formation of calcium silicate hydrate (CSH) phases. Secondary CSH phases of varying morphologies were observed throughout the column (Fig. 2), in some cases these phases were seen to be cementing the grains together, leading to a reduction in porosity.

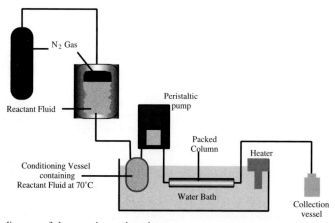

Fig. 1. Schematic diagram of the experimental equipment.

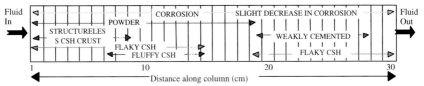

Fig. 2. Summary diagram of experimental observations showing the distribution and nature of the secondary reaction products in the quartz column reacted with the 'young' fluid.

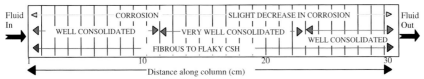

Fig. 3. Summary diagram of experimental observations showing the distribution and nature of the secondary reaction products in the quartz column reacted with the 'evolved' fluid.

Electron microprobe analyses of the CSH phases showed a considerable range in Ca:Si ratios. These ranged from approximately 0.5 close to the inlet, rising to 1.6 further into the column. The Ca:Si ratios are equivalent to the CSH(I) and CSH(II) group of minerals (Lea 1970) (i.e. hillebrandite, Ca:Si = 0.5; foshagite, Ca:Si = 0.75; tobermorite, Ca:Si = 1.2).

Similar results to those observed in quartz columns were seen in the columns reacting albite with the simple 'young' (Na–K–Ca–OH) fluid, where the Ca:Si ratios of the CSH phases varied from approximately 0.6 to 1.3, with increasing distance from the inlet of the column. The main difference in the albite columns relative to the quartz columns was that all analyses of the CSH phases showed they contained up to 1 wt% Al (i.e. calcium aluminium silicate hydrates (CASH) phases were formed).

The columns reacting the muscovite/quartz mixture with the 'young' (Na–K–Ca–OH) fluid showed a similar pattern to that of the albite columns.

All of the experiments using the simple 'evolved' fluid (Ca(OH)$_2$-saturated) generally showed the same pattern of reaction observed in the experiments using the Na–K–Ca–OH fluids, with CSH and CASH (when Al present) phases being precipitated throughout the columns but with less dissolution of the starting materials (Fig. 3).

Modelling: validation strategy and acceptance criteria

These experiments provided an opportunity to conduct 'blind' predictive modelling to test the capabilities of current coupled chemistry and flow models. Such models have to be used to estimate the growth and extent of the chemical disturbed zone surrounding a repository for radioactive waste as it is impossible to conduct experiments over the very long timescales that need to be considered. It is important that any model(s) used is validated, and that the model(s) are shown to be an adequate representation of the physical system.

There are various definitions of the term 'validation'. The IAEA (Radioactive Waste Management Glossary 1988) definition is: 'Validation is a process carried out by comparison of model predictions with independent field observations and experimental measurements. A model cannot be considered validated until sufficient testing has been performed to ensure an acceptable level of predictive accuracy. (Note that the acceptable level of predictive accuracy is judgmental and will vary depending on the specific problem or question to be addressed by the model.)'

The process of developing and validating numerical models of natural processes is one of repeated iteration between model predictions and field and experimental observations (although in this paper it is only experimental observations that are considered). At each stage, testing of the theoretical predictions against observational data yields refinements of our understanding of the processes involved. These refinements may be minor adjustments to the parameter values used in a particular model, or they may be major re-considerations of the fundamental conceptual basis of our understanding. Thus, the model validation process can be represented as follows:

(a) Review and select conceptual models based on current knowledge and observations, propose a new model if necessary;
(b) Calibrate the model by comparison with experiment(s);
(c) Define acceptability of the model (with due regard to its purpose);
(d) Predict the outcome of experiments and compare with the results;
(e) Analyse discrepancies and consider implications (e.g. refinements);
(f) Recommend further experiments if necessary, to provide additional data for validation or to address further issues.

It may be necessary to iterate through the sequence of these steps a number of times before a model can be considered to be validated. The aim of the modelling described in this paper is to predict the effect of the cementitious porewater from a repository on the surrounding rock. It should be noted that several of the quantities that the models predict (such as porosity, mineral masses dissolved or precipitated as a function of position in the column) are not readily measured accurately, and may be only available as a qualitative description. Other quantities that can be measured relatively accurately, such as the chemical composition of the out-flowing fluids, are values that represent an integration of the interactions taking place along the whole column.

Conceptual model and assumptions for column experiments

Models of the type used for this study require values for a large number of parameters. Many of these can be determined before any experiment begins or may be set by the choice of experimental conditions. The models used in the study assumed that dissolution of the minerals in the packed columns would occur and that, once dissolved, the aqueous components would be transported by the fluid. Within the aqueous phase, the various chemical components may interact and under certain conditions form secondary precipitates within the column.

The parameters set by the experimental conditions included the physical dimensions of the columns (internal diameter 7.5 mm, length 300 mm), the flow rates through the columns, and the temperature (70°C) at which the apparatus was maintained. In addition, the mineralogical composition of the columns (quartz, albite, calcite and 25wt% quartz/75wt% muscovite) and the chemistry of the input fluids (Table 1) were specified. Surface area analyses of

the mineral components were carried out before the columns were assembled. Inert tracer tests were performed on the unreacted columns to provide values for dispersivities, which were used in the modelling.

Two conceptual models were tested in this study and were implemented in the programs PRECIP (Noy 1990) and CHEQMATE (Haworth & Smith 1994). The main difference between these models was whether the dissolution/precipitation processes occurred fast enough for the aqueous phase to be considered to be in equilibrium with the primary and secondary minerals (as implemented in CHEQMATE), or whether a kinetic description of the system was necessary (as implemented in PRECIP).

Acceptance criteria

As discussed above, models cannot be considered validated until sufficient testing has been performed and demonstrated an acceptable level of predictive accuracy. It is therefore necessary to define suitable acceptance criteria for the models. The acceptance criteria are discussed below with regard to the experimental observations.

(a) Distances of the dissolution/precipitation fronts.
   The distances, from the column inlet, at which the dissolution of the primary minerals and precipitation of secondary minerals occur should give a clear indication of the differences between the local equilibrium and kinetic assumptions. If the distances observed in the experiments are close to the distances predicted assuming equilibrium and the fronts relatively sharp, then the rate of reaction of the minerals must be relatively fast. If fronts that are more diffuse were observed in the experiments, this would suggest that the rates of reaction of the minerals are slow.
   The distances over which the mineralogical changes occur are thought to be important indicators of the rates of reaction. In the experiments with columns of length 300 mm, uncertainty in the distances of a few cm would be reasonable.

(b) Identification of secondary minerals.
   If minerals are observed in the experiments that are not predicted by the model, it would suggest that the thermodynamic data used were either wrong or that the model database was incomplete. The absence, in the experiments, of minerals that were predicted might indicate either incorrect thermodynamic data in the models, or that the

effect of kinetics of mineral dissolution reactions are important, or that there was kinetic inhibition of secondary phase precipitation. Distinguishing between these different possible effects is difficult.

While prediction of the exact secondary mineral phase may not be crucial, the models should be capable of predicting the broad types of mineral (e.g. zeolites or CSH phases). In this context, minerals of the same type will have broadly similar properties in terms of sorption behaviour and the tendency to block or increase porosity. The models will have been considered to perform acceptably if the minerals predicted are of the same type as those observed.

(c) Qualitative/semi-quantitative changes in porosity.

Changes in porosity in the experiments may only be identifiable qualitatively, since it is only possible to make detailed measurements at the end of the experiments. Thus, there may not be sufficient information to distinguish between the models. However, it is an important factor for the assessment of the suitability of the models. The models should therefore be able to predict broad trends in porosity change, i.e. a tendency to close porosity or to increase it.

(d) Qualitative indications of the form of the reaction fronts.

The widths of secondary mineral precipitation bands and the extent to which these bands are segregated will indicate the validity of the assumption of rapid precipitation starting as soon as the fluid becomes supersaturated with respect to any given mineral. The detailed form of the mineral fronts may not be significant in terms of performance assessment. The comparison with experiment here is chiefly of interest in identifying whether the kinetics of precipitation are important.

PRECIP modelling

PRECIP (Noy 1990) was developed to examine changes in the mass-transfer properties of a system resulting from mineral reactions. The rates of these mineral reactions are explicitly included by using a kinetic formulation of dissolution and precipitation. The conceptual model is of a one-dimensional flow path along which the flow is Darcian and in which precipitation and dissolution reactions can take place amongst a number of components. The flow field is defined by fixed values of head at each end of the profile; the chemical reactions are described by kinetic rate laws and the aqueous components are transported by advection, dispersion, and diffusion. As the chemical reactions proceed in response to the transporting action of the flow field, the masses of the precipitates change at each point on the profile with consequent changes in the porosity. These changes may be related to changes in the permeability of the flow path and consequently affect the flow field. The transport and reaction equations are fully coupled and solved simultaneously. By restricting the calculation to just the major aqueous components, the computing times are kept relatively short, allowing the use of refined grids and the simulation of long times.

The dissolution rates for quartz and albite were taken from Knauss & Wolery (1986, 1988), and that for muscovite from Knauss & Wolery (1989). The rate for calcite dissolution was based on the review work in Rochelle et al. (1998). Precipitation rates for all secondary phases were set to be at least an order of magnitude greater than that of the mineral dissolution rate, for all conditions likely to be found in the experiments. Thermodynamic data for the potential secondary phases were taken from Sarkar et al. (1982)

PRECIP predictions for calcite columns

The predictions for the reaction of calcite with both 'young' (Na–K–Ca–OH) and 'evolved' ($Ca(OH)_2$-saturated) fluids showed dissolution to occur only immediately adjacent to the inlet end of the column, with the accompanying formation of small amounts of portlandite. Changes in porosity were predicted to be negligible except close to the inlet.

PRECIP predictions for quartz columns

The predictions for reaction of quartz with the 'young' fluid show quartz being dissolved along the entire length of the column, though with much more dissolution near the inlet than towards the outlet. Product minerals, however, were predicted to be restricted to close to the inlet end of the column, hillebrandite forming in the first 10 mm and a smaller quantity of tobermorite forming between 10–20 mm. Beyond this point, all available calcium was removed from the fluid and no further product minerals were predicted. There was a large increase in porosity over most of the column except very close to the inlet, where all available pore space was predicted to be filled within about 100 days.

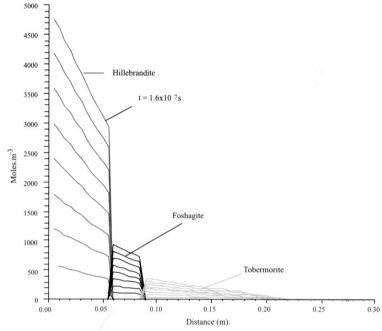

Fig. 4. Predicted profiles of secondary minerals at time intervals of 1×10^6 s (≈ 11.5 days) during the simulated reaction of quartz with the 'evolved' fluid. (PRECIP modelling).

Predictions involving the reaction of quartz with the 'evolved' (Fig. 4) fluid indicated reduced quartz dissolution compared to the simulation with the 'young' fluid. The calcium silicate hydrate (CSH) product phases, were predicted to form a succession down the column, with hillebrandite precipitating in the first 60 mm, foshagite between 60–90 mm, and tobermorite between 90–250 mm. There was a net reduction of porosity along much of the profile with particularly large changes in the first 60 mm where hillebrandite formed.

PRECIP predictions for albite columns

The predictions for the reaction of albite with the 'young' fluid anticipated strong dissolution close to the inlet end of the column. Tobermorite was predicted to form in the first centimetre together with a very small amount of foshagite. Mesolite was predicted in small quantities in the second centimetre of the column, with trace quantities up to 40 mm from the inlet of the column.

Calculations involving the reaction of albite with the 'evolved' fluid (Fig. 5) also showed vigorous reaction close to the inlet end of the column. In this case, hillebrandite was predicted to be formed in large quantities in the first two centimetres of the column. Foshagite was pre-

dicted to form between 5–30 mm, and tobermorite between 25–90 mm. Mesolite was predicted to precipitate between 10–90 mm, overlapping the zones of foshagite and tobermorite formation. The porosity was predicted to be reduced over the whole of the reaction zone with the hillebrandite completely filling the available pore space in the first centimetre of the column after approximately 35 days simulated reaction time.

PRECIP predictions for muscovite/quartz columns

Predictions for the reaction of the muscovite/quartz mixture with the 'young' fluid showed uniform dissolution of quartz and muscovite along the length of the column. Hillebrandite was predicted to be found in large quantities over the first 35 mm of the column. A small amount of foshagite was predicted to be found at 30 mm and a similar amount of tobermorite at around 35–40 mm. Between 40–65 mm, mesolite was predicted. A complex pattern of changes in the porosity was predicted in this experiment. Within the region in which hillebrandite, formed there was a marked reduction in porosity (about 10%). Over the rest of the column, a general increase in the porosity was predicted.

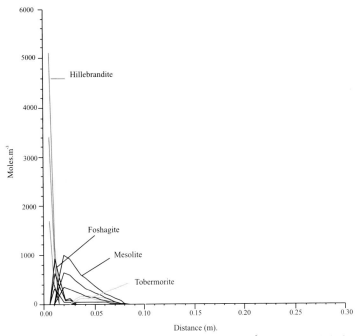

Fig. 5. Predicted profiles of secondary minerals at time intervals of 1×10^6 s (≈ 11.5 days) during the simulated reaction of albite with the 'evolved' fluid. (**PRECIP** modelling).

The predicted reaction of the muscovite/quartz mixture with the 'evolved' fluid was much simpler than the above. Dissolution was predicted to take place along the entire length of the column but at about half the rate found with the 'young' fluid, and with more variation between the inlet and outlet ends. The only products expected to be found in this experiment were hillebrandite, which was predicted to form over the first 240 mm of the column, and foshagite, predicted to form from then on. It was predicted that there would be small net reductions in the porosity throughout the column.

CHEQMATE modelling

Details of the CHEQMATE model can be found in Haworth & Smith (1994). A schematic figure showing the CHEQMATE grid used is given in

Fig. 6. A 31-cell grid was used for the calculations. The first cell contained the incoming alkaline solution and no minerals. The remaining cells contained demineralized water and a finite amount of the appropriate mineral. Demineralized water was present within these cells rather than mineral-equilibrated water because the columns were flushed with demineralized water prior to commencement of the experiment.

CHEQMATE predictions for calcite columns

In the predictions for the reaction of calcite with both 'young' and 'evolved' fluids, a small amount of calcite dissolution was predicted at the inlet end of the column. In the 'evolved' fluid case, this was associated with a porosity increase where no secondary mineral precipitation was

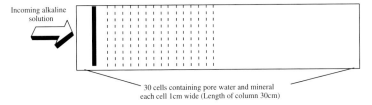

Fig. 6. Schematic representation of grid used for the CHEQMATE calculations.

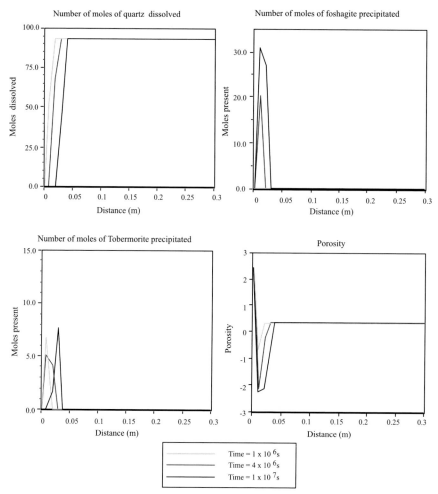

Fig. 7. Predictions for quartz column experiments reacting with the 'evolved' fluid up to 10^7 s (≈ 115 days) using CHEQMATE.

predicted. However, in the experiments with the 'young' fluid, there was a predicted porosity decrease due to precipitation of $Ca(OH)_2$.

CHEQMATE predictions for
quartz columns

The predictions for reactions of quartz with the 'young' fluid indicated that by end of the experiment quartz would be completely dissolved in the first half of the column, with a very sharp dissolution front. It was predicted that tobermorite would precipitate at the dissolution front, with foshagite precipitating behind the dissolution front replacing the quartz. Only small quantities of the CSH phases were predicted, together with an increase in porosity. Additionally, $Ca(OH)_2$ was predicted to precipitate at the

inlet of the column.

Predictions involving the reaction of quartz with the 'evolved' fluid (Fig. 7) indicated dissolution was slower compared to the simulation with the 'young' fluid. The dissolution front was predicted to be very sharp but to reach only a few centimetres into the column. Large quantities of CSH phases were predicted to precipitate, with an associated reduction in the porosity of the column.

CHEQMATE predictions for
albite columns

The predictions of albite reaction with the 'young' fluid anticipated complete albite dissolution in the first half of the column by end of the experiment. Tobermorite and mesolite were

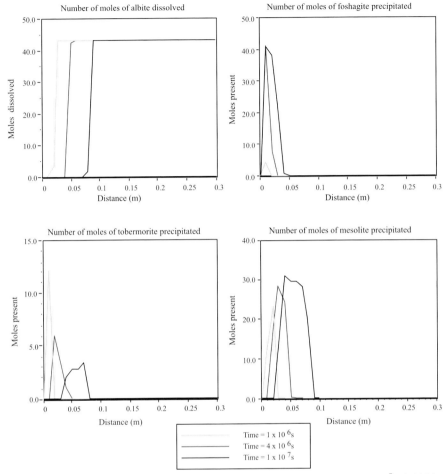

Fig. 8. Predictions for albite column experiments reacting with the 'evolved' fluid up to 10^7 s (≈ 115 days) using CHEQMATE.

predicted to precipitate at the dissolution front, with foshagite predicted to precipitate behind the dissolution front. There was a large predicted increase in the porosity in the first half of the column (i.e. where dissolution had occurred).

In the predictions with the 'evolved' fluid, albite (Fig. 8) dissolution was slower than for the 'young' fluid, and was limited to the initial 5–10 cm of the column. A complicated series of secondary phases were predicted to form. These phases were foshagite, tobermorite, $CaSi_2O_5.H_2O$, mesolite dachiardite, yugawaralite and laumontite. Most of these phases appeared at the dissolution front. Foshagite and mesolite, however, were predicted to precipitate in large amounts behind the dissolution front. At short timescales, the predicted porosity decreases and then increases in the first few cm

of the column. At longer timescales, there was a predicted decrease in porosity over the first 50–100 mm of the column followed by a sharp increase at the dissolution front.

CHEQMATE predictions for muscovite/quartz columns

The predictions for the muscovite/quartz mixture reacting with the 'young' fluid showed quartz dissolving quickly. The quartz dissolution front was predicted to be sharp and appeared to be associated with a brief period of zeolite precipitation. The muscovite dissolution front was also predicted to be sharp. However, the muscovite took longer than the quartz to dissolve. A region of muscovite precipitation in front of the dissolution front was predicted.

Tobermorite was predicted to form at the dissolution front together with the zeolite phase, mesolite. Foshagite was predicted to precipitate behind the dissolution front. There was a predicted increase in porosity for the entire column. At short timescales, a sharp decrease in porosity was predicted, associated with the zeolite precipitation at the quartz dissolution front. This disappeared over longer timescales following removal of quartz from the column. In the first 100 mm of the column, there was predicted to be a large increase in porosity corresponding to the muscovite dissolution front. For the remainder of the column, there was also a predicted increase in porosity but this was minor by comparison.

As with the quartz column modelling, it was predicted that the reaction of the muscovite/quartz mixture with the 'evolved' fluid would produce larger amounts of the secondary phases than with the 'young' fluid. The predicted development of porosity was complex, over the first 150–180 mm of the column a reduction in the porosity was predicted, as a result of secondary phase precipitation. Then there was a short length of column where there was an increase in porosity. This appeared to correlate with the quartz dissolution front.

Comparison of modelled and experimental results

In comparing the results of the predictive modelling with the column experiments it should be noted that a restricted set of mineral phases were included in the models used namely the reactant minerals, CSH phases; and some zeolite phases (PRECIP – mesolite; CHEQMATE – dachiardite, erionite, faujasite, heulandite, laumontite, mesolite and phillipsite). The rationale behind this choice was twofold. Firstly, to limit the number of minerals active in the model so as to allow the simulation of long time periods, without requiring excessive computational time. Secondly, because of the lack of availability of good kinetic and thermodynamic data for the CSH, CASH and zeolite phases of interest. The predictions of the models are discussed below with regard to the validation strategy and acceptance criteria set out earlier.

Distances of the dissolution/precipitation fronts

The distances, from the column inlet, at which the dissolution of the primary minerals and precipitation of secondary minerals occur should give a clear indication of the differences between the local equilibrium and kinetic assumptions.

Generally, the PRECIP model predictions showed the primary minerals being *significantly*, but not totally, dissolved along the length of the column, whilst the CHEQMATE modelling predicted *total* dissolution of the primary minerals in the first half of the columns, (except for the predictions of the muscovite/quartz mixture). The experiments differ from both these predictions in that *partial* dissolution of primary minerals was observed, mainly in the first part of the columns, although there was some evidence for limited dissolution further into the columns. Predictions of product minerals for both models are similar, although the zone of precipitation extends only partially into the columns. In the experiments, product minerals were observed throughout the columns, closely associated with the dissolution of the primary minerals.

The experimental observations showed that zones of dissolution/precipitation overlapped rather than being sharp and this suggests that the kinetics for the mineral reactions are slower than those used in the models. Therefore, although the rates of reaction and precipitation implemented in PRECIP do involve kinetic functions they would appear to be too fast. In addition, the CHEQMATE equilibrium model leads to the total dissolution of the primary minerals in parts of the columns.

Identification of secondary minerals

Electron probe microanalysis data for the CSH phases observed in the columns give Ca : Si ratios ranging from approximately 0.5 close to the inlet, rising to 1.6 further into the column. These ratios are representative of both the CSH(I) and CSH(II) groups of minerals (Lea 1970), and are consistent with the composition of the phases used in the model predictions (i.e. hillebrandite, Ca : Si = 0.5; foshagite, Ca : Si = 0.75; tobermorite, Ca : Si = 1.2). Many analyses showed the presence of up to 1 wt% Al (i.e. CASH phases). Although the predicted CSH phases did not contain all the variations of Ca : Si produced in the experiments, as a result of the restricted number of CSH (and CASH) phases in the models, they do demonstrate evolution of CSH phase compositions with time and with distance along the columns.

Secondary zeolites were predicted by both PRECIP and CHEQMATE for the albite, and muscovite/quartz columns. However, no experimental evidence was found for zeolite formation. This suggests that both models were not representing accurately this part of the system, possibly due to either a kinetic inhibition on precipitation of the zeolites in the experiments, or that the equilibrium constants used in the models were incorrect, or that competition for Al by the CSH phases (to form CASH phases) prevented zeolite formation. It is not clear which, if any, of these is the case and it is therefore not possible to resolve whether the PRECIP or CHEQMATE models were more appropriate.

Qualitative/semi-quantitative changes in porosity

The changes in porosity are difficult to identify quantitatively in the experiments but qualitative comments can be made. The models were able to predict the broad trends in porosity change, i.e. a tendency to close porosity or to increase it. However, the length scales of zones of predicted porosity changes were not in total agreement with the experimental observations.

Generally in the PRECIP predictions, porosity was reduced over the whole of the reaction zone but with an increase of porosity associated with the dissolution of the primary minerals. CHEQMATE generally gave similar predictions but with larger increases in porosity due to the prediction of the total dissolution of the primary minerals in some cases. The experimental observations showed a reduction in porosity close to the inlet of the columns whilst close to the outlet of the column an increase in porosity was observed. This was due to the dissolution of primary minerals coupled with little secondary mineral precipitation. In general, the predicted porosity variations obtained with the PRECIP model are closer to the observations. The CHEQMATE model predicts the total dissolution of the primary minerals. This results in large porosity increases, which were not observed in the experiments.

Qualitative indications of the form of the reaction fronts

The widths of secondary mineral precipitation bands, as predicted by the models, were generally shorter than the experimental observations. This challenges the validity of the assumption of rapid precipitation, starting as soon as the fluid becomes supersaturated with respect to any given mineral. It also suggests that the kinetics of mineral precipitation is important, and should be considered in the modelling.

The differences between the predictions and the experiments, apart from the variety of secondary CSH phases, are mainly of scale. The PRECIP modelling predicted the formation of secondary phases only part way into the columns, whereas the experiments showed the CSH phases to be formed along the entire length of the columns.

These differences can probably be accounted for by a combination of factors. One possible explanation for the differences was that the dissolution rate for the reacting minerals was too fast. A reduction in magnitude of the dissolution rate would not be unreasonable given that the initial rates used are only valid for far from equilibrium conditions, a condition that was not true for the experiments. The rate of reaction of most minerals is known to decrease with approach to equilibrium. PRECIP uses a linear function to correct for this change such that the rate decreases as equilibrium was approached. However, for most minerals the change in dissolution rate on approach to equilibrium does not vary linearly. Therefore the simple model used here may overestimate the dissolution rate as the fluids become more saturated with respect to the dissolving minerals.

A possible explanation for the differences between the CHEQMATE modelling predictions and the experiments, is the use of a thermodynamic equilibrium approach (rather than kinetics) which leads to the total removal of primary minerals in some parts of the columns.

In both models, the surface area available for reaction was assumed to remain constant throughout the duration of the experiments. In practice however, as the reactant mineral dissolves, the surface area will change, which will affect the rate of dissolution of mineral into the fluid. This would result in the spreading out of the zone of precipitation of secondary phases. In addition, the models cannot take into account any reduction in surface area available for reaction due to the precipitation of product phases covering the reacting surface. Again, this would have the effect of changing the surface area available for reaction as the experiments progressed.

Finally, for the mineral phases of interest, there is limited availability of good quality kinetic and thermodynamic data, particularity for the conditions expected in the alkaline plume from the cementitious engineered barriers of a geological disposal facility for radioactive wastes.

Conclusions

The coupled models PRECIP and CHEQMATE were used to provide 'blind' predictive calculations based upon known parameters and data available from the literature. In parallel, laboratory column experiments were undertaken to provide 'test cases' against which the models could be compared. The predictions did not replicate all the variations in mineralogy observed in the experiments, primarily due to restrictions in the availability of kinetic and thermodynamic data for the range of CSH, CASH and zeolite phases of interest. In the models, the CSH phases were only predicted to form part way into the columns, whereas the experiments showed them to be formed throughout. However, the model predictions did reproduce the general variation of Ca : Si ratios with time and distance along the columns, with more Ca-rich phases forming first followed by phases containing increasing amounts of Si. Further experimental and modelling work is currently underway to refine the capabilities of the currently available coupled models, to predict more accurately the spatial and temporal variations in solid phase and output fluid compositions. A better understanding of the above processes will lead to greater confidence in predictive modelling of the migration of alkaline plumes away from the cementitious barriers of radioactive waste disposal facilities.

This work was funded jointly by United Kingdom Nirex Limited, as part of the Nirex Safety Assessment Research Programme (NSARP), Nagra (Switzerland) and SKB (Sweden). This paper is published with the permission of the Director of the British Geological Survey and the funding bodies.

References

ATKINSON, A. 1985. *The Time Dependence of pH within a Repository for Radioactive Waste Disposal.* UKAEA Report AERE-R 11777.

HAWORTH, A. & SMITH, A. C. 1994. *Improvements to the Coupled Chemical Equilibria and Transport Code CHEQMATE.* Nirex Report NSS/R324.

KNAUSS, K. G. & WOLERY, T. J. 1986. Dependence of albite dissolution kinetics on pH and time at 25°C and 70°C. *Geochimica et Cosmochimica Acta*, **50**, 2481–2497.

——1988. The dissolution kinetics of quartz as a function of pH and time at 70°C. *Geochimica et Cosmochimica Acta*, **52**, 43–53.

——1989. Muscovite dissolution kinetics as a function of pH and time at 70°C. *Geochimica et Cosmochimica Acta*, **53**, 1493–1501.

LEA, M. 1970. *The Chemistry of Cement and Concrete.* Arnold, London.

NOY, D. J. 1990. *PRECIP: A Program for Coupled Groundwater Flow and Precipitation/Dissolution Reactions.* Nirex Report NSS/R300.

RADIOACTIVE WASTE MANAGEMENT GLOSSARY, 2nd Edition. 1988. International Atomic Energy Agency Technical Document, IAEA-TECDOC-447.

ROCHELLE, C. A., BATEMAN, K. MACGREGOR, R., PEARCE, J., SAVAGE, D & WETTON, P. 1998. *The Evaluation of Chemical Mass Transfer in the Disturbed Zone of a Deep Geological Disposal Facility for Radioactive Wastes. IV. The Kinetics of Dissolution of Chlorite and Carbonates at Elevated pH.* Nirex Report NSS/R368.

SARKAR, A. K., BARNES, M. W. & ROY, D. M. 1982. Office of Nuclear Waste Isolation Technical Report ONWI-201.

SAVAGE, D. (ed.) 1995. *The Scientific and Regulatory Basis for the Geological Disposal of Radioactive Waste.* Wiley, New York.

——, HUGHES, C. R. MILODOWSKI, A. E., BATEMAN, K., PEARCE, J.. RAE, E. & ROCHELLE, C. A. 1998a. *The Evaluation of Chemical Mass Transfer in the Disturbed Zone of a Deep Geological Disposal Facility for Radioactive Wastes. I. Reaction of Silicates with Calcium Hydroxide Fluids.* Nirex Report NSS/R244.

——, BATEMAN, K., HILL, P., HUGHES, C. R., MILODOWSKI, A. E., PEARCE, J. & ROCHELLE, C. A. 1998b. *The Evaluation of Chemical Mass Transfer in the Disturbed Zone of a Deep Geological Disposal Facility for Radioactive Wastes. II. Reaction of Silicates with Na–K–Ca-Hydroxide Fluids.* Nirex Report NSS/R283.

The mineralogy and geochemistry of cement/rock reactions: high-resolution studies of experimental and analogue materials

EMILY S. HODGKINSON & COLIN R. HUGHES

Department of Earth Sciences, University of Manchester, Oxford Road, Manchester M13 9PL, UK

Abstract: Cements are commonly used to engineer the environment around waste disposal sites. As groundwaters move through these sites, cement gradually dissolves and a reactive 'hyperalkaline plume' forms downstream. Published experimental and modelling studies of cement/rock reactions suggest that host-rock mineralogy will dissolve and be replaced initially by high-volume calcium silicate hydrate gels, and later by zeolites. Results of mineral/alkali experiments are presented. Examination of the reaction products, at the high analytical resolution afforded by ATEM, show new disordered phases which are structurally related to calcium silicate hydrates but are highly aluminous (relative to the published results of previous work). A review of archaeological analogue studies shows that old cement pastes easily become carbonated in the presence of CO_2 or carbonate waters. If protected from carbonation, however, cement gels can survive unchanged for thousands of years. New results of an ATEM examination of 1700 year-old Roman mortar from Hadrian's Wall confirm that cement gels remain poorly ordered and reveal an aluminous CSH gel phase to be the chief product of mortar/dolerite reaction. These findings are discussed in the context of barrier engineering.

Cement is used in a variety of applications in which it comes into contact with the geological environment. In hydrocarbon well-bores and mine-shafts it is a principal component of the engineered environment adjacent to the host geology. In many radioactive waste (or 'radwaste') concepts and some landfill and contaminated land sites, cementitious material forms part of the system of barriers between waste and the biosphere. Its barrier properties include high sorption capacity, potentially low permeability, and the ability to condition its porewaters to a highly alkaline pH. This is particularly important around radwaste repositories where high pH lowers the solubility of a number of the key radionuclides and also slows the corrosion of the metal canisters in which the waste may be placed.

Atkinson (1985) modelled the leaching action on cement by groundwater moving through a radwaste repository to predict the duration of pH buffering in the 'near field'. The model predicted that, with a groundwater flux density of $10^{-10} \, \mathrm{m \, s^{-1}}$, for the first hundred thousand years a pH of 12.5 or above will be maintained, chiefly by the dissolution of portlandite (calcium hydroxide). When the portlandite has dissolved, the main constituents of cement, calcium silicate hydrate (or 'CSH') gels, will undergo dissolution to condition the pH to at least 10.5. This stage was predicted to last for well over a million years. During this time, the fluids produced by cement dissolution will migrate through the repository and into the host geology. This body of fluid is commonly described as a 'hyperalkaline plume'. When the plume encounters the host geology, significant dissolution, primarily of aluminosilicates, is expected, followed by precipitation of high volume, calcium aluminosilicate hydrate (or C-A-S-H*) phases. The first phases to form are likely to be amorphous or poorly ordered 'gels' which, although metastable, can be very long lived. The composition of the plume will evolve with both time and distance from the repository as alumino-silicates dissolve and as calcium hydroxide is removed from solution, so that zeolites are expected to precipitate from the more highly evolved fluids. Reactions between minerals and alkaline fluids will affect the barrier properties of the host geology. Both the transport properties (including permeability) and the chemical properties (e.g. sorption potential and Cation Exchange Capacity) of the host-rock are likely to be altered by the dissolution/precipitation reactions.

---

* Standard cement chemistry nomenclature is used, whereby $C = CaO$; $A = Al_2O_3$; $S = SiO_2$; $H = H_2O$.

From: METCALFE, R. & ROCHELLE, C. A. (eds) 1999. *Chemical Containment of Waste in the Geosphere.* Geological Society, London, Special Publications, **157**, 195–211. 1-86239-040-1/99/$15.00 © The Geological Society of London 1999.

A typical approach to studying cement/rock reactions is through experimental investigation, with experiments usually being based in the laboratory where parameters such as temperature and fluid chemistry can be controlled and monitored. The question of what happens over longer periods of time cannot, however, be answered by laboratory-based experiments alone. Fortunately, analogues of long-term cement/rock interaction exist both in archaeology (e.g. in ancient Roman mortars) and in the natural environment (e.g. hyperalkaline groundwater systems). The study of archaeological and geological analogues can therefore extend the timescale of investigation to thousands and even millions of years. The chief disadvantage of this approach is that in these environments many geochemical and physical parameters, and their variance with time, are not usually well known. A third approach is to use geochemical and/or transport models to predict phase formation and solution chemistry and then to compare these predictions with the results of experiments and analogues for validation. Difficulties with this approach arise from a lack of reliable thermodynamic and kinetic data for the relevant phases, and from the inability of most models to match the complexity of geological systems.

This article reviews the relevant literature on cement/rock and mineral/alkali interaction, and presents the results of experimental and archaeological analogue studies in which Analytical Transmission Electron Microscopy (ATEM) has been used to characterize C-A-S-H phases.

Previous work

Experimental and modelling studies

A number of experimental studies have been carried out which are concerned with the effects of hydrocarbon well-bore operations on the properties of reservoirs. There is also a growing body of literature on the expected interactions of alkaline fluids with 'host rocks' around radioactive waste repositories. Tables 1 and 2 present a summary of experimental and modelling work which has mostly been carried out by the radwaste disposal or hydrocarbon industries. Experimental studies have frequently been used to validate coupled geochemical and transport computer models of the migration and evolution of the hyperalkaline plume, and the accompanying mineralogical alterations. Special attention has been given to the stability of bentonite at high pH because bentonite is a frequent choice of barrier material. A number of issues which

arise from this body of work regarding the reactions observed or predicted, and the predicted effects on the host rock, are discussed below.

Most of the experiments listed in Tables 1 and 2 show that the dominant mechanisms of host-rock alteration are the dissolution of the primary aluminosilicates and the precipitation of secondary phases, although ion exchange may also occur in clays and may play a significant part in the alteration of smectitic materials (Jefferies et al. 1988; Savage & Rochelle 1993). Where calcium hydroxide was present in the reactant fluid, the precipitating phases are CSH gels and other cement minerals such as hydrogarnet (Bateman et al. 1998; Braney et al. 1993; Hughes et al. 1995; Rochelle et al. 1992; Savage & Rochelle 1993; van Aardt & Visser 1977a, b). The gels are all amorphous to poorly crystalline but are structurally and stoichiometrically related to fully crystalline C-S-H minerals such as tobermorite and jennite, and are similar to the gels found in cement. In a few cases, zeolites were tentatively identified forming at a late stage (Rochelle et al. 1992; Savage et al. 1992). These calcium hydroxide experiments were all conducted at relatively low temperatures (up to 110°C). The experiments in which only alkali hydroxides (NaOH and KOH) were present in the reactant fluid have mostly been conducted at higher temperatures (greater than 150°C). In these cases, zeolites, and sometimes feldspars, formed rapidly and ubiquitously (Chermak 1992, 1993; Inoue 1983; Komarneni & White 1981, 1983; Johnston & Miller 1984). Johnston & Miller (1984) also found that at high temperatures, in addition to zeolite and feldspar formation, high pH caused the alteration of smectite to illite or illite/smectite interlayers.

The mathematical models (e.g. EQ3, PHREEQE) are limited by a lack of data, in particular thermodynamic data for many zeolites and cement minerals and kinetic data for feldspar formation (Braney et al. 1993; Rochelle et al. 1992; Savage & Rochelle 1993; Savage 1996). To circumvent these problems, certain simplifications have to be made. For example, purely crystalline C-S-H phases such as tobermorite and foshagite may be chosen to represent the more poorly ordered C(-A)-S-H gels, and a few specific zeolites may represent the zeolites generically. Despite these difficulties, there is broad agreement between the results of models and experiments as regards changes in fluid chemistry, pH and plume evolution. Initially, CSH minerals such as tobermorite and foshagite are predicted to precipitate, but most of the model simulations run for much longer than lab

Table 1. *Summary of experimental and modelling studies of cementitious fluid/rock interaction*

Authors	Method	Conditions	Solid reactants	Alkali	Secondary products	Effect on host rock/other comments
Bateman et al. (1998)	Experiment/model	6 month, 70°C	Single minerals	$Ca(OH)_2 \pm$ (Na,K)OH	CSH in experiments; C-S-H minerals & zeolites in models	Porosity mostly decreases; porosity increase predicted in quartz model
Braney et al. (1993)	Experiment/model	280 day, 25°C	Sandstone	$Ca(OH)_2$	CSH	Experiment and model broadly agree. Slight porosity decrease
Chermak (1992)	Experiment	150–200°C, 42 day	Shale	NaOH	Zeolite; replaced by clay	
Chermak (1993)	Experiment	150–200°C, 50 day	Shale	KOH	Zeolite; replaced by feldspar and clay	
Cunningham & Smith (1968)	Experiment	160°C, 3000 psi	Sandstone	Cement filtrate		No major permeability reduction
Eikenberg & Lichtner (1992)	Model	50 000 a	Muscovite – chlorite marl	Cement fluids	Mostly feldspars & zeolites up to 3 km from site	Plume front propagates slowly and retains a steep pH gradient
Hughes et al. (1995)	Experiment	110°C	Single minerals, sandstone	Cement filtrate and leachate	CSH; hydrogarnet; hydrocalumite	Sandstone lost cohesiveness
Krilov et al. (1993)	Experiment	25°C, 0.6 MPa	Sandstone	Cement filtrate	Ca carbonates and sulphates	Permeability decrease
Krueger (1986)	Experiment	Ambient	Sandstone	Cement filtrate	Ca carbonate	No major permeability reduction
Madé et al. (1994)	Model		Sandstone	NaOH		Validated by experiment
Novosad & Novosad (1984)	Experiment		Sandstone	NaOH		CEC increase
Rochelle et al. (1992)	Experiment	70°C, 4 month	Single minerals; crystalline rock	$Ca(OH)_2 \pm$ (Na,K)OH	CSH; hydrogarnet; one late zeolite?	Large permeability drop within decades.
	Model	>1000 years	Crystalline rock		Zeolites ± feldspars	
Savage et al. (1992)	Experiment/model	70°C, 100 day	Single minerals	$Ca(OH)_2 \pm$ (Na,K)OH	Na,K,Al-CSH; one zeolite? (zeolites ± feldspars in model)	Net volume decrease; results agree with experiments
Savage & Rochelle (1993)	Model	71 a	Granite	$Ca(OH)_2 \pm$ (Na,K)OH	CSH, hydrogarnet; replaced by zeolite	Permeability decreases perpendicular to fracture and increases parallel to it; matrix sealed off from fracture
Steefel & Lichtner (1994); Lichtner et al. (1998)	Model	>500 a	Marl; tuff	Cement leachate	Mainly calcite & brucite, C-S-H and C-A-H phases	Free alkali released from feldspars
van Aardt & Visser (1977a)	Experiment	95°C	Various rocks; feldspar	$Ca(OH)_2$	Hydrogarnet; tobermorite; 10 Å phase; possible CSH and/or silica gels	
van Aardt & Visser (1977b)	Experiment	20–40°C, 250–730 day	Feldspars; clays; sandstone	$Ca(OH)_2$	Alkali silicate gel; CSH; hydrogarnet; C_4AH_n	Possible expansion of alkali silicate gels; analogous to 'concrete cancer'
Yang & Sharma (1991)	Experiment	700 psi	Sandstone	Cement filtrate	Mainly $CaCO_3$	Permeability decrease

Table 2. *Summary of experimental and modelling studies of cementitious fluid/clay interactions*

Author(s)	Method	Conditions	Solid reactants	Alkali	Secondary products	Effect on host rock/other comments
Ewart et al. (1985)	Experiment		Natural clays	$Ca(OH)_2$		Multiple ion exchange. pH buffered to 9
Haworth et al. (1988)	Model	~1000 a	Clays	$Ca(OH)_2$		Ion exchange reduced pH significantly over this timescale
Inoue (1983)	Experiment	25–300°C, pH 6–10.7, 64 day	Pure clays	KOH	Mixed layer illite/smectite (from smectite at 300°C); Ca-zeolites (from Ca-smectite at 300°C)	Ion exchange
Jefferies et al. (1988)	Experiment/model	Several months	Mixed and pure clays.	Mainly $Ca(OH)_2$	Calcite	Smectite and smectitic clays buffer pH to 9. Diffusivity reduced. Model and experiment agree
Johnston & Miller (1984)	Experiment	150–275°C	Na and Ca smectite	KOH	Zeolites and feldspars (above 200°C); illite and illite/smectite	
Komarneni & White (1981)	Experiment	100–300°C, 300 bar, 1–4 months	Clay minerals and shales	CsOH	Cs-aluminosilicates	
Komarneni & White (1983)	Experiment	200–300°C, 30 MPa, 12 week	Clay minerals and shales	$Sr(OH)_2$	Sr compounds including Sr-zeolite	
Lentz et al. (1985)	Experiment		Mg smectite	NaOH	Brucite	Permeability drop
Mohnot et al. (1987); Mohnot & Bae (1989)	Experiment	49–82°C, 2 month	Clays	NaOH	Zeolites; chlorite; amorphous material	
Pusch (1982)	Experiment Field test Bore-hole test	1 year 70 year 3 month, 85°C	Na bentonite Natural clay Natural clay	NaOH Concrete fluids Concrete fluids	Calcite	Quartz dissolution Possible ion exchange. Possible quartz dissolution

experiments (as long as 50 000 years in the case of Eikenberg & Lichtner 1992) and they predict that zeolites and/or feldspars will be the final products of reaction.

When smectitic material such as bentonite comes into contact with hyperalkaline fluids, cation exchange on the clays is also expected to occur. Studies of clay/alkali reaction are summarised in Table 2. Ewart *et al.* (1985) investigated the reaction between natural clay and calcium hydroxide and found that the clay buffered a pH of 9 due to substitution of calcium for more than one ion in the clay. Haworth *et al.* (1988) modelled this ion exchange process and found it to have a significant effect on hyperalkaline plume composition, decreasing the pH over a one thousand year timescale. Jefferies *et al.* (1988) compared models and experiments of ion exchange in several types of clay and alkaline solution. Pure Na smectite and a natural smectitic clay were both found to buffer a pH of 9 by ion exchange while Ca smectite showed a small buffering capacity and pure kaolinite showed very slight resistance to a pH rise. Thus ion exchange may be a significant process around a repository if it is situated within a smectite-rich host rock which has a large enough buffering capacity to lower the pH of the hyperalkaline plume. Otherwise, ion exchange is unlikely to be significant in the long term because smectite, like other alumino-silicates, is eventually expected to dissolve and be replaced (e.g. Mohnot & Bae 1989).

A third important alteration process in cement/rock interaction is the precipitation of calcite and other carbonates and sulphates. This happens when calcium hydroxide-rich cement fluids come into contact with various sources of carbonate and sulphate, such as local groundwaters or carbonate and sulphate phases already present in the rock which may dissolve when the highest pH fluids are encountered. Calcite precipitation due to these processes has been predicted by models (Jefferies *et al.* 1988; Lunden & Andersson 1989; Eikenberg & Lichtner 1992; Steefel & Lichtner 1994). Calcite has also precipitated in some experiments, even if attempts were made to keep carbonate and CO_2 out of the system (e.g. Pusch 1982; Jefferies *et al.* 1988).

Salt precipitation from cement filtrates around hydrocarbon well-bores occurs by a different mechanism which is unlikely to be significant in the radwaste repository environment. These cement filtrates contain dissolved calcium carbonate and sulphate at near-supersaturation levels. When the cement is pumped down a well-bore, filtrates are forced into the surrounding rocks at high differential pressure,

and salts can precipitate. Experimental studies replicating this scenario have shown the precipitation of calcium carbonates and sulphates to be the chief alteration process in the short term, resulting in a (usually minor) permeability decrease (Krueger 1986; Yang & Sharma 1991; Krilov *et al.* 1993).

Effects of alteration on transport and chemical properties

Zeolites and cement minerals have higher molar volumes than most of the primary silicates in a host-rock (Lea 1970, p. 270; Savage 1996), but it is not known whether alteration will result in net dissolution or net precipitation, and therefore it is not known how host-rock porosity and permeability will change. The conclusions that have been made on the basis of the experimental and modelling studies do not concur in this matter and are largely speculative. If there is a net solid volume decrease, due to dissolution (as predicted by the model of Savage & Rochelle 1993), then porosity and permeability will increase. This is more likely to occur close to a repository where fluids will be particularly aggressive to rock-forming minerals, due to their high alkalinity. Net precipitation of secondary phases is more likely to occur further away and may cause a porosity decrease, as observed in experiments by Braney *et al.* (1993) and modelled by Eikenberg & Lichtner (1992). This may be accompanied by a permeability decrease as fractures gradually become sealed up; Rochelle *et al.* (1992) predicted that precipitation may seal up a fracture within decades. Alternatively, the precipitation of high-volume phases may cause expansion and micro-cracking, thereby *increasing* permeability. This possibility was suggested by Hughes *et al.* (1995) who found a reacted sandstone to have lost its 'cohesiveness' but could not determine whether this was due to the dissolution of primary phases or to micro-fracturing. Van Aardt & Visser (1977*a, b*) identified two further possible mechanisms of expansion. They found that alkali feldspars, on reacting with calcium hydroxide, alter to form a suite of phases including tetracalcium aluminate hydrates (or 'AFm' phases) and also possibly alkali silicate gels. Alkali silicate gels are known to cause 'concrete cancer' by taking in water, expanding and thereby fracturing the surrounding concrete matrix. AFm phases are also known to cause expansion in concrete by altering to sulphate/carbonate-bearing phases such as ettringite.

Steefel & Lichtner (1994) highlighted the need to take flow geometries into account when assessing the effects of host-rock alterations. They modelled diffusive and advective transport processes along a hyperalkaline fluid-filled fracture in marl and also perpendicular to it between fracture and matrix. Dolomite dissolution was found to result in increased permeability parallel to the fracture, and diffusion was responsible for the precipitation of a 'calcite front' in the wall rock, thus isolating the fracture physically and chemically from the rock matrix. This may reduce the effective buffering and sorption capacity of the rock. The mechanisms which affect the transport properties of a host rock are shown in this work to depend on many different factors and may be far more complex than can easily be modelled or simulated in a laboratory.

The chemical properties of the host rock will also be affected by mineral/alkali reactions. Barrier properties, such as sorption capacity for radionuclides, have been well studied for the primary mineralogies of the unaltered host rock and bentonite barriers (e.g. Ames et al. 1983; Carroll 1991; Berry et al. 1993). The effect of alkali reactions on these properties has been less widely considered and is largely based on what is known of the chief expected reaction products, zeolites and CSH gels. The chemical barrier properties of zeolites are well known. They have high cation exchange capacities (CECs) and sorption capacities, although they do not sorb tri- and tetra-valent elements as well as iron oxyhydroxides (Sand & Mumpton 1977; Johnston & Miller 1984; Savage 1996). Zeolitization may significantly alter the chemical barrier properties of bentonites: zeolites have a higher CEC than smectites but their cation selectivity for multivalent ions is much poorer (Johnston & Miller 1984; Heimann 1985). The properties of CSH gels and associated cement minerals have been the subject of comparatively few studies. CSH gels have high surface areas, making them especially good sorbents for many actinides and lanthanides (e.g. U Th and Eu), although less so for the alkalis and alkaline earths (e.g. Cs and Sr; Allard et al. 1985; Faucon et al. 1998; Wieland et al. 1998).

Analogue studies

Geological and archaeological materials have been studied to assess the effect of alkaline fluids on rocks or geological systems and to document the longevity of CSH gels (which occur rarely in nature). For example at Maqarin, Jordan, a unique hyperalkaline groundwater system is buffered by naturally occurring cement minerals (mainly portlandite) and so mimics hyperalkaline plume chemistry very closely. Reactions with the local geology have resulted in the precipitation of gels, zeolites and other cement minerals. An extensive site characterization project has been running for a number of years and is described in detail by Alexander et al. (pers. comm.). Another classic site is at Scawt Hill, Northern Ireland, where CSH gels have formed by retrograde hydrothermal alteration of contact-metasomatized flints near a Tertiary dolerite intrusion (McConnell 1954, 1955; Tilley & Alderman 1934). Although these gels formed in very different conditions to those expected in the cement–rock environment, their presence does show that gels may survive for tens of millions of years.

The study of archaeological cements, mortars and concretes can provide information about the behaviour of cementitious materials over tens to thousands of years. There has long been interest in the durability of old structures themselves; early studies mainly involved mechanical tests and determinations of bulk chemical composition (Mallinson & Davies 1987). More recently, the radioactive waste disposal industry has begun to look more closely at the mineralogical and microstructural factors affecting durability of cements. There has been relatively little interest in using archaeological materials to study cement/rock reactions, partly because the environmental conditions (temperature, pore fluid composition, degree of fluid saturation and presence of CO_2) are not always well known and may differ from the conditions expected in waste repositories.

The history of calcareous cement and concrete is briefly summarized by Lea (1970) and also by Stanley (1979) who describes a number of individual structures significant to the development of concrete technology. A much more in-depth history is given by Mallinson & Davies (1987) who also provide a thorough review of the early literature. Jull & Lees (1990) give a brief ancient history of lime mortars and pozzolanic cements.

The simplest, lime-based mortars came into common use in Europe in the first century BC (Stanley 1979). These were made by first calcining limestone at temperatures high enough to produce lime (CaO), and then hydrating it to form portlandite ($Ca(OH)_2$) which would later have become carbonated, i.e. converted to calcium carbonate. Aggregates (mostly of quartz) in these mortars have often been found to be etched and pitted, presumably by the alkaline conditions established by hydration, although it is possible that this dissolution occurred during natural weathering of the aggregate materials prior to their emplacement in mortar (Znaczko-Janworski 1958; Malinowski et al. 1962; Roy & Langton 1982, 1983). Calcite is the only replacing phase seen in these cases.

Limestones are rarely pure and often contain alumino-silicate components. When these impure limestones are calcined, dicalcium silicate ($2CaO.SiO_2$) may form as well as lime. This hydrates on contact with water to form CSH gel. Many archaeological lime-based mortars are therefore likely to have originally contained small amounts of gel.

The ancient Greeks and Romans developed 'pozzolanic' cement by mixing lime mortar with a source of reactive silica such as volcanic ash, tuff or crushed bricks and tiles. The silica reacts slowly but continuously with calcium hydroxide after initial hydration to form large amounts of CSH gel.

The essentials of modern cement manufacture have been followed for the last 200 years (Lea 1970). A calcium source (usually limestone) and an aluminosilicate source (usually shale or clay) are mixed together and fired at high temperatures (1300 to 1500°C) to produce calcium silicates, aluminates and alumino-ferrites. The calcium silicates hydrate on contact with water to produce CSH gels which are the chief binding agents in the cement. Cements containing CSH gel (including pozzolanic cements) are termed 'hydraulic' because they set and harden on reaction with water and once set, continue to harden if placed underwater (Taylor 1990).

Many studies of the mineralogy of ancient lime-based mortars have shown them to be entirely carbonated (Malinowski et al. 1962; Jedrzejewska 1967; Roy & Langton 1982, 1983; Perander & Raman 1984; Mallinson 1986; Mallinson & Davies 1987). It is likely that all of these materials were of pure lime, that is, non-hydraulic, with no initial CSH gel having formed.

Ancient hydraulic mortars, in contrast, may remain uncarbonated and CSH gel has been identified in some Roman and Greek mortars (Malinowski 1979; Roy & Langton 1983; Rayment & Pettifer 1987; Rassineux et al. 1989; Jull & Lees 1990) and a sixteenth century Dominican mortar (Luxán & Dorrego 1996). Rayment & Pettifer (1987) examined 1700 year-old samples of Roman mortar from Hadrian's Wall, UK and found large amounts of CSH gel in the more compact, low porosity mortar, whereas elsewhere the cement paste had become fully carbonated. This suggests that the preservation of the gel depended on the extent to which it

could be sealed off from sources of carbonate or CO_2. Rassineux et al. (1989) found a similar phenomenon in 1800 year-old Gallo-Roman materials. They also noted that a continuous crust of calcite coated the surfaces of the more compact mortars and concretes, thereby protecting the interior from the further ingress of CO_2 or carbonate waters. They attribute the compactness of Roman cement pastes to the thoroughness of mixing and ramming, also noted by Lea (1970) as being the chief reason for the unusual durability of Roman cementitious structures. It has been suggested (Thomassin & Rassineux 1992) that early failures to find CSH gel in Roman cements were largely due to the lack of availability of modern analytical techniques, including scanning electron microscopy (SEM), electron microprobe analysis (EMPA), X-Ray diffraction (XRD), thermal analysis and infrared spectrometry (IRS).

The long-term survival of CSH gels (and perhaps other cement phases) at surface and near-surface conditions thus appears to depend on their being protected from sources of carbonate (i.e. atmospheric gases and carbonate-bearing groundwaters); either by the low permeability of the matrix or by a carbonated surface. Complete carbonation in the surface zone of concretes is also known to occur within a few decades in modern structures (Kobayashi et al. 1994). These observations imply that around a landfill or waste disposal site, a layer of calcite may coat CSH gels wherever they are exposed to a high CO_2/carbonate flux (e.g. adjacent to fractures with a high flux of carbonate-rich groundwater). This applies to both gels in cement and gels formed by cement/rock reaction. The calcite may act as a low permeability physical barrier and may make CSH gel less available as a potential sorbent, ion exchanger or high pH buffer. This possibility warrants further study. The physical isolation of rock matrices due to calcite precipitation was also predicted by Steefel & Lichtner (1994; described earlier) to result from cement/rock interaction in carbonate rocks.

If CSH gels escape carbonation, they might be expected to become increasingly well ordered and hence more thermodynamically stable with time. However, the CSH gels characterized in the ancient and historical materials described above were found to have no significant differences in structure, morphology or composition from modern cement gels. Roy & Langton (1983), for instance, conclude that such amorphous binders are effectively 'kinetically inert' and remain unchanged for at least three thousand years. A similar conclusion is reached by

Jull & Lees (1990). The only exception to this is the work of Mchedlov-Petrosyan et al. (1968), who identified a number of crystalline C-S-H phases by XRD including hillebrandite, gyrolite, afwillite and tobermorite in an ancient mortar. The authors suggested that initially amorphous CSH gels may have altered to more crystalline phases with time. This process has been observed to occur within laboratory timescales at 90°C or more (e.g. Atkinson et al. 1995) but has not been documented at low temperatures, even over long times.

Numerous studies of historical cements and concretes (from 68 to 136 years old) have found the CSH gels in them to be almost identical in composition, ordering and morphology to modern gels, when analysed by XRD (Grudemo 1982), optical microscopy (Idorn & Thaulow 1983), NMR (Glasser et al. 1989), SEM (Gebauer & Harnik 1975; Jull & Lees 1990; Mallinson & Davies 1987; Scrivener 1986) and EMPA (Rayment 1986).

ATEM characterization of C-A-S-H phases

Rationale

The analytical techniques commonly used to characterise experimental and analogue materials are far from ideal for characterising CSH gels. For instance, Grudemo's (1982) XRD study of gels in a 68 year-old concrete found that the method was not suitable for assessing increases in crystallinity over this time scale. XRD is a bulk method which is best suited to characterization of phases that are highly crystalline, present as a large fraction of the sample and structurally dissimilar to any other phases present. CSH gels, whether formed in mineral/alkali experiments or present in old mortars, tend to be poorly ordered, present in small quantities and structurally similar to other phases present. They are also often too finely grained and intermixed with other phases for their chemical composition to be resolved by SEM or EMPA.

In none of the archaeological analogue studies described, and in only very few of the experimental studies, were analyses made at the resolution afforded by analytical transmission electron microscopy (or ATEM). Use of bulk or low resolution methods alone can lead to unreliable phase identification in these materials, and only limited information can be obtained about gel phases. ATEM allows compositional, morphological and crystallographic information to be obtained at a much higher resolution than more standard methods, and is thus well suited to the

characterization of CSH gels, as demonstrated by Richardson & Groves (1993) and Viehland et al. (1996). It was therefore decided to examine both experimental products and archaeological materials by a combination of more standard methods (e.g. XRD and SEM) with ATEM, with the dual aims of characterizing mineral/alkali reaction products and obtaining a better understanding of the transformation of amorphous gels to more well-ordered and stable states.

Experimentally produced C-A-S-H phases

Method

A selection of common rock-forming and fracture-coating minerals were reacted with calcium hydroxide solution in accelerated tests for up to 800 hours. Since the focus of the experiment was to generate and characterize secondary products, closed-system batch tests were carried out to encourage extensive precipitation. The tests were accelerated by raising the temperature to 85°C, somewhat higher than would be expected around a radioactive waste repository.

Calcium hydroxide was chosen as a simple representative of cementitious fluid: a saturated solution was made by filling tubes of semipermeable membrane with calcium hydroxide slurry and equilibrating them with deionized water for two weeks. (The tubes of slurry were removed before the experiment, leaving calcium hydroxide solution.) The mineral samples used were anorthite, albite, quartz, chlorite (chamosite), muscovite, haematite, magnetite and a forsterite/enstatite mixture. These were powdered, grain-size separated and placed in polymethyl-pentene flasks. The calcium hydroxide solution was then filtered and added to the flasks while under nitrogen gas to limit the ingress of CO_2. The flasks were placed in agitating water baths at 85°C for up to 800 hours.

At the end of each test the pH of the hot fluid was measured immediately using a pH meter and combination electrode which had been calibrated up to pH 10. The meter automatically corrected for temperature, giving equivalent readings for solutions at 25°C. The pH was observed to fall slightly during the time it took to take these measurements, presumably due to contamination from atmospheric CO_2 and the subsequent precipitation of calcite, and so the measured pH values are likely to be lower than the 'true' values. The solids were then filtered, dried and analysed by XRD, SEM and ATEM. The fluids were analysed by atomic absorption spectrophotometry (AAS) for Ca, Mg, Fe, Na and K and inductively-coupled plasma optical emission spectrometry (ICP-OES) for Al and Si.

Solid products

In all the tests, some calcite and/or portlandite had precipitated (identified by XRD, morphology and chemical analysis). Some precipitation of portlandite was expected to occur on heating because its solubility decreases with increasing temperature. The additional presence of calcite in nearly all the products, despite attempts to limit the ingress of CO_2, shows how readily calcium hydroxide becomes carbonated.

The quartz, albite and anorthite samples underwent dissolution, often along preferred crystallographic directions to form etch pits visible by SEM. The phyllosilicate reactants were too fine grained for any dissolution textures to be visible by SEM. New calcium aluminosilicate phases precipitated abundantly in the anorthite, chlorite and muscovite samples. The resolution of the SEM analytical facility was not sufficiently high to obtain quantitative chemical data for these phases, whereas good quality chemical data were obtained for them by ATEM (Tables 3 and 4).

The major new precipitates seen in the anorthite tests can be described in terms of an evolving metastable product assemblage:

100 hours: katoite (a siliceous hydrogarnet)
200 hours: katoite; poorly-ordered fibrous C-A-S-H phase
800 hours: katoite; well-ordered fibrous C-A-S-H phase.

The katoite was identified by XRD, morphology (SEM and ATEM) and chemical analysis (ATEM). Table 3 presents the ATEM-derived compositions of all these phases. The earlier-forming fibres had a poorly crystalline morphology and did not give any visible XRD or electron diffraction pattern. The late-forming fibrous phase produced a 'diffuse ring' electron diffraction pattern characteristic of CSH gel and its formation coincided with the appearance of a 9.88 Å reflection on the XRD trace. These phases may both be intermediate in ordering between amorphous cement gels and fully crystalline C-S-H phases (such as tobermorite and jennite). Notably, both of them had high Al contents, equivalent to 19 and 23 mol% Al for Si substitution respectively.

In the muscovite tests, large amounts of a highly aluminous 'crumpled sheet' C-A-S-H phase precipitated (shown in Fig. 1). The ATEM-derived composition (see Table 4) shows that its Al content is equivalent to 37 mol% substitution for Si. A number of minor unidentified reflections were present in the XRD traces. Diffuse ring patterns obtained by electron diffraction gave d-spacings which are all

Table 3. *ATEM compositional data of the products of anorthite/calcium hydroxide experiments*

Wt% oxide	Katoite ($n = 17$)		Fibres at 200 h ($n = 5$)		Fibres at 800 h ($n = 15$)	
	Mean	Std deviation	Mean	Std deviation	Mean	Std deviation
CaO	52.3	4.6	25.9	2.5	51.9	2.9
Al_2O_3	23.6	2.8	12.1	1.5	9.8	1.1
Fe_2O_3	1.0	1.4	0.0		0.0	
SiO_2	23.1	3.0	62.0	2.3	38.3	2.2
Total	100.00		100.00		100.00	
Formula*						
Ca	3.40	0.28	0.45	0.04	1.45	0.08
Al	1.69	0.19	0.23	0.03	0.30	0.03
Fe^{3+}	0.04	0.06	0.00		0.00	
Si	1.41	0.21	1.00	0.04	1.00	0.06
O	8.81		2.80		3.90	
H_2O	3.19		n†		n†	
Molar ratios						
Ca/Si			0.45		1.45	
Ca/(Al + Si)			0.37		0.66	
Al/(Al + Si)			0.19		0.23	

*The katoite formula is recalculated to fit the ideal hydrogarnet formula: $3CaO.(Al,Fe^{3+})_2O_3.xSiO_2$. $(6 - 2x)H_2O$. The formulae for the fibrous phases are recalculated to 1.00 silicon as is standard in cement chemistry.
† H_2O content not measured by ATEM. CSH phases have high, and highly variable, hydration states.

Table 4. *ATEM compositional data of the products of phyllosilicate/calcium hydroxide experiments*

Wt% oxide	Muscovite experiment: Crumpled sheet ($n = 19$)		Chlorite experiment: Katoite ($n = 18$)	
	Mean	Std deviation	Mean	Std deviation
CaO	31.2	3.7	46.2	5.1
K_2O	1.6	1.4	0.0	
Al_2O_3	21.5	2.1	16.5	3.5
Fe_2O_3	2.5	0.9	23.0	4.8
SiO_2	43.2	2.1	14.43	2.7
Total	100.00		100.00	
Formula*				
Ca	0.77	0.09	2.83	0.28
K	0.05	0.04	0.00	
Al	0.59	0.06	1.11	0.21
Fe^{3+}	0.04	0.01	1.00	0.24
Si	1.00	0.05	0.83	0.05
O	3.74		7.67	
H_2O	n†		4.33	
Molar ratios				
Ca/Si	0.77			
$(Ca + K_2)/(Al + Fe + Si)$	0.50			
$Al/(Al + Si)$	0.37			

* The crumpled sheet formula is recalculated to 1.00 silicon as is standard in cement chemistry. The katoite formula is recalculated to fit the ideal hydrogarnet formula: $3CaO.(Al,Fe^{3+})_2O_3.$ $xSiO_2.(6 - 2x)H_2O$.
† H_2O content not measured by ATEM. CSH phases have high, and highly variable, hydration states.

characteristic either of CSH gel or of the semi-crystalline phases C-S-H (I) and (II) (Taylor 1990). This precipitate is thus intermediate in ordering between fully crystalline CSH phases and amorphous gels. It has a composition far

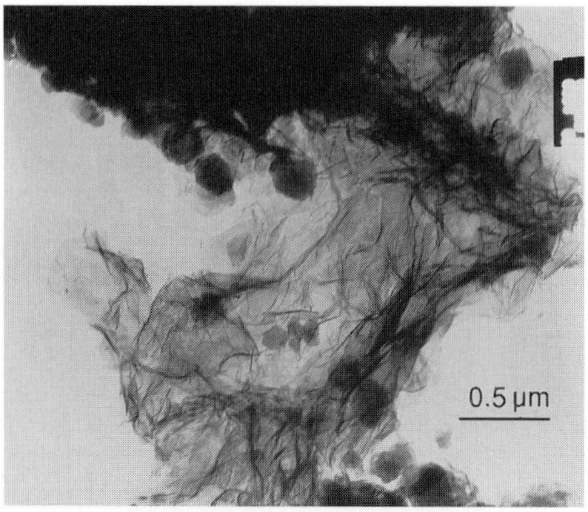

Fig. 1. TEM image of the solid products of the muscovite/calcium hydroxide experiment after 800 h. The 'crumpled sheet' phase is a highly aluminous calcium silicate hydrate. The hexagonal plates are the reactant muscovite. Magnification ×28 000.

closer to zeolites than C-S-H phases, but it is clearly not a zeolite.

Reactions with chlorite resulted chiefly in the formation of katoite, identified by XRD, morphology (SEM and ATEM) and chemical analysis (ATEM). The ATEM-derived chemical data for this phase are presented in Table 4.

In the quartz and albite tests, only very minor precipitation of new phases was seen. Fibrous material, a sheet-like phase and some amorphous gel all formed in the albite tests and a 'crumpled sheet' C-S-H phase was observed in one of the quartz tests (by ATEM and/or SEM). These were all in such small quantities that it was not possible to obtain fully quantitative chemical data from the products by ATEM. Both of these reactant samples were unusually clean, that is with very few fines adhering to grain surfaces. It is likely that this limited the dissolution rate (by lowering surface area) and so precipitation of secondary material may also have been inhibited.

Fluid products

The concentration of Ca in the reactant solution (i.e. the initial fluid prepared before the

Table 5. *Fluids data for the mineral/calcium hydroxide experiments*

Time (h)	Concentration (mgl^{-1})						
	Ca	Al	Si	Fe	Mg	K	pH
Anorthite							
0	554	0.097	0.0167				11.0
100	763	0.472	0.105				10.6
200	679	5.44	0.378				10.6
400	707	5.54	0.406				11.0
800	384	10.1	0.489				11.1
Chlorite							
0	554	0.097	0.0167	0.00	0.000		11.0
100	40.0	33.4	0.00	0.000	0.051		8.9
200	35	27.6	0.271	0.000	0.032		8.2
400	35.4	19.2	0.339	0.000	0.032		8.5
800	38	19.7	0.319	0.000	0.032		9.11
Muscovite							
0	554	0.097	0.0167			0.312	11.0
100	0.68	55.6	0.00			1230	10.4
200	0.92	58.3	2.36			1380	9.7
400	0.60	59.6	4.13			1560	9.8
800	1.31	61.6	2.72			1340	10.1

experiments were carried out) was found to be compatible with a fluid close to calcium hydroxide saturation. The measured pH, however, was significantly lower than a saturated solution should be (close to 11 instead of 12.5). There may have been problems with all the pH measurements, because attempts to charge balance the fluids data, by using pH to calculate the OH$^-$ concentration, were not successful.

The fluids data are however in general agreement with observations of the solid products of the anorthite, chlorite and muscovite tests, in that they indicate dissolution of Al- and Si-bearing material and precipitation of Ca-bearing material (see Table 5). In the chlorite and muscovite tests, the fluids data also indicate a decrease in pH and the buffering of Al to a roughly constant level, possibly caused by the precipitation of aluminous material.

Comments

XRD, morphological and electron diffraction data show that three phases precipitated which have degrees of ordering intermediate between fully crystalline C-S-H phases and amorphous gels. They are more well ordered than reaction products would presumably have been at lower (ambient) temperatures. Hydrothermal ageing of gels is known to cause their alteration to crystalline phases (Lea 1970, 197–201; Atkinson *et al.* 1995), but it is not clear from these experiments whether longer periods of ageing at

lower temperatures (e.g. in the natural environment) would have the same effect.

The C-S-H-type phases formed in these experiments are highly aluminous (8–24 mol%), suggesting Al for Si substitution from 19 to 37 mol% – significantly more than has been recorded for crystalline C-S-Hs such as tobermorite (e.g. Kalousek 1957; Komarneni *et al.* 1982, who synthesized tobermorites with up to 20% Al for Si substitution). The product of the muscovite experiments contains so much Al that it lies almost within the zeolite compositional field.

C-A-S-H phases in archaeological analogues

Roman mortar from Hadrian's Wall, UK

Hadrian's Wall was built to secure the British frontier of the Roman Empire. The material examined comes from cores taken by the National Trust in 1986 from a section which was rebuilt under Emperor Severus from around 200 AD. The wall here consists of a stone outer facing enclosing a rubble core bonded with mortar. The cores were originally examined by the Building Research Establishment using bulk methods (XRD, thermal analysis and wet chemistry) and low resolution microscopy (transmitted light, SEM and EMPA). The results were reported by Rayment & Pettifer (1987) and Jull & Lees (1990). Rayment & Pettifer (1987) found the binding phase to be a calcium silicate hydrate gel. They suggested that the gel may

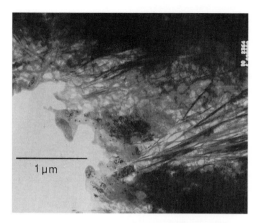

Fig. 2. TEM image of CSH gel in 1800 year-old cement paste from Hadrian's Wall; showing both a fibrous aspect and a typical 'outer CSH' porous honeycomb aspect in its morphology. Magnification ×28 000.

structures which have survived the ages intact: the mechanical integrity of these structures is by definition undisturbed by cement/rock or cement/aggregate reactions (such as the Alkali-Aggregate-Reaction known to cause 'concrete cancer').

In this study the cement paste was found to be variably preserved, being entirely carbonated in some regions and uncarbonated in others. The binding phase is a calcium alumino-silicate hydrate gel. It appears by SEM to have a typical 'cement gel' morphology. By TEM, it can be seen that some of the paste is a typical amorphous gel giving no electron diffraction pattern, but much of it has developed a fibrous morphology. A typical image is shown in Fig. 2. The fibres are poorly ordered but do sometimes give a ring pattern by electron diffraction. Less commonly, a 'crumpled sheet' morphology is developed, giving the same ring diffraction patern. The patterns have reflections characteristic of CSH gel as well as reflections characteristic of numerous structurally-related crystalline and semi-crystalline C-S-H phases. This suggests that the material is more well ordered than modern cement paste CSH gel. These fibrous and crumpled sheet morphologies were not distinguishable from one another by their chemistry but the chemical data set as a whole has a clearly defined compositional range (as shown in Fig. 3.) with a Ca/Si molar ratio between 0.34 and 1.25, 7–15 mol% Al, and minor Fe. The $Ca/(Al + Fe^{3+} + Si)$ molar ratio ranged from 0.26 to 1.07.

There is some evidence for cement/aggregate reaction. Some quartz, potassium feldspar and calcareous material has undergone dissolution

have formed by the hydration of calcium silicates. The calcium silicates may have formed in a local siliceous limestone, either where it had undergone contact metamorphism, or when it had been fired by the Romans whose intention would have been to produce lime.

A core sample was loaned by the Building Research Establishment and has been examined by optical microscopy, BSEM and ATEM. The objective was to characterize cement paste gels and the products of cement/rock or cement/aggregate reactions in greater detail than has been done before in materials of this age, principally by ATEM. It should be noted that an inherent sampling bias is incurred by sampling

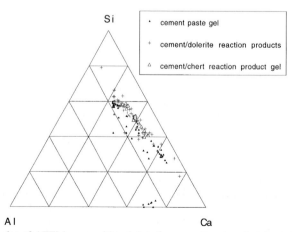

Fig. 3. Mol% element plot of ATEM compositional data for cement paste gel and cement/mineral reaction products in Hadrian's Wall.

and chert aggregate has clearly acted as a pozzolan, having been dissolved and replaced by calcium silicate hydrate material. The paste around the cherts is Si-enriched and their boundaries have become diffuse. The replacing material has filled the porosity of the cherts but there has been no cracking or expansion in most cases. The Si-enriched haloes around the cherts are more resistant to carbonation than cement paste elsewhere, possibly due to decreased porosity. ATEM showed the replacement phase to be a poorly ordered gel with a 'honeycomb' morphology, typical of the 'outer product' CSH of modern cement pastes (Viehland *et al.* 1996). The compositional range is well defined with a Ca/Si molar ratio varying from 0.48 to 1.23 and with 9–13 mol% Al; the chemical data are plotted in Fig. 3. Selected area electron diffraction occasionally produced ring patterns.

The material examined includes an interface between the mortar and a block of quartz dolerite. Analysis by BSEM together with element maps of the interface (made by wavelength-dispersive electron microprobe analysis) show an extensive reaction zone in the dolerite. A zone of Ca enrichment and Al and Si depletion extends roughly 0.5 mm into the dolerite, caused by replacement with calcium aluminosilicate hydrate phases. Beyond this, extending up to 2 mm into the dolerite, is a zone of increased porosity, caused principally by anorthite dissolution. Feldspar is clearly the most reactive of the dolerite minerals and is not usually replaced. Pyroxene is less reactive and is almost always replaced. The degree of reaction undergone by quartz was hard to assess because it is anhedral and is usually surrounded by the highly reacted feldspar. Fe-Ti oxides appeared largely unreactive. Olivine grains were protected by an unreactive layer of iddingsite, a primary alteration product. Using the SEM EDX facility, it was possible only to obtain qualitative chemical data of the replacing phases. This showed them to be calcium (alumino)silicates, also containing Fe and Mg if replacing ferromagnesian minerals. TEM imaging showed that the replacing phases are, for the most part, intimately intermixed with remnants of reactant mineralogy (as shown in Fig. 4.). Most of this material, replacing feldspar and pyroxene, is a typical porous CSH gel, too poorly ordered to give any electron diffraction pattern. In composition it is similar to the cement paste gel except that the Ca/Si molar ratio has a wider distribution (from 0.42 to 2.8), as shown in Fig. 3. The Al content varies from 6 to 16 mol%. A possible tetra-calcium aluminate hydrate (C_4AH_n) phase was also identified replacing pyroxene.

Fig. 4. TEM image of reacted dolerite at a mortar/dolerite interface in Hadrian's Wall. Feldspar has been partially altered to a gel which is intimately intermixed with the feldspar. Magnification ×28 000.

Discussion

The experimental reaction products and historical analogue gels formed and aged at different temperatures and over vastly different lengths of time. This is the chief difference between their environments and is the main factor dictating the degree of ordering of the C-A-S-H phases. The CSH gels in Hadrian's Wall have a similar degree of ordering to modern cement paste gels (based on morphology and electron diffraction). Thus, a long period (thousands of years) of ageing at low temperatures has not in this case increased C-S-H phase crystallinity as much as temperature-accelerated ageing of experimental reaction products, which produced more well ordered phases.

All the C-S-H phases characterized by ATEM, whether high temperature experimental products or low temperature analogue phases, were highly aluminous, with up to 37 mol% Al for Si substitution. Currently, theoretical models cannot include such aluminous C-S-H phases – all Al is assumed to go into other phases, principally zeolites. Thus, models may overestimate zeolite precipitation.

Hydrogarnets were a major reaction product in the mineral/alkali experiments, as some models have predicted (e.g. Savage & Rochelle 1993). However, the analogous phase forming at the cement/dolerite interface in Hadrian's Wall was a possible tetra-calcium aluminate hydrate (C_4AH_n). This is known to be metastable with respect to hydrogarnet and in fact converts slowly to hydrogarnet during cement curing (Lea 1970, p. 182). Its persistence in a 1700 year-old mortar implies that it may form and persist as a low temperature product of cement/ rock reaction in host rocks around waste disposal sites.

In Hadrian's Wall, mineral/cement fluid reactions in the dolerite produced a zone of increased porosity (caused by dissolution) and a deeper, parallel zone in which macroporosity was unchanged due to replacement by reaction products. The combined effect is to increase the surface area of solid material (dolerite or reaction product). This could increase the extent of any subsequent matrix/fluid interaction.

Conclusions

Published experimental studies of mineral/calcium hydroxide reactions show that at low temperatures (below 110°C), the chief reaction products are calcium silicate hydrate (CSH) gels, while zeolites and feldspars are formed at higher temperatures and in the presence of alkalis NaOH and KOH. The phase identifications have however often been made by low resolution or bulk methods, neither of which are ideal for such material. Published results of numerical simulations are in broad agreement with those of experimental studies of cement/ rock interaction. These models predict that CSH gels will be replaced by zeolites and maybe feldspars as plume chemistry evolves.

In the literature, experiments, models and archaeological analogue studies all highlight the importance of carbonate fluids or CO_2 in the cement/rock system. CSH gels and portlandite show a pronounced tendency to become carbonated (i.e. alter to calcium carbonate) and calcite readily precipitates wherever calcium hydroxide fluid comes into contact with a carbonate source. If protected from carbonation, the CSH gels found in archaeological and geological analogues can survive for thousands and even millions of years largely unchanged.

In the new work presented here, ATEM has been used to accurately identify and characterize the reaction products of alkaline fluid/mineral experiments at 85°C. Semi-crystalline C-A-S-H phases were formed which are structurally related, and morphologically similar, to amorphous CSH gels and fully crystalline C-S-H phases. The products are highly aluminous, in one case with analyses close to the compositional field of zeolites.

An ATEM examination of 1700 year-old mortar from Hadrian's Wall has shown that the cement paste gel has remained quite poorly ordered. The results confirm previous studies made at lower analytical resolution and also provide more detailed information about the morphology, structure, ordering and chemistry of the gel. At a mortar/dolerite interface in the wall, a reaction zone extends up to 2 mm into the dolerite, in which the primary mineralogy has been dissolved and partially replaced chiefly by poorly-ordered CSH gel. This appears to have produced a net increase in porosity and to have increased the solid surface area within the dolerite.

The ATEM characterizations of experimental reaction products and archaeological analogue material have shown that highly aluminous C-S-H phases precipitate due to cement/rock interaction. This contrasts with the results of theoretical models, in which all Al is assumed to go into other phases, principally zeolites. Zeolite precipitation may therefore be overestimated by the models.

Chemical reactions between alkaline cement fluids and host rocks have implications for the containment properties of cementitious barriers and surrounding geology. A number of mechanisms have been identified which will alter the chemical and physical barrier properties of the host rocks. Chemical reaction between cement fluids and aluminosilicate rock components leads to the precipitation of new phases (e.g. CSH gels and zeolites) with highly sorptive surfaces which may help attenuate a non-conservative pollution plume. The dissolution of host rock minerals may in places increase porosity and permeability, while replacement by high-volume reaction products may either decrease or increase permeability, depending on the geometry and mechanical properties of the host rock. Both mechanisms may increase the surface area adjacent to fracture flow paths, which would allow greater interaction (e.g. sorption and cation exchange) between matrix and subsequent pore fluids. However, in the presence of carbonate, there is a tendency for fracture porosity to become physically isolated from the reacted host matrix by the precipitation of calcite. This might limit the attenuation properties of the barrier system.

This work was funded by a Natural Environment Research Council Studentship, grant no. GT4/93/220/G. National Heritage and the Building Research Establishment are thanked for the loan of material from Hadrian's Wall.

References

ALLARD, B., PERSSON, G. & TORSTENFELT, B. 1985. *Radionuclide sorption on concrete.* Nagra Technical Report No. **NTB 85-21**, Nagra, Wettingen, Switzerland.

AMES, L. L., MCGARRAH, S. E. & WALKER, B. A. 1983. Sorption of trace constituents from aqueous solutions onto secondary minerals. I. Uranium. *Clays and Clay Minerals*, **31**(5), 321–334.

ATKINSON, A. 1985. *The time dependence of pH within a repository for radioactive waste disposal.* UKAEA, Report No. **R11777**.

——, HARRIS, A. W. & HEARNE, J. A. 1995 *Hydrothermal alteration and ageing of synthetic calcium silicate hydrate gels.* Nirex Safety Studies, Report No. **NSS/R374**.

BATEMAN, K., COOMBS, P., NOY, D. J., PEARCE, J. M. & WETTON, P. D. 1998. Numerical modelling and column experiments to simulate the alkaline disturbed zone around a cementitious radioactive waste repository. *Materials Research Society Symposium: Proceedings*, **506**, 605–611.

BERRY, J. A., BISHOP, H. E., COWPER, M. M., FOZARD, P. R., MCMILLAM, J. W. & MOUNTFORT, S. A. 1993. Measurement of the sorption of actinides on minerals using microanalytical techniques. *The Analyst*, **118**(10), 1241–1246.

BRANEY, M. C., HAWORTH, A., JEFFERIES, N. L. & SMITH, A. C. 1993. A study of the effects of an alkaline plume from a cementitious repository on geological materials. *Journal of Contaminant Hydrology*, **13**, 379–402.

CARROLL, S. A. 1991. Mineral-solution interactions in the $U^{6+}–CO_2–H_2O$ system. *Radiochimica Acta*, **52/53**, 187–193.

CHERMAK, J. A. 1992. Low-temperature experimental investigations of the effect of high pH NaOH solutions on the Opalinus shale, Switzerland. *Clays and Clay Minerals*, **40**(6), 650–658.

——1993. Low temperature experimental investigation of the effect of high pH KOH solutions on the Opalinus shale, Switzerland. *Clays and Clay Minerals*, **41**(3), 365–372.

CUNNINGHAM, W. C. & SMITH, D. K. 1968. Effect of salt cement filtrate on subsurface formations. *Journal of Petroleum Technology*, March 1968 issue, 259–264.

EIKENBERG, J. & LICHTNER, P. C. 1992. Propagation of hyperalkaline cement pore waters into the geologic barrier surrounding a radioactive waste repository. *In*: KHARAKA, Y. K. & MAEST, A. S. (eds) *Water–Rock Interaction*. Balkema, Rotterdam, 377–380.

EWART, F. T., SHARLAND, S. M. & TASKER, P. W. 1985. The chemistry of the near-field environment. *In*: WERME, L. (ed.) *Scientific Basis for Nuclear Waste Management IX. Materials Research Society*, **50**, 539–546.

FAUCON, P., CHARPENTIER, T., HENOCQ, P., PETIT, J-C., VIRLET, J. & ADENOT, F. 1998. Interaction of alkalis (Cs^+) with calcium silicates hydrates. *Materials Research Society Symposium Proceedings*, **506**, 551–559.

GEBAUER, J. & HARNIK, A. B. 1975. Microstructure and composition of the hydrated cement paste of an 84 year old concrete bridge construction. *Cement and Concrete Research*, **5**, 163–170.

GLASSER, F. P., MACPHEE, D., ATKINS, M., POINTER, C., COWIE, J., WILDING, C. R., MATTINGLEY, N. J.

& EVANS, P. A. 1989. *Immobilisation of radwaste in cement based matrices.* Dept of the Environment Report No. **DOE/RW/89/058**.

GRUDEMO, Å. 1982. X-ray diffractometric investigation of the state of crystallisation in the cement matrix of a very old concrete. *Nordic Concrete Research*, The Nordic Concrete Federation, Section **7.1**.

HAWORTH, A., SHARLAND, S. M., TASKER, P. W. & TWEED, C. J. 1988. Evolution of the groundwater chemistry around a nuclear waste repository. *Materials Research Society Symposium Proceedings*, **112**, 645–651 or 425–434.

HEIMANN, R. B. 1985. Stability and dissolution of Ca- and Na-bentonites. *Proceedings of the 19th Information Meeting of the Nuclear Fuel Waste Management Program*, September 1985, Pinawa, Manitoba. Atomic Energy of Canada Ltd, **2**.

HUGHES, C. R., HILL, P. I. & ISAAC, K. P. 1995. An experimental evaluation of the formation damage potential of cement filtrates and leachates. *Society of Petroleum Engineers*, SPE **30087**, 87–99.

IDORN, G. M. & THAULOW, N. 1983. Examination of 136 years old Portland cement concrete. *Cement and Concrete Research*, **13**, 739–743.

INOUE, A. 1983. Potassium fixation by clay-minerals during hydrothermal treatment. *Clays and Clay Minerals*, **31**(2), 81–91.

JEDRZEJEWSKA, H. 1967. New methods of investigation of ancient mortars. *In*: LEVEL, M. (ed.) *A Symposium on Archaeological Chemistry*, University of Philadelphia Press, Philadelphia, PA, 147–165.

JEFFERIES, N. L., TWEED, C. J. & WISEBY, S. J. 1988. The effects of changes in pH within a clay surrounding a cementitious repository. *In*: APTED, M. & WESTERMANN, R. F. (eds) *Scientific Basis for Nuclear Waste Management*, Materials Research Society, **11**, 43–52.

JOHNSTON, R. M. & MILLER, H. G. 1984. *The effect of pH on the stability of smectite.* Atomic Energy of Canada Ltd, Report No. **AECL 8366**.

JULL, S. P. & LEES, T. P. 1990. *Studies of historic concrete. CEC, Management and storage of radioactive waste. Part A Task 4; Geological disposal studies*, Luxembourg.

KALOUSEK, G. L. 1957. Crystal chemistry of hydrous calcium silicates: I Substitution of aluminium in lattice of tobermorite. *Journal of the American Ceramic Society*, **40**(3), 74–80.

KOBAYASHI, K., SUZUKI, K. & UNO, Y. 1994. Carbonation of concrete structures and decomposition of C-S-H. *Cement and Concrete Research*, **24**, 55–61.

KOMARNENI, S. & WHITE, W. B. 1981. Hydrothermal reactions of clay minerals and shales with cesium phases from spent fuel elements. *Clays and Clay Minerals*, **29**(4), 299–308.

—— & ——1983. Hydrothermal reactions of strontium and transuranic simulator elements with clay minerals, zeolites and shales. *Clays and Clay Minerals*, **31**(2), 113–121.

——, Roy, D. M. & Roy, R. 1982. Al-substituted tobermorite: shows cation exchange. *Cement and Concrete Research*, **12**, 773–780.

Krilov, Z., Romic, L., Celap, S. & Cabrajac, S. 1993. Permeability damage due to precipitation of insoluble salts from cement slurry filtrate. *Society of Petroleum Engineers*, SPE **25218**, 631–640.

Krueger, R. F. 1986. An overview of formation damage and well productivity in oilfield operations. *Journal of Petroleum Technology*, **38**(2), 131–152.

Lea, F. M. 1970. *The Chemistry of Cement and Concrete*. (3rd edn) Arnold, London.

Lentz, R. W., Horst, W. D. & Uppot, J. O. 1985. The permeability of clay to acidic and caustic permeants. *In:* Johnson, A. I., Froebel, R. K., Cavalli, N. J. & Pettersson, C. B. (eds) *Hydraulic barriers in soil and rock*, American Society for Testing and Materials. ASTM STP **874**, 127–139.

Lichtner, P. C., Pabalan, R. T. & Steefel, C. I. 1998. Model calculations of porosity reduction resulting from cement-tuff diffusive interaction. *Materials Research Society Symposium Proceedings*, **506**, 709–718.

Lunden, I. & Andersson, K. 1989. Modelling of the mixing of cement pore water and groundwater using the PHREEQE code. *Materials Research Society Symposium Proceedings*, **127**, 949–956.

Luxán, M. P. & Dorrego, F. 1996. Ancient XVI century mortar from the Dominican Republic: its characteristics, microstructure and additives. *Cement and Concrete Research*, **26**(6), 841–849.

McConnell, J. D. C. 1954. The hydrated calcium silicates riversideite, tobermorite and plombierite. *Mineralogical Magazine*, **30**, 293–305.

——1955. The hydration of larnite (β-C_2S) and bredigite (α_1-C_2S) and the properties of the resulting gelatinous mineral plombierite. *Mineralogical Magazine*, **30**, 672–680.

Madé, B., Jamet, P. & Salignac, A-L. 1994. Modeling of hydro-geochemical processes in waste disposal systems with a coupled chemical reactions-transport code. *Mineralogical Magazine Goldschmidt Conference,* Edinburgh, **58A**, 551–552.

Malinowski, R. 1979. Concretes and mortars in ancient aqueducts. *Concrete International*, January 1979, 66–76.

——, Slatkine, A. & Ben Yair, M. 1962. Durability of Roman mortars and concretes for hydraulic structures at Caesarea and Tiberias. *Final Report of the International Symposium on the durability of Concrete 1961*. RILEM, Prague, Czechoslovakia.

Mallinson, L. G. 1986. *An historical examination of concrete*. CEC Report No. **WAS-U.K.-432–84-7**. Taylor Woodrow Construction, Southall (UK).

—— & Davies, I. L. 1987. *A historical study of concrete*. CEC Report No. **EUR 1093 EN**, Luxembourg.

Mchedlov-Petrosyan, O. P., Vyrodov, I. P. & Papkova, L. P. 1968. Written discussion. *Proceedings of the Fifth International Symposium on the Chemistry of Cement*, Part **2**, Tokyo, Japan, 32–34.

Mohnot, S. M. & Bae, J. H. 1989. A study of mineral/alkali reaction – part 2. *Society of Petroleum Engineers Reservoir Engineering*, August issue, 381–390.

——, Bae, J. H. & Foley, W. L. 1987. A study of mineral/alkaline reactions. *Society of Petroleum Engineering and Reservoir Engineering*, November 1987 issue, 653–663.

Novosad, Z. & Novosad, J. 1984. Determination of alkalinity losses resulting from hydrogen ion exchange in alkaline flooding. *Journal of the Society of Petroleum Engineers*, **24**(1), 49–52.

Perander, T. & Raman, T. 1984. Durability of ancient mortars as a basis for development of new repair mortars. *Proceeding of the 3rd International Conference on the Durability of Building Materials and Components*, Espoo, Finland, **1**, 280.

Pusch, R. 1982. *Chemical interaction of clay buffer materials and concrete*. SKBF/KBS, Report No. **SFR 82-01**.

Rassineux, F., Petit, J. C. & Meunier, A. 1989. Ancient analogues of modern cement: calcium hydrosilicates in mortars and concretes from Gallo-Roman thermal baths of Western France. *Journal of the American Ceramic Society*, **72**(6), 1026–1032.

Rayment, D. L. 1986. The electron microprobe analysis of the C-S-H phases in a 136 year old cement paste. *Cement and Concrete Research*, **16**, 341–344.

—— & Pettifer, K. 1987. Examination of durable mortar from Hadrian's Wall. *Materials Science and Technology*, **3**, 997–1004.

Richardson, I. G. & Groves, G. W. 1993. Microstructure and microanalysis of hardened ordinary Portland cement pastes. *Journal of Materials Science*, **28**, 265–277.

Rochelle, C. A., Bateman, K., Milodowski, A. E., Noy, D. J., Pearce, J., Savage, D. & Hughes, C. R. 1992. Reactions of cement pore fluids with rock: implications for the migration of radionuclides. *In:* Kharaka, Y. K. & Maest, A. S. (eds) *Water–Rock Interaction*. Balkema, Rotterdam, 423–426.

Roy, D. M. & Langton, C. A. 1982. *Longevity of borehole and shaft sealing materials: 2. Characterisation of cement-based ancient building materials*. Office of Nuclear Waste Isolation, Columbia, Ohio. Report No. **ONWI 202**.

—— & ——1983. Characterisation of cement-based ancient building materials in support of repository seal materials studies. Office of Nuclear Waste Isolation, Columbia, Ohio. Report No. **BMI/ONWI-523**.

Sand, L. B. & Mumpton, F. A. (eds) 1977. *Natural Zeolites: Occurrence, Properties, Use*. Pergamon Press.

Savage, D. 1996. *Zeolite occurrences, stability and behaviour: a contribution to phase III of the Jordan Natural Analogue Project*. QuantiSci. DoE Report No. **DOE/HMIP/RR/95.020**.

—— & Rochelle, C. A. 1993. Modelling reactions between cement pore fluids and rock: implication for porosity change. *Journal of Contaminant Hydrology*, **13**, 365–378.

——, BATEMAN, K., HILL, P., HUGHES, C., MILO-DOWSKI, A., PEARCE, J., RAE, E. & ROCHELLE, C. 1992. Rate and mechanism of the reaction of silicates with cement pore fluids. *Applied Clay Science*, **7**, 33–45.

SCRIVENER, K. L. 1986. A study of the microstructure of two old cement pastes. *Proceedings of the 8th International Conference on the Chemistry of Cement*, Rio de Janeiro, 389–393.

STANLEY, C. C. 1979. *Highlights in the History of Concrete*. Cement and Concrete Association.

STEEFEL, C. T. & LICHTNER, P. C. 1994. Diffusion and reaction in rock matrix bordering a hyperalkaline fluid-filled fracture. *Geochimica et Cosmochimica Acta*, **58**, 3595–3612.

TAYLOR, H. F. W. 1990. *Cement Chemistry*. Academic, London.

THOMASSIN, J. H. & RASSINEUX, F. 1992. Ancient analogues of cement-based materials: stability of calcium silicate hydrates. *Applied Geochemistry, Supplementary Issue*, **1**, 137–142.

TILLEY, C. E. & ALDERMAN, A. R. 1934. Progressive metasomatism in the flint nodules of the Scawt Hill contact-zone. *Mineralogical Magazine*, **23**, 513–518.

VAN AARDT, J. H. P. & VISSER, S. 1977a. Formation of hydrogarnets: calcium hydroxide attack on clays and feldspars. *Cement and Concrete Research*, **7**, 39–44.

& ——1977b. Calcium hydroxide attack on feldspars and clays: possible relevance to cement–aggregate reactions. *Cement and Concrete Research*, **7**, 643–648.

VIEHLAND, D., LI, J-F., YUAN, L-J. & XU, Z. 1996. Mesostructure of calcium silicate hydrate (C-S-H) gels in Portland Cement Paste: short range ordering, nanocrystallinity, and local compositional order. *Journal of the American Ceramic Society*, **79**(7), 1731–1744.

WIELAND, E., TITS, J., SPIELER, P. & DOBLER, J-P. 1998. Interaction of Eu(III) and Th(IV) with sulphate-resisting Portland cement. *Materials Research Society Symposium Proceedings*, **506**, 573–578.

YANG, X. M. & SHARMA, M. M. 1991. Formation damage caused by cement filtrates in sandstone cores. *Society of Petroleum Engineers Production Engineering*, November 1991 issue, 399–405.

ZNACZKO-JANWORSKI, I. L. 1958. Experimental research on ancient mortars and binding materials. *The Quarterly of History of Science and Technology*, Warsaw, Poland, no. **3**, 377–407.

Chemical containment of mine waste

R. J. BOWELL,[1] K. P. WILLIAMS,[2] R. J. CONNELLY,[1]
P. J. K. SADLER[1] & J. E. DODDS[3]

[1] *Steffen, Robertson & Kirsten (UK) Limited, Summit House, Cardiff CF1 3BX, UK*
[2] *Department of Materials and Minerals, School of Engineering,*
University College of Wales, Cardiff, CF2 3TA, UK
[3] *JDIA, 20 The Pinfold, Newton Burgoland, Coalville, Leicestershire LE67 2SP, UK*

Abstract: The safe disposal of material considered hazardous is a natural part of good housekeeping for any industrial development. This is particularly so for the mining industry which has historically not always been so well managed in this aspect and as such has a high political profile today. Typical problems associated with mineral processing and mine waste are:

- poor water quality, through the release of sulphate, acidic waters, metals, metalloids and other undesirable substances
- poor air quality, through the release of mineral dust and gases such as SO_2, As_2O_3, Hg vapour and CO_2 amongst others
- contamination of land by the above with elevated levels of metals, sulphate and metalloids.

When the impact has been allowed to occur, control can be difficult and remediation costly. Consequently great efforts have been made over the last ten to fifteen years in understanding the nature of the potential pollutants, their pathways, environmental tolerances and solutions to their impact either by 'dilute and disperse' methods, recovery or containment.

In discussing the chemical containment of mining waste three broad categories can be defined: source control, migration control and dispersion control. Source control refers to measures employed to prevent the release of contaminants from solid phases. Migration controls restrict the interaction between environmental agents (principally water and humid air) and the deleterious element solid hosts. Dispersion control involves the collection and treatment of contaminated material such that they can be dispersed into the environment without significant impact. This last option, although the most common practice is also the most expensive and also may be required long term.

The increased awareness of the environmental impact associated with mining and mineral processing has increased the interest in developing effective processes for the treatment of mine waste. Presently there are a variety of physical, chemical and biological processes for the removal and recovery of metals, sulphate, cyanide and other by-products of mining and mineral processing. The goal of any treatment is to develop the most economical process or combination of processes which will be acceptable to environmental regulators and not pose a threat to the receiving environment.

In the selection and development of containment and treatment options a thorough knowledge of site geology, hydrogeology, hydrology, ore geochemistry, metallurgy and potential waste sources and characteristics are essential.

Any chosen option must be reliable and flexible enough to maintain a consistently high quality effluent throughout the active life of the mine and to provide sufficient safeguards against pollution after closure.

Greater emphasis is being placed on the early consideration of waste disposal as a part of the environmental management plan for new mining operations. In many countries it is now established practice to submit waste disposal and mine closure plans within the feasibility study of a new mine before any site development can take place.

Mining can generally be considered to be a short-term operation at most of the order of up to 100 years. Only in some unique areas such as the tin mines of Cornwall, the pyrite and copper mines of Rio Tinto in Spain or the lead, zinc and

From: METCALFE, R. & ROCHELLE, C. A. (eds) 1999. *Chemical Containment of Waste in the Geosphere.* Geological Society, London, Special Publications, **157**, 213–240. 1-86239-040-1/99/$15.00 © The Geological Society of London 1999.

gold mines in Northern Greece, has mining been active for thousands of years. Environmental concerns about mining are almost as old with the emperor Augustus Caesar introducing legislation to protect the forests around Rome from the iron industry (Fetter 1986). The deleterious consequences of pyrite oxidation and the production of acid salts were documented by Agricola (1556). Mining and processing methods have changed significantly with more concern about the environmental impact. As the twenty-first century approaches, the pressure to ensure protection of the natural environment is becoming more intense, heightened by the quality of modern communications and fostered by the media which brings the environment to the attention of society. All industries, which produce toxic waste, are affected, but the mining industry has received particular focus (Kraicheva 1996; Ricks & Connelly 1996; Sides 1996). Not only is it a large industry with a very visible profile but it suffers from a legacy of poor historical husbandry of the environment.

Therefore it is appropriate to review the general status of toxic waste disposal in the mining industry, the natural processes which exacerbate or mitigate the environmental impact and what analogues can be drawn for related industries which deal with toxic waste.

Material for waste disposal in a mine can be the marginal ore or the waste rock or chemically weathered equivalents of the precursor, *tailings* (processed ore material) or chemicals used in the mining or mineral processing operation. A discussion of the main features of mining and geochemical considerations in waste disposal on a mine is presented below.

Types of mining

A brief description of various mining operations is given to put waste disposal issues into context (Fig. 1).

Open pit mining

Open pit mining comprises surface excavation which can range from a small shallow pit of tens of metres across and 5 m deep to large pits which may be of the order of 2000 m across and 400 m deep. A special case would be strip mining of coal where a relatively narrow but long 'strip' is open at any one time but the face is continuously moving forward and waste material piled on the other side of the excavation. The waste pile also moves forward and is progressively rehabilitated as the mining progresses. In the conventional open pit, the ore will generally form only a part of the excavated material and large volumes of waste are generated which have to be placed in waste piles outside of the pit.

In the case of industrial minerals and aggregates, little waste is generated, therefore a large hole remains at the end of working. Backfilling of large open pits is not often undertaken due to the high cost of this procedure. However, excavated pits are used as landfill sites with backfilling of domestic waste if close to an urban area. Open pit mining is becoming a more common practice in the metal mining industry, particularly in the last ten years. It has the advantage that mine development is much more rapid and is often used as an initial step in developing underground resources, such as at the Meikle or Getchell gold mines in northern Nevada (Berentsen *et al.* 1996).

Underground mining

This covers mining of an underground ore body accessed either by drifts or tunnels into hillsides or inclined or vertical shafts to deeper ore bodies. Depth of mining can be from very close to surface, such as small-scale mining at depths of 20 m, to the deepest mines in the world, such as the Western Deep mine in South Africa which reaches a depth of around 4000m.

Some methods of underground mining, such as block caving induce collapse of the overlying material into the mine. This collapse is progressive as the mine develops and probably because of effects at surface represents the worst case of impact. Other methods involve leaving pillars of intact rock to support the roof either permanently or temporarily. The amount of movement of surrounding rock is a function of the geology, mine geometry and mining method but there will be some movement which may or may not be reflected in ground surface subsidence. Long wall mining of coal is designed to totally extract the coal from a panel. This induces total collapse of the mining horizon which will reflect as surface subsidence. The amount of surface subsidence is a function of lithology of the roof rocks, bulking and separation and the depth below surface.

Solution mining

The extraction of metals by chemical reagents on either *in situ* or stockpiled ore has been termed solution mining. The difficulties in controlling

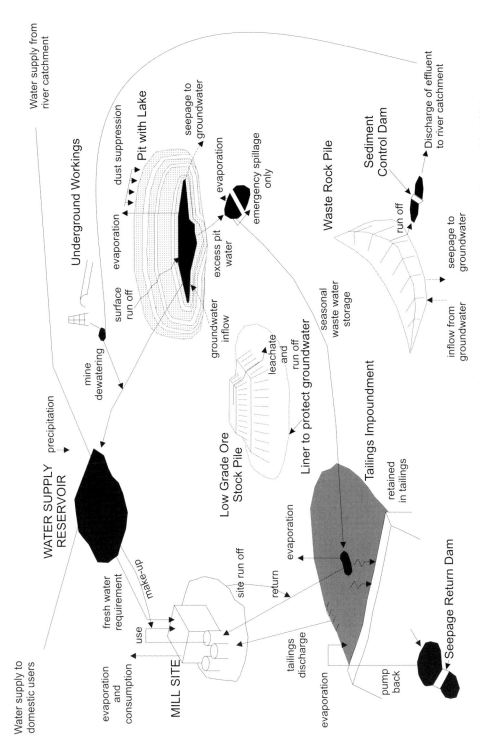

Fig. 1. Schematic diagram showing potential routes of contamination by water around a mine site (modified from Smith & Mudder 1991).

fluid flow in fractured rocks, often a characteristic of ore deposits, lead to problems of
environmental contamination by reagents and
dissolved ore elements.

A good example is the contamination of
groundwater by mineral acids from *in situ* leach
projects. In the Stratz mining area of the Czech
Republic, over a 25-year period over 3.8 million tons of sulfuric acid, 100 000 tons each of
nitric and hydrofluoric acid and 1 million tons
of ammonia was injected into a fracture and
litholgically controlled uranium orebody. Considerable leakage of the system has resulted in
major contamination which is currently being
re-addressed (Thomas 1996).

Types of processing

The recovery of a valuable commodity from its
mineral source requires a number of stages in the
process route which are dependent on the metal
or mineral to be extracted. A general review of
the subject is given by Woollacott & Eric (1994).
Crudely this includes one or more of the
following.

Concentrating

The ore is initially crushed from rock matrix to
mineral grain size or smaller. Often this rock
powder is mixed with water and/or chemical
reagents to form slurry. This mixture is injected
with air and agitated by stirring. The mineral
particles then attach themselves to air bubbles
and float off as a concentrate, while the wet
grains of rock (waste) sink to the bottom.

Pre-treatment

Very often prior to recovery of the ore metal
pre-treatment is required to make the ore
minerals amenable to leaching. For example in
the case of gold ores, gold can occur as a
chemically bound trace element in sulphides
such as arsenopyrite (Cabri *et al.* 1991). In order
to recover the gold the sulphide requires oxidation. Traditionally this required roasting in a
furnance, leading to the environmental release of
SO_2 as well as trace substances such as As_2O_5
and PbO. More environmentally acceptable
methods include autoclave or biological oxidation such as the BIOX® process developed by
Gencor and utilized at the Sansu gold mine in
the Ashanti region of Ghana (EMJ 1989).

Recovery

The method of recovery is dependent on the
chemistry of the metal to be recovered. Gold is
recovered very efficiently as a cyanide complex
from solution and can then be exchanged for a
more reactive metal such as zinc and coprecipitated (Smith & Mudder 1991). In the
case of base metals, strong mineral acids which
form highly soluble complexes are used. From
solution the metal is recovered by reduction by
either co-precipitation or electrolysis. Further,
for base metals, an additional refining step is
required. For example, in the case of copper
processing leaching can be carried out by nitric
acid forming soluble copper nitrate. The copper
is recovered by passing an electrical current
between plates of impure copper (anodes) and
plates of pure copper (cathode) immersed in the
copper nitrate solution. Copper is dissolved
from anodes and transferred to the cathodes
producing a very pure copper.

Types of mining waste

The key types of waste can be classified as:

- waste rock
- low grade ore
- tailings
- contaminated water
- chemical reagents

The physical and chemical nature of the wastes
and typical methods of disposal are summarized
below.

Waste rock

Waste rock is material excavated as overburden
or rock outside the ore body that must be moved
to gain access to the ore but would normally not
contain any ore material of significance. The
rock would be placed in specific sites on surface
around the mining operation. Some waste rock
is taken underground to provide backfill and
roof support. Waste rock has also been used as
a source of construction material if suitable.
The potential of contaminant release from
waste rock has been addressed by several studies
(SRK 1989; Blowes *et al.* 1991; Fuge 1993;
Bowell *et al.* 1996; Levy *et al.* 1997).

Low grade stock piles

These will be similar to the waste rock piles but
kept separate for possible ore processing at a
later stage, usually towards the end of the mining

operation. They therefore represent a source of contaminants due to prolonged exposure of this marginal ore to chemical weathering.

Tailings impoundment

Tailings and slimes dams represent the waste product after crushing, grinding and processing of the ore. Deposition can be in a dry form or hydraulically emplaced in the impoundment. Traditionally these 'dams' are a major source of potential problems from escape of contaminants into the environment. For example a poly-metallic sulphide ore can generate elevated anions, cations and metals together with chemical reagents from the process.

Contaminated water

Water, which has been contaminated by mining, can come from a number of sources (Fig. 1):

- Underground or open pit dewatering. Water typically has high Total Dissolved Solids (TDS) in excess of $1000 \, mg/l^{-1}$, in metal and in some coal mines very often sulphate is in excess of $2000 \, mg/l^{-1}$. Some of this water may be highly acidic and contain high levels of dissolved metals, particularly iron, such as in the Butte Pit, Montana (Robins et al. 1997). Pit lakes, developed in abandoned open pits are influenced by exchange with groundwater and can also impact local and regional groundwater. Surface runoff and reaction with soluble salts, precipitation and evaporation as well as the chemistry of any water discharged directly into the pit affect the chemistry of the lake also.
- Tailings effluentwaters. These can be highly alkaline as well as highly acidic. For example water in effluent ponds at Parys Mountain copper mine in North Wales is typically pH 2 with high levels of iron, copper, zinc, cadmium and sulphate (Fuge et al. 1993). However, tailings water from gold cyanide leach operations are typically pH > 10 and high sulphate but low levels of metals although oxyanions such as arsenate, anti-monate and molybdenate are high (Smith & Mudder, 1991).
- Water contaminated by natural interaction with minerals. For example at the Levant mine, Cornwall, groundwater is acidic, of the order pH 2–4 and has high levels of iron, aluminium, copper, zinc and sulphate (Bow-ell & Bruce 1995), whereas in the Yerrington pit, Nevada, mine water is neutral and has low

metal content despite reacting with sulphide-bearing host rocks (Murphy 1997; Price et al. 1995). Similarly some surface waters can be highly acidic by so-called 'natural contamination' (Runnells et al. 1997).

- Saline groundwater which is released into the mine from regional groundwater storage. For example the Beatrix gold mine in South Africa has a groundwater with high Fe, Mn, Na and Cl and sulphate (Juby & Pulles 1990).

Chemical reagents

Most forms of mineral processing utilize inorganic and organic reagents in the separation of economic elements and minerals from waste rock; additionally drilling fluids and petroleum products are major waste components on a mine site. Part of the chemical containment required at a mine site will involve collection, treatment and disposal of waste chemicals. These are very often the most hazardous materials on site and include. A major problem is the release of strong mineral acids, such as sulphuric acid from TiO_2-plants in the Ukraine (Schuiling & van Gaans 1997; van Gaans & Schuiling 1997).

In rare cases cyanide-rich contaminants can be released leading to alkaline metal-rich waters in which metals are stabilized as cyano-complexes. These species are quickly buffered by changes in water Eh-pH so the extent of impact is limited. Such examples include the Globe and Phoenix mine in Zimbabwe (McGill & Comba 1990; Smith & Mudder 1991; Williams & Smith 1994) and Omai mine in Guyana (Harcourt & Wickham 1996).

Dispersion mechanisms

Mineral pollutants in the mining industry are generally of two types.

Solid pollutants

These include asbestos and other fibrous or fine-grained minerals (Bish & Gutherie 1993). Coal and rock dust emission can lead to several health-related problems (Kagey & Wixson 1983). These problems are generally dealt with by controlling dust emissions and are usually restricted to the immediate mine environment. Dust emissions either from processing operations or rock piles or from dry tailings beaches can lead to dispersion of significant concentrations of potentially toxic elements including Pb, Cd and Zn (Kuo *et al.* 1983; Cotter-Howells & Thornton 1991). However, on most processing operations dust emissions are reduced by collection of dust. The mechanisms of dust dispersion and inhalation are outside the scope of this paper.

Hydro-dispersed pollutants

The transport, mobilization and precipitation of an element is a balance between processes which release the element from its precursor through dissolution or desorption to processes which scavenge or fix the element through mineral precipitation or adsorption (Appelo & Postma 1993; Stumm & Morgan 1995). The actual mechanisms involved are complex and reviews have been published elsewhere (Lowson 1982; Nordstrom 1982; Morse 1983; Trescases 1992; Bigham 1994; Bowell *et al.* 1996).

The most common cause of acid generation in metal or coal mines is the oxidation of FeS_2, although other minerals can also generate H^+ (Thornber 1992, 1993). Pyrite and/or marcasite generate the acidity of the mine waters and simultaneously supply large quantities of Fe and sulphate and consequently produce large volumes of ochres. Where alkalinity–acidity is balanced, acid generation and neutralization is localized. Where this does not occur or where acid generation exceeds the acid neutralizing capacity of the mineral lode an acidic discharge will be emitted, in some cases even leading to supersaturation of H^+ such as recorded for Iron Mountain drainage (Alpers & Nordstrom 1991). The dispersal of metal-rich acid waters is influenced by the ability of the surrounding environment to neutralize the drainage. Where neutralization occurs metals and sulphate are precipitated forming a range of mineral

Table 1. *Oxidation reactions of iron sulphide and sulphate minerals generating acidity*

Reaction 1

(a) $FeS_2 + \frac{3}{2}O_2 + H_2O = Fe^{2+} + 2SO_4^{2-} + 2H^+$

(b) $2FeS_2 + 7O_2 + 2H_2O = 2Fe(SO_4) + 2H_2SO_4$

Reaction 2

(a) $Fe^{2+} + \frac{3}{2}H_2O + 3O_2 = Fe(OH) + 2H^+$

(b) $2Fe(SO_4) + H_2SO_4 + 2O_2 = Fe_2(SO_4)_3 + H_2O$

Reaction 3

$Fe^{2+} + 2O_2 + H^+ = Fe^{3+} + 2H_2O$

Reaction 4

$FeS_2 + 14Fe^{3+} + 8H_2O = 15Fe^{2+} + 2SO_4^{2-} + 16H^+$

Stage 1
Reaction 1: proceeds abiotically and by bacterial oxidation (reaction b more
    common with bacterial oxidation)
Reaction 2: proceeds abiotically, slows as pH falls (reaction b more common with
    bacterial oxidation)
pH approximately 4.5 or higher, high sulphate, low Fe, low pH

Stage 2
Reaction 1: proceeds abiotically and by bacterial oxidation (reaction b more
    common with bacterial oxidation)
Reaction 2: proceeds at rate determined primarily by activity of bacteria such as
    T.ferrooxidans
pH approximately 2.5–4.5, high sulphate, Fe and low pH. Low Fe^{3+}/Fe^{2+} ratio

Stage 3
Reaction 3: proceeds at rate determined by activity of T.ferrooxidans
Reaction 4: proceeds at rate determined by rate of reaction 3
pH generally below 2.5, high sulphate, high Fe and low pH. High Fe^{3+}/Fe^{2+} ratio

After Kleinmann & Pacelli 1991; Thornber 1992.

precipitates dominated by iron oxyhydroxides and oxysulfates such as ferrihydrite $(Fe(OH)_3)$, goethite $(FeOOH)$ and iron sulphate polymorphs. Secondary ferrous sulphate salts may act as a temporary store for acidity which is released with the destabilization of the salts (Cravotta 1994). Acidity from these minerals is generated by the hydrolysis of ferrous iron to ferric ion and subsequently to ferric hydroxide (Table 1) and is often termed 'mineral acidity'. Thus three essential mechanisms are operative, primary and secondary acid generation and alkaline generation or acid buffering (Bowell et al. 1996).

Hydrogeology and hydrology

The water-related aspects of mining are critical to a discussion on pollution control and mining waste management. Water is actively involved in geochemical reactions and provides a transport medium for both dissolved and solid reaction products.

Water forms an integral part of any mining operation whether it is as a result of groundwater inflow, direct precipitation and storm water flow or as a need for water resources for operations (Fig. 1). The water balance on a mine can be simplified to an in/out balance comprising input from rainfall, stream flow and groundwater and output as evaporation, direct water discharges and seepage losses from ponds and tailings management facilities. In addition to the direct in/out balance the mine and its surrounding environment will experience both groundwater and surface water fluxes.

The nature of the impact of each phase of a project on the water environment is different in terms of location, quality and quantity of discharges. Typical risk areas associated with each phase are given in Table 2. In order for potential contaminants to get from the mine to the environment a pathway must exist. This pathway can be considered as a vector, with both a value and a direction. During the life of a mining project the pathways will change both in terms of their ability to transport contaminants and their direction. For example, consider an underground pathway. During exploration the size of the pathway will be limited by the permeability of the rock mass or a particular feature. The direction will be governed by a regional flow field. During mining the size of an underground pathway will be controlled by the physical openings created by mining and the direction will be toward the mine. After closure, the size of the pathway will be dependent on the manner in which the mine has been restored, by underground or open pit backfilling, collapse, flooding and what openings remain such as shafts, adits etc. The direction of the pathway will return to something similar to

Table 2. *Summary of hydrogeological assessment*

Stage	Risks (Examples)
Mineral exploration	Point source pollution of surface and groundwater from drilling Erosion due to land clearing leading to impact on surface water
Design	Field investigations resulting in impacts on surface and groundwater from drilling, testing, pumping, river studies Point source pollution from plant
Construction	Erosion from land stripping Point source pollution from plant
Operations	Impact on surface and groundwater resources from mine water management Impact on ground and surface water quality from mineral oxidation in mine, rock dumps and tailings
Closure	Mobilization of $2°$ minerals on re-flooding Re-emergence of ground water Cessation of pumping Long-term seepage Deep fluxes Ground water rebound

the pre-mining condition, but will be modified, particularly locally by the presence of high and low permeability zones created by mining and backfilling/sealing. In areas with a long history of extensive mining, such as the European concealed coal fields, the post-closure influences and pathways may be regionally extensive and cross what would normally be considered as surface or groundwater catchments.

Environmental requirements

The legislative requirements for environmental protection related to a mining operation are dependent on the status of the operation and the geographical location with respect to sensitivity of the local ecosystem and the proximity of urban developments and agriculture to the mining venture.

Emissions into the air are perhaps the most visible aspect, and the majority of mineral processing operations employ technology which involves gaseous discharges. Most operations generate liquid streams which have to be returned to the natural environment. Contaminated solids are also subject to stringent guidelines (Table 3) which if violated require that the solid phase be stabilized, treated or appropriately contained to prevent further impact.

Nationally and internationally developed guidelines based on water quality and quantity are now in place in most countries. These typically are based on USEPA or WHO guidelines as to the maximum concentration of a particular chemical constituent in drinking water or water utilized by spawning salmon (Table 4). There exist also the requirements to maintain the area of a mineral operation, both during and after the

Table 3. *International guidelines for the maximum level of solid contaminants for safe disposal. Concentrations in $mg\,kg^{-1}$*

Parameter	European Community	South Africa*	USA/Canada
pH		6–8	6–8
Arsenic	10	10	5
Cadmium	3	10	0.5
Chromium (VI)	25	60	2.5
Manganese	50	100	10
Mercury	1	1	0.1
Nickel	70	200	20
Lead	500	500	25
Sulphate	2000	2000	500
Zinc	300	500	100

USNAS 1977; WHO 1984; Angus Environmental 1991; Visser 1993; USEPA 1995.
* Recommendations only

Table 4. *Typical water quality guidelines for surface discharge and mine waters*

Parameter	EC	South Africa	USA
pH	6–8	5.5–8	6–8
Total dissolved solids	1000	1000–1500	1000
Sulphate	500	500	400
Cyanide (free)	0.1	0.5	0.1
Arsenic (total)	0.05	0.5	0.05
Cadmium	0.05	0.05	0.01
ChromiumVI	0.05	0.05	0.015
Copper	1	2	0.2
Iron	2	2	0.2
Manganese	0.1	0.4	0.05
Mercury	0.001	0.02	0.0025
Nickel	0.05	2	0.2
Lead	0.05	0.1	0.01
Zinc	3	5	0.05

Flanagan 1990; Angus Environmental 1991; DWAF 1993; USEPA 1995.
All parameters except pH in $mg\,l^{-1}$.

life of the operation, in a condition sympathetic to the surrounding environment.

Criteria for selection of containment option

Criteria for selecting the most appropriate containment strategy for mine waste can be defined in terms of the contaminant load mobilized from the waste that can be acceptably accommodated by the receiving water body. In order to define this load a Water Quality Objective (WQO) should be set for the receiving course downstream of the discharge. At present in the UK, Statutory Quality Objectives have not been set for surface water courses, but some pilot studies are currently ongoing. The WQO should therefore be set specifically for the site in question and should take account of local factors such as downstream effluent discharges.

To define the acceptable load or treatment objective for the discharge, a simple mixing model can be employed to calculate the mean downstream contaminant concentration for a particular discharge contaminant loading (Sadler 1997). Two approaches can be taken depending on the availability of data.

- Establish the time of year when the problem is worst, estimate worst case conditions for each parameter and combine these to calculate the acceptable conditions for that period. This period is often likely to be late summer, but if the discharge is very responsive to recharge conditions it may be low or absent at this time.
- Probabilistic analysis to combine the frequency distributions of the analysed variables to predict the probability of exceeding the limit in the downstream receiving water.

In this way a remediation objective can be calculated such that the impact on the receiving water course is acceptable.

The assessment of environmental stability for any product created is essential in assessing the role of a particular treatment option. Protocols for the geochemical assessment of material have been published elsewhere and cover acid generating capacity, leachable and total metals through a variety of procedures (SRK 1989; COSTECH 1990). The tests are used as the major criteria for long-term disposal of waste products and highlight the need to take geochemical processes which will modify the physical and chemical characteristics of the compounds into account. As an example the influences on several products are summarized in Table 5.

Finally there are special situations created where potential operations exist on sites where high priorities are placed on the area for other reasons, such as nature conservation or cultural or religious grounds. The response of these environmental considerations to a mining venture depends on status, giving rise to three or four different scenarios based on the level of

Table 5. *Geochemical considerations in the long-term stability of contaminant waste products (after Broadbent & Warner 1985; SRK 1989; Swash & Monhemius 1995)*

Product	Consideration	
Ferrihydrite	Dehydration leads to instability goethite Precipitate at ambient temperature Fe (III) to Fe (II)	Recrystallization Biogeochemical reduction
Ca-salts/cements	High intrinsic solubility Ca-salts converted to $CaCO_3$. Release of contaminants (influence of CO_2) High CaO levels – high pH (11–12) CaO converts to $CaCO_3$	
Crystalline products (e.g. scorodite)	Low solubility Geological stability High production cost	
Slags	Recrystallize glass-devitrification Unknown stability long term Quenched slags, low solubility Low efficiency, high volume High production cost; can theoretically get energy back (Broadbent & Warner 1985)	

activity from planning stage of a future operation to active extraction to planning closure or dealing with a closed mine.

Chemical containment methods

Chemical containment strategies fall into the following categories (Fig. 2):

- source control; includes sulphide oxidation and infiltration control (**A**)
- migration control; chemical and physical barriers (**B**)
- discharge control; dispersion and treatment (prior to drainage)

Before investigation of potential remediation options begins, however, remediation objectives should be defined. The discussion below

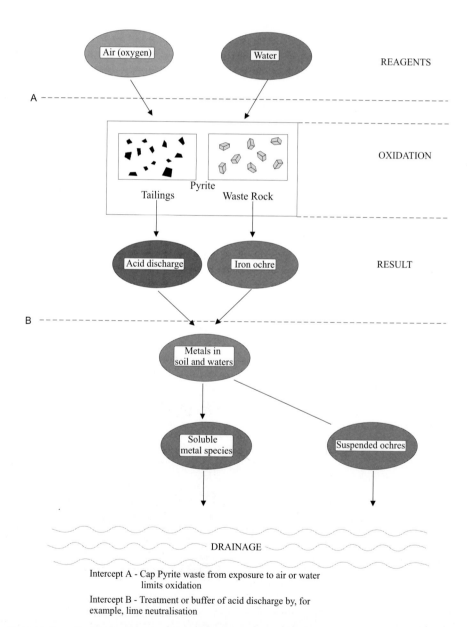

Intercept A - Cap Pyrite waste from exposure to air or water limits oxidation

Intercept B - Treatment or buffer of acid discharge by, for example, lime neutralisation

Fig. 2. Oxidation of sulphide minerals and generation of acid rock drainage.

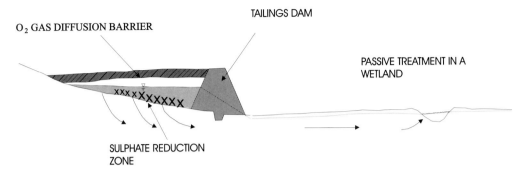

Fig. 3. Schematic diagram showing the use of a combination of treatment techniques including low oxygen gas barrier, sulphate reduction barrier and a passive treatment wetland system at the point of discharge (after Blowes *et al.* 1994).

introduces the setting of containment objectives and details potential containment and remediation options from the three categories above. It is likely that the selected option in many cases will be a combination of strategies from the different categories. For example in the case of a closed mine the containment scheme involved arrangement of controls to limit contaminant transfer from a tailings impoundment through (Fig. 3):

- the use of oxygen diffusion barriers (source control)
- a chemical membrane to hold contaminants or a liner to act as a barrier (migration control)
- treatment cells or dilution of effluent or wetland prior to discharge (discharge control)

Source control

Sulphide oxidation control. Sulphide oxidation can be controlled by excluding sulphides, water bacteria or oxygen. In most cases to exclude water is impractical as even in low humidity environments sulphide oxidation can occur (Lowson 1982).

To exclude sulphides can be achieved by mechanical removal of the sulphides by traditional or innovative mineral processing techniques (Thornhill & Williams 1995). However, this still presents the problem of the disposal of the sulphides in a different manner elsewhere and is not applicable where sulphides are still present in large quantities underground. A more practical barrier is to exclude the sulphides by precipitation of an insoluble non-reactive precipitate thereby isolating the sulphide from oxidants. Ferric phosphates and oxyhydroxides have been proposed and the rate of sulphide oxidation has been observed to decrease in samples amended with phosphate (Achmed 1991; Huang & Evan

gelou 1994). However, in field studies the presence of alteration products on sulphides has not led to a significant decrease in acid generation (Blowes & Jambor 1990; Bowell & Bruce 1995). Exclusion of bacteria will also significantly reduce sulphide oxidation and this involves the use of bactericides either as an intimate mixture with tailings or backfill material or applied directly onto sulphide surfaces (Erickson & Ladwig 1985). However, such treatment requires continual reapplication based on currently available bactericides and is only suitable as a short-term option.

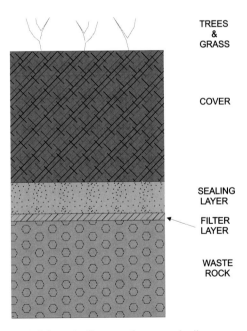

Fig. 4. Schematic diagram of waste rock pile cover with sealing layer and filter layer. This system is designed to reduce sulphide oxidation in the rock pile, thereby limiting the impact of acid drainage.

Placing a diffusion barrier to limit atmospheric oxygen entry into tailings impoundments or underground workings would greatly impede oxidation of reactive sulphides. Such a barrier could include a geomembrane or geofabric cover which actively consumes moisture and/or oxygen (Fig. 4). Underground, the simplest solution is flooding of the workings. With tailings material a number of approaches have been proposed including dry covers composed of fine-grained materials which maintain high water and low oxygen diffusion (Nicholson *et al.* 1989), synthetic oxygen diffusivity materials (Malhotra 1991), and oxygen consuming materials (Tasse *et al.* 1994).

Fine-grained covers rely on the moisture-retaining characteristics of these materials to maintain high moisture contents above the water table. Naturally formed covers can be encouraged by intentional formation of 'hard pans' (Blowes *et al.* 1991). These are analogues to lateritic iron accumulation in tropical soils (Mann 1983). The hard pan can comprise a ferricrete (iron-rich cement), silicrete (silica-rich cement), a gypcrete (gypsum-cemented zone) or calcrete (calcite cemented zone). The zone acts as a barrier for water and oxygen movement thus limiting sulphide oxidation (Fig. 5).

Synthetic covers are expensive and susceptible to cracking after installation due to desiccation or subsidence. The effectiveness of such covers relies on long term integrity and because they are located on the tailings surface they are most susceptible to erosion (Blowes *et al.* 1994). The strategy is only applicable to rock waste, spoil or backfill and not to *in situ* rock (Robertson 1987).

At the end of stabilization the waste rock piles can be covered by top soil, landscaped and revegetated (Epps 1993). Often landscaping and revegetation are undertaken as a staged process in which areas are reclaimed as the active face of the mine or impoundment or waste rock pile is no longer needed.

Infiltration controls. The most effective strategy to control the movement of mine drainage is to

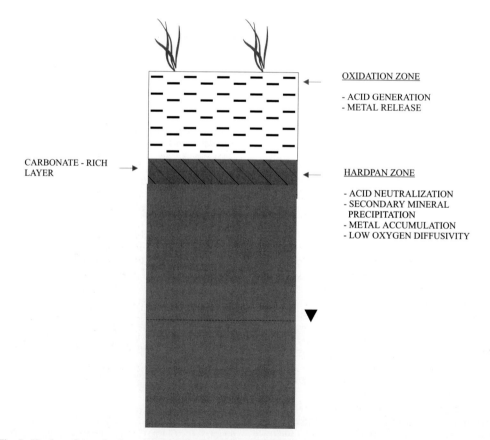

Fig. 5. Hardpan formation in a waste rock pile (Blowes *et al.* 1991).

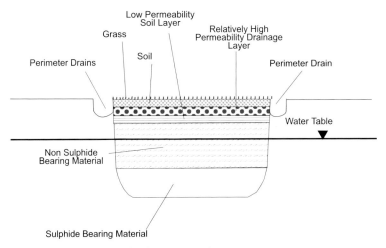

Fig. 6. Conceptual model of a closure plan for an open pit or quarry.

restrict entry of water which will in turn reduce the quantity of low quality drainage. The use of liners, cut-off walls and diversion trenching may be considered in controlling drainage. The use of these barriers can be used to direct flowing ground water or drainage toward an *in situ* treatment system, for example a funnel and gate system.

Flooding may also be used to prevent oxygen filtration when:

- the diffusive transfer of oxygen to the reactive sulphides is reduced, as water has a low oxygen diffusivity and/or
- the barometric pumping of air into the void spaces (which results in the flushing of air through the acid generating rock) is prevented by water occupying the void spaces in the mine and between the solid rock particles

The combination of these two effects results in the effective exclusion of oxygen from the sulphides, resulting in control of oxidation and acid generation. This measure is effective where the primary mechanism of acid generation or ferruginous water generation is active oxidation of sulphides under unsaturated conditions. For example at the Roquis mine, St Salvy, France, flooding of the underground workings has limited sulphide oxidation (Sadler 1997). However, flushing of underground workings has led to initially low water quality due to the dissolution of stored secondary products.

Consideration should always be given to the potential geotechnical impacts of flooding and the potential for uncontrolled leakage (Younger 1995). These may be a particular problem where

coal seams have been worked from surface and where the extent of old workings is not well documented. Flooding of underground workings can be achieved by installing plugs underground to block drainage paths.

Covers over mine piles have often been ineffective in reducing air contact with sulphide minerals and oxidation of sulphides, as a result of barometric pumping and convective transport of air through and around the covers. At some abandoned coal mines it may be possible to significantly reduce the amount of air, and therefore of sulphide oxidation, by sealing old open adits to the passage of air. This would only be effective where mining has not significantly impacted on the ground surface, as subsidence zones are likely to provide alternative points of air entry to the mine which could not so easily be sealed.

In the closure of an open pit, confinement of the environmentally unstable material, such as pyrite or radioactive materials can be achieved by storage below the water table (Fig. 6). For example at the Wismut uranium mine in Germany radioactive and sulphide-rich waste is stored below the water table preventing extensive further breakdown. Above this is a buffer zone and low-level waste and a clay barrier which acts to reduce oxygen and water permeability and also absorb any radium emission (Chapman & Hockley 1996; Hockley *et al.* 1997).

Migration control

Diversion of water flows. Production of contaminated effluents can be reduced significantly if the flow of comparatively clean water which contacts potentially polluting solids is reduced. In addition to isolating such solids by means of barrier systems as discussed below, this can be achieved for surface water flows by construction of suitable cut-off walls and diversion trenches. Placement of wastes in positions ('high and dry') where exposure to water is minimized is also an option both during production and, less commonly, after closure.

In an underground mine, isolation of pollution sources can be engineered through the use of plugs (bulkheads) to reduce the flow of water through parts of the mine where contamination is high. This may have the effect of:

- diverting cleaner waters around known, more contaminating zones, reducing the overall rate of flushing and therefore contaminant migration to the environment.
- reducing mine through-flow and therefore the underdraining effect of the mine. This can in turn increase near-surface circulation of groundwaters so that the total contaminated discharge flow is reduced.

Bulkheads must be sited where the rock strength is sufficient to withstand the required pressure and where by-pass flows are minimized. It would generally be necessary to grout the rock around the plug to reduce permeability.

Hydraulic balancing. Hydraulic balancing may be achieved by installing spillways (either direct via adits or via boreholes) and by closing existing discharge points such as adits. This can only be of benefit if it allows decanting of relatively clean recharge water, or reduces the drainage path length and limits the iron load that can be picked up by the water moving through the workings. It is therefore essential to understand the hydraulic and geochemical regimes as well as mine geometry.

Barrier systems. Isolation of the source of pollution by encapsulating in an impermeable liner is an attractive solution to waste management in general, and has been the technical area of landfill engineering which has probably received the greatest amount of attention (see e.g., Bentley 1996; Arnould *et al.* 1993). In the mine environment it is seldom possible to consider such complete isolation but partial solutions to the migration problem may be accomplished during production by providing an impermeable base onto which waste is placed and by the use of covers and vertical barriers once production has ceased. Grouting to provide horizontal barriers beneath an existing

waste is also technically feasible but usually extremely costly.

Seldom is any single barrier capable of preventing contaminant transfer, and deterioration with time must also be considered. For these reasons it is normal to use composite liner systems which may incorporate one or more geomembranes whose function is to prevent any advective transport together with components designed to reduce chemical transport whilst possibly allowing passage of water. Even with an efficient barrier system it may still be necessary to consider water treatment, albeit at a much reduced capacity. Considerable development has occurred in the synthetic materials used for geomembranes (Koerner 1990) and these can now be tailored to a specific environment. The remaining components of a composite or reactive barrier are usually inorganic and may include compacted clays (possibly modified by addition of metal cations or organics such as amines), *in situ* soils with or without admixture with bentonite, fly ash, zeolites, bentonite slurry and bentonite cement mixtures. It is possible, therefore, to engineer a barrier which would have the following zones:

- Sorptive zones
- Neutralizing zones
- Sulphate reduction zones
- Sulphate precipitation zones.

These processes all have natural analogues, for example:

- sorption of metals to clay minerals, iron hydroxides and organics
- neutralization of acid waters by reaction with carbonate minerals
- sulphate reduction and precipitation of sulfide minerals in organic substrates in peat bogs and wetlands
- sulphate precipitation as a product of sedimentary evaporative processes

Constructed sorption barriers involve the emplacement of materials with high sorption capacities in the contaminated flow path. Zeolites have a very high sorptive capacity and have received attention in the mining industry in North America for this reason. Essentially the zeolites act to absorb heavy, less reactive constituents and release lighter, more mobile elements, for example the absorption of metals such as Cu^{2+} and the release of Ca^{2+}. In the case of oxyanions and organic molecules absorption occurs by the expansion of the zeolite structure (Tsitsishvili *et al.* 1992; Bish & Guthrie 1993).

The use of local clays or fly ashes may also be effective. However, the difficulty in constructing

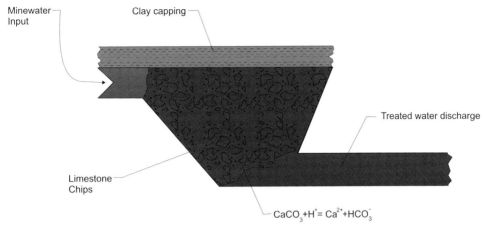

Minewater Input

Clay capping

Treated water discharge

Limestone Chips

$CaCO_3 + H^+ = Ca^{2+} + HCO_3^-$

Fig. 7. Cross-section of an anoxic limestone drain.

a sorptive barrier would limit potential applications to sites beneath new surface emplacements and across groundwater discharge areas. Even in these applications, the use of neutralizing barriers is likely to prove more cost-effective.

Constructed neutralization zones immobilize contaminants by causing their precipitation. In general, neutralizing zones are constructed by placing alkaline material in the contaminated flow path. As the pH of the contaminated waters rises, contaminants are precipitated from solution. Normally limestones are used for this purpose and contaminants are precipitated as either hydroxides (ferric hydroxide) or carbonates (zinc, cadmium, manganese).

In order to ensure continued neutralization it is necessary to avoid precipitation of blinding minerals on the surface of the alkaline material. This can be avoided if the water is anoxic and ferric iron and aluminium are not present in the water. Anoxic limestone drains are designed to achieve this (Fig. 7). These work by reacting $CaCO_3$ with the drainage acidity in an anaerobic environment such that ferric hydroxide precipitation will not occur. If it does then this leads to mantling of the limestone and effectively reduces the efficiency of the drain. This effect occurred at the Cwmstywth drain in central Wales and led to a complete failure of a limestone drain (Fuge 1993).

Sulphate reduction zones utilize the fact that under anoxic conditions metals may be removed from the mine waters as stable sulphide precipitates. Under these conditions sulphide minerals remain stable and have extremely low solubilities. Flooded underground mine workings and open pits can be anoxic, and as such provide a suitable environment for the implementation of a sulphate reduction system. In tailings and

waste/ore rock piles left exposed to the atmosphere the utilization of sulphate reduction barriers requires the emplacement of an oxygen consuming material over the spoil (Fig. 8). This acts to create the anoxic environment required within the body of the spoil to actively reduce sulphate to sulphide.

The biological reduction of sulphate to hydrogen sulphide is brought about by specialized strictly anaerobic bacteria and is accomplished primarily by two genera: *Desulfovibrio* (five species) and *Desulfotomaculum* (three species). These organisms have a respiratory metabolism in which sulphates, sulphites and/or other reducible sulphur compounds serve as the final electron acceptors, with the resulting production of hydrogen sulphide. The organic substrates for these bacteria are generally short chain acids such as lactic and pyruvic acid. In nature these substrates are provided through fermentative activities of other anaerobic bacteria on more complex organic substrates. Thus when oxygen is depleted where organic matter is available in the presence of sulphate, hydrogen sulphide production can be expected as follows:

$$SO_4^{2-} + 2CH_2O + H^+$$
$$\xrightarrow{\text{bacteria}} HS^- + 2H_2O + 2CO_2 \quad (1)$$

where CH_2O represents organic matter.

The production of sulphide ions results in the sulphide mineralization of a number of heavy metals. Precipitation of the corresponding metal sulphides proceeds as follows:

$$Fe^{2+} + HS^- \rightarrow FeS + H^+ \quad (2)$$

Due to the natural occurrence of the sulphate reducing bacteria (SRB), sulphate reduction can

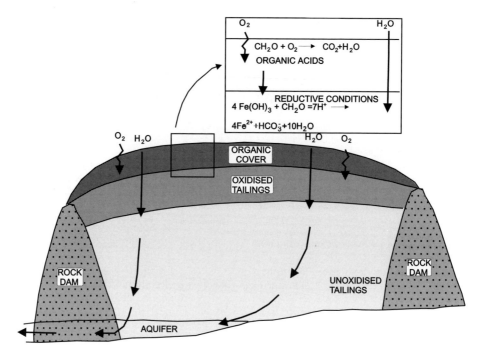

Fig. 8. Schematic diagram of a sulphate-barrier used in the chemical containment of tailings (from Blowes *et al.* 1994). Sulphate reduction occurs by reaction with organic carbon. Released sulphide reacts with metals to precipitate metal sulphides in the impoundment.

Table 6. *Biological sulphate reduction rates*

Reference	$g(SO_4^{2-})l^{-1}day^{-1}$	$g(SO_4^{2-})/gVSS\,day^{-1}$	Temperature (°C)	Nutrient source
Middleton & Lawerence 1977	2.25	0.11	35	acetic acid
Oleszkiewicz *et al.* 1986	1.5	–	35	cheese whey
Marree 1989	6.4	0.11	30.5	molasses
McIntire *et al.* 1990	0.029	–	–	wetland
Reynolds *et al.* 1991	0.23	–	18	sodium lactate
	0.413	–	18	hay extract
	0.087	–	18	mushroom compost
	0.115	–	18	manure & soil

be utilized *in situ* for the treatment of acid mine drainage provided the correct conditions can be maintained to sustain bacterial activity. Sealing shafts, adits and air vents may enhance anaerobic conditions. For conditions to be sufficiently anaerobic, however, it is likely that a significant depth of water will be required.

A wide variety of organic substrates have been investigated for this purpose including molasses, sewage sludge, straw, newspaper, sawdust and manure. Other possibilities are wastes from the chemicals industry such as short chain organic acids. The overall rate of sulphate reduction is dependent on various parameters; amongst these is the rate dictated by the concentration of active biomass. In turn, the bacterial concentration will be dependent on the availability from the substrate of nutrients and the efficiency by which bacteria can extract them. Some rates of sulphate reduction in the presence of different nutrient sources have been published (Table 6).

Bacterial growth can be inhibited by the presence of metals. The effect is highly variable with as little as $2\,mg\,l^{-1}$ copper found to inhibit

Desulfovibrio by at least 50% (Widdel 1987). However sulphate reduction has been found to be highly active in the Løkken mine, Norway despite zinc levels in excess of $2000 \, mg \, l^{-1}$ (Arnesen *et al.* 1991).

Because the bacteria consume the organic material, constant addition would be necessary. In underground mines this may be added by injection via shafts or boreholes (Arnesen *et al.* 1991). It is necessary to have a maximum retention time in the workings and good mixing for the treatment to be effective.

Sulphate precipitation zones may be constructed using an inorganic source which will produce a low-solubility sulphate phase. Both lime and barium salts have been proposed (Trusler 1988; Marree 1989; Marree & Bosman 1989; Bosman *et al.* 1991; Everett *et al.* 1994; Marree & DuPlessis 1994). A number of barium and calcium salts can be used but the most commonly proposed are carbonate and sulphide by reactions such as:

$$BaS\,(s) + SO_4^{2-} + 2H^+ \quad (3)$$
$$\rightarrow \; BaSO_4\,(s) + H_2S\,(g)$$

$$BaCO_3\,(s) + SO_4^{2-} \; \rightarrow \; BaSO_4\,(s) + CO_3^{2-}. \quad (4)$$

Because of the cost of barium and its environmental toxicity it is advantageous to have a barium recovery plant to recycle barium salts. A typical design would be to reduce the formed barium sulphate to barium sulphide by roasting with coal by a reaction such as:

$$BaSO_4\,(s) + 2C \; \rightarrow \; BaS\,(s) + 2CO_2\,(g). \quad (5)$$

This is then carbonated with CO_2 to produce a recycled batch of $BaCO_3$ by a reaction such as:

$$BaS\,(s) + CO_2\,(g) + H_2O \quad (6)$$
$$\rightarrow \; BaCO_3\,(s) + H_2S\,(g).$$

Liberated hydrogen sulphide can be used elsewhere in effluent treatment by precipitating insoluble metal sulphides. Because of its lower

Table 7. *Treatment of sulphate-rich effluents by barium and calcium salt precipitation, shown as % removal*

pH	Precipitating agent			
	CaO	BaCO$_3$	BaS	Ba(OH)$_2$
2.9	62.3	24.2	95.6	>100
7.9	80.5	>100	>100	>100
12	51	90.1	90.1	>100

Products are gypsium ($K_{sp} = 10^{-2.3}$) with CaO and barite ($K_{sp} = 10^{-9}$) with Ba compounds (Bowell *et al.* 1997).

sulphate solubility barium is a more useful cation than calcium for precipitation and barium sulphide is more widely applicable, especially when treating acidic drainage, than other salts (Table 7). The results in Table 7 show in some cases removal at >100% sulphate due to co-precipitation effects of the cation.

Discharge control

This refers to both active chemical treatment and passive treatment systems such as wetlands. There are numerous technologies and strategies available for water treatment for metals removal. They tend to be expensive in terms of capital and operating costs and should be considered a last resort. These include:

- dilute and disperse
- hydroxide precipitation
- reverse osmosis
- ion exchange
- chemical precipitation and adsorption
- biological systems (including wetlands).

These systems generally require temporary storage and regulation because the rate of discharge from mines and run off from mine waste tips is usually variable and water treatment systems have limited operating flexibility. Temporary storage capacity is required to retain extreme precipitation event flows until the water can be treated and so that a regulated feed rate to the system can be maintained.

In addition to storage it may be necessary to blend several streams of different flow rates and water quality so that steady operating conditions can be maintained. Storage facilities may require lining to prevent contaminated seepage.

Selected systems which may be relevant to mine water treatment are described briefly below.

Dilute and disperse. Commonly, in natural water systems, the peak flows of groundwater discharges and surface stream flows (rivers) do not coincide (Fetter 1986). Groundwater flows normally lag behind surface flows because of the flow retardation that occurs during infiltration from a rainfall event. Consequently, peak contaminant loadings from groundwater sources do not coincide with peak surface flows. By actively managing the discharge flows, limiting discharge to high flow periods only, contaminant levels in the receiving waters may be controlled to within the water quality objective. While this is not strictly a water treatment approach, the impact on the receiving environment could be reduced.

Chemical precipitation and adsorption. Most metals can be precipitated as their respective hydroxides or hydrated metal oxides. During neutralization a slurry can be aerated to oxidize reduced metals to oxidized forms to improve recovery of metals as more stable hydroxides.

In practice, the most frequently applied method for the treatment of iron-rich waters is oxidation of ferrous iron to ferric iron and simultaneous pH adjustment and hydroxide precipitation. Typical process configurations for hydroxide precipitation comprise a neutralization-precipitation stage, and a settling or clarification stage.

The pH adjustment is most commonly achieved by calcium hydroxide (added as a slurry). Sodium-based reagents preclude the formation of gypsum but result in higher total dissolved-metals levels for the treated discharge water stream. The neutralization-precipitation stage is typically completed in a single process step, complemented with pre-aeration or direct aeration. Final clarification is achieved by flocculation and settling of the iron hydroxide precipitates.

Settling equipment typically includes high rate thickeners or settling lagoons. While sludge densities of approximately 10% (w/w) are observed, high-density processes may achieve sludge densities in the order of 15% or higher (w/w). These processes incorporate sludge recycling to the thickener to promote precipitation and increase dewatering of the sludge. Filtering may be required to reduce the volume of the sludge further and so reduce disposal costs. Capital and operating costs for these can be high, however.

The use of hydrothermal precipitation with iron to remove oxyanions such as those of arsenic, antimony, vanadium and titanium, particularly ferric arsenate has been proposed (Swash & Monhemius 1995, 1996). This disposal option is based on geochemical considerations in that these minerals are stable, for example over 60% of arsenic minerals are arsenates of which

the most common is scorodite, $FeAsO_4.2H_2O$. A number of elaborate schemes have now been proposed, including MIRO's *Scorodite process* (MIRO 1994; Swash & Monhemius 1995, 1996). In these high temperature ($c.\,300°C$) precipitates arsenic is chemically incorporated into the structure of the crystalline material. This is much more stable than arsenic which has been 'stabilized' through surface adsorption with ferrihydrite.

Ferrihydrite is a disordered high surface area, iron hydroxide formed by ferrolysis at ambient temperatures. With time ferrihydrite becomes unstable and converts to goethite (Waychunas *et al.* 1993). Arsenical ferrihydrite are produced by the neutralization of effluents which contain As together with ferric iron. Low solubilities are associated with products with an Fe : As ratio of $>3:1$. Lower Fe : As ratios tend to produce a highly soluble product (Krause & Ettel 1985, 1989; Harris & Krause 1993; MIRO 1994). Adsorption of oxyanions in mildly acidic conditions is an effective method of attenuating some elements such as arsenic (Bowell 1994). However any environmental changes will lead to liberation of the contained species.

Fixation can also be achieved within a slag matrix (Broadbent *et al.* 1994; Machinganuta & Broadbent 1994; Kontopoulos *et al.* 1996). Recent work has attempted to stabilize up to 10 wt% As in slags. Solidification and stabilization is used for many hazardous materials. This involves the mixing of sludges with a cement binder to produce a solid which is structurally sound and relatively impermeable (Diel & Tolppi 1991; Emmett & Khoe 1994). For example it has been experimentally shown that up to 70% As_2O_3 (as Ca-arsenates) can be incorporated into cement, but high loadings reduce structural strength and solidification of the cement. In cement-stabilized materials iontransfer is reduced as porosity and permeability are reduced. The high levels of lime added increase Ca content and pH of test solutions

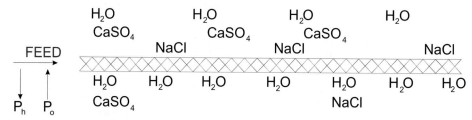

P_h = Hydraulic Pressure
P_o = Osmotic Pressure

Fig. 9. Schematic diagram of reverse osmosis mechanism.

leading to erroneous measurements of solubility. Cements can deteriorate rapidly and the overall characteristics of the waste need to be considered before this disposal option can be chosen. The cement requires a dry, low humidity, CO_2-free environment at constant temperature for long-term stability (Diel & Tolppi 1991).

The recovery of gold by the carbon-in-pulp (CIP) method utilizes the strong adsorption of $[Au(CN)_2^-]$ onto the surface of activated carbon in alkaline ($pH \geq 11$) regimes (Adams & Fleming 1989; Adams 1993). The same mechanisms can be utilized to adsorb contaminants from a water column. At the present time only experimental studies have been carried out (Hegenberger *et al.* 1987; McEnamey 1988; van der Merwe & van Deventer 1988). The technique has also been applied to volatile and dense organic solvents (Pankow & Cherry 1996).

Reverse osmosis. This is one of several membrane processes for the removal of ions from aqueous solutions. It involves applying a pressure in excess of the osmotic pressure to a concentrated solution on one side of a semipermeable membrane thus driving pure solvent through the membrane to a solvent stream on the other side (Fig. 9). In this way a stream of more concentrated solution is obtained together with an increased flow of the clean water. It must be noted that it is not possible to achieve complete separation of solute from solvent and careful consideration must be given to the disposal or reuse of the concentrated solution stream. It is also usual to treat the feed in order to remove all particles larger than about 0.25 mm, remove turbidity, control pH and temperature as these can have a significant influence on membrane life. Periodic cleaning of the membrane is also required. The values of important operating parameters for reverse osmosis systems are summarized in Table 8. The parameters are self-explanatory apart from the recovery factor which is the ratio of clean water flow rate (i.e. that which permeates the membrane) to the feed flow rate. The concentrated solution flow is then the difference between the feed and permeate flow.

Use of reverse osmosis is not widespread in the mining sector due to high operating costs in comparison to other options. Its use in plating systems for recovery of Cd, Ni and Cr is common where both the metals and the clean water are recycled. Feed streams with up to $10\,000\,\text{mg}\,\text{l}^{-1}$ total dissolved solids can be treated in modern systems where the membrane is usually in fine tubular form with water flowing from inside the tube to the outside.

Electrolysis. This process uses direct electric current across a stack of alternating cation and anion selective membranes. In the effluent, anions are attracted to the anode but cannot pass through anion-impermeable membranes and are thus concentrated. Cations move in the opposite direction and are impeded by cation-impermeable/anion-permeable barriers. The initial container has thus been depleted of salts and the cleaned water can be extracted. By the use of current reversal the process is greatly improved. The anode and cathode can be periodically changed as can the effluent and clean water channels. This reduces potential for membrane fouling and facilities regeneration of the membrane by self-cleaning. A major advantage of electrolysis over other RO techniques is that the system is not sensitive to effluent temperature or pH. Capital costs are reduced as are working costs due to lower working pressures.

Ion exchange. In classical wastewater treatment systems, ion exchange involves the removal

Table 8. *Summary of system operating parameters for RO plant*

Parameter	Range	Typical
Pressure (MPa gauge)	2.5–6.5	4
Temperature (°C)	15–38	21
Packing density ($m^2\,m^{-3}$)	160–1650	–
Flux ($m^3\,(\text{day}\,m^2)^{-1}$)	0.5–3.5	0.6–1.6
Recovery factor (%)	75–95	80
Salt rejection factor (%)	85–99.5	95
Membrane life (a)	–	2
pH	3–8	4.5–5.5
Turbidity (JTU)	–	1
Feedwater velocity ($m\,s^{-1}$)	0.01–0.8	–
Power utilization $kWh\,m^{-3}$	2–4.5	–

After Eckenfelder (1989).

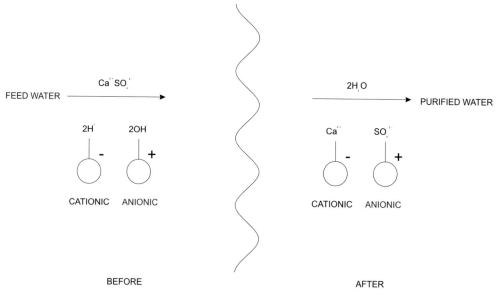

Fig. 10. Schematic diagram of ion exchange mechanism.

of cations from the water by exchange with H^+ or Na^+ and anions by exchange with OH^- ions on specially designed resins. These are in columns which require regeneration once breakthrough (i.e. appearance of the undesirable ion in the effluent) occurs (Fig. 10). The cost-effectiveness of the semi-batch process depends on the exchange capacity of the resin and the degree of regeneration required.

A continuous ion exchange system (see Fig. 11) nominally for the removal of calcium sulphate from water has been developed in South Africa by Chemical Effluent Treatment (Pty) Ltd (CHEMEFFCO) but although it has wide potential for application in the mining industry it has not been used at full scale to date (Everett *et al.* 1994). The process involves cascading the feed water through a series of tanks where the cationic resin is fluidized and transported counter currently by means of airlifts. The loaded resin is regenerated by means of 10% sulfuric acid in a batch reactor, creating gypsum for ease of disposal. The decationized water is contacted with air in a degassing tower to remove CO_2 before passing through the anion exchange fluidized bed cascade in a counter-current direction to the anionic resin. Anionic resin regeneration is performed batch wise with lime slurry, again creating gypsum for disposal. The product water should meet effluent disposal criteria and will be at neutral pH. The advantages claimed for the process are:

- low cost operation
- no need for feed pre-treatment
- simple chemical suite
- degree of desalination controlled by varying resin flow rate
- slurry discharge of waste for ease of disposal
- high water recoveries
- low resin loss
- other salts can be removed with slight modification of the flowsheet

Biological controls. Where metals are the problematical component of the contaminated mine water discharge, rather than salts *per se*, a number of bioadsorption aids are available to remove metals for subsequent disposal or recovery, including:

- Bio Fix Process: The US Bureau of Mines has developed porous polymeric beads containing immobilized biological materials, (non-living sphagnum moss) called Bio-Fix beads, which readily remove Cd, Zn, Mn, Cu and other metals from acidic waters (Wood 1995).
- Bioclaim: Another biological metal removal system is the Advanced Mineral Technologies AMT-Bioclaim. This is a granulated, non-living biomass product with a high capacity for soluble metal removal from wastewaters. Effluent is fed into a fluidized bed reactor containing the granules water under controlled conditions. Essentially, Bioclaim acts

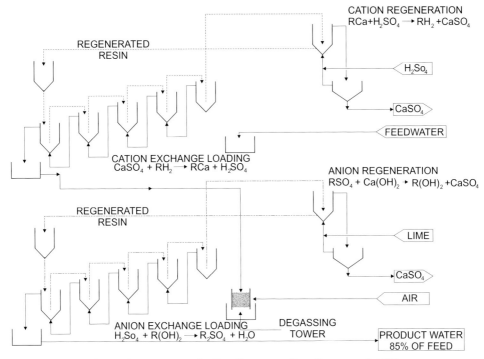

Fig. 11. Continuous ion-exchange system. The Gyp-Cix process (from Everett *et al.* 1994).

as a conventional expanded activated carbon, where the granules have a high capacity for accumulation of metal cations.

Constructed wetlands. Peat bogs and wetlands are known accumulators of metals (Kalin *et al.* 1991; Kleinmann & Pacelli 1991). Constructed wetlands, employ similar techniques to conventional treatment for the removal of metals and buffering of acidic drainage. Treatment of coal and metal mine drainage may involve aerobic or anoxic conditions. An important requirement of a wetland system is its ability to withstand both current and future metal loadings. Of major concern is the gradual infilling of the wetland basin and the potential for remobilization of contaminants.

Drainage in deep mines usually has a low dissolved oxygen content (typically $< 0.5\,\mathrm{mg\,l^{-1}}$) with prevalent reducing conditions. As a consequence sulphides are exposed to negligible oxidation and any dissolved iron is in the ferrous state. Increase of oxygen levels results in the hydrolysis of iron to the ferric form. This rate is controlled by dissolved oxygen levels, pH and iron speciation. Below pH 3 oxidation is slow and dominated by bacterial processes (Colmer

& Hinkle 1947; Kleinmann & Pacelli 1991; Gould *et al.* 1994). Above pH 5 oxidation is more rapid as ferric iron is present and can act as an oxidizer. Consequently, aerobic wetlands are designed to encourage oxidation and so are shallow ($< 0.3\,\mathrm{m}$), vegetated (reed beds) with predominantly surface flow (Fig. 12). Pretreatment of very acidic discharge is often advisable and also reduces the area of land required by the wetland because the rate of oxidation increases. A common method of pre-treatment is through an Anoxic Limestone Drain (ALD) consisting of lime or limestone sealed from air (Fig. 9). Alkalinity of the water is increased by calcite buffering of pH:

$$CaCO_3 + H^+ \quad \rightarrow \quad Ca^{2+} + HCO_3^- \qquad (7)$$

The need for anoxic conditions is to prevent hydrolysis of ferric salts which would mantle the limestone chips reducing the ALD efficiency. Elevated dissolved oxygen levels ($> 2\,\mathrm{mg\,l^{-1}}$) or presence of Fe^{+3} will cause precipitation of ferric salts, mantling limestone and reducing ALD efficiency. Removal of other metals will require an anaerobic wetland to encourage precipitation of insoluble sulphides.

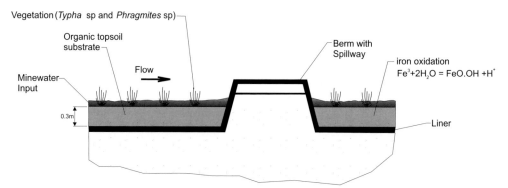

Fig. 12. Section through an aerobic wetland.

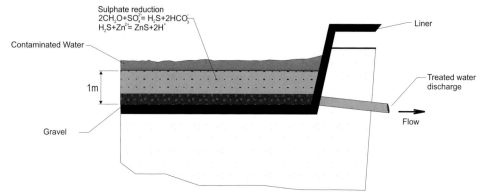

Fig. 13. Section through an anaerobic wetland.

Anaerobic systems require metal-rich waters to flow through a layer of organic material under anaerobic conditions (Fig. 13). The organic layer can consist of sewage, manure, sawdust or compost. It must contain the sulphate-reducing bacteria, *Desulfovibrio*. The bacterial acceleration of sulphide oxidation can be significant with contributions from many other different species including *Thiobacilli* (Gould *et al.* 1994). *Desulfovibrio desulfuricans* readily reduce sulphur species to sulphide and can participate in secondary sulphide production (Kleinmann & Pacelli 1991).

The mechanisms resulting in dissolved metal immobilization are two stage:

$$2CH_2O + SO_4^{2-} \rightarrow H^+ + HS^- + 2HCO_3^- \quad (8)$$

where CH_2O is the labile organic material. Hydrogen sulphide is produced which reacts with dissolved metals (Me represents a divalent metal cation) by a mechanism such as:

$$Me^{2+} + HS^- \rightarrow MeS(s) + H^+. \quad (9)$$

Organic carbon is replenished by annual growth of wetland plants and degradation products. The production of alkalinity as a result of sulphate reduction is an additional pre-treatment for an aerobic system and may be useful if the efficiency of an ALD is reduced. It is important to maintain anaerobic conditions in this process.

Aluminium is unlikely to form a sulphide but on sulphate reduction may be precipitated as a hydroxide or adsorbed by the organic substrate. Similarly MnS has a higher solubility than other sulphides and is unlikely to precipitate unless other metals are removed first or are present in low concentrations. In aerobic systems Mn is not rapidly oxidized, unlike iron, and may pass almost entirely through an aerobic wetland. A separate passive treatment would be required where Mn concentration is high enough to precipitate the metal by a mechanism such as:

$$Mn^{2+} + HCO_3^- = MnCO_3(s) + H^+ \quad (10)$$

$$2MnCO_3(s) + O_2 = 2MnO_2(s) + 2CO_2. \quad (11)$$

Precipitation would increase with pH. When dealing with a closed mine with few or no records preserved it is not possible, in general, to prevent or contain the toxic waste. In this case

Table 9. *Performance results, Whitworth Number 1 adit wetland (10/95–8/97)*

Parameter		Inlet	Cell 1	Cell 2	Cell 3	Cell 4	Outlet
pH	range	5.68–7.24	6.76–7.95	6.62–7.84	6.43–7.92	6.75–7.92	6.67–8.13
	average	6.41	7.28	7.28	6.86	7.25	7.26
Fe	range	11.7–37.4	0.51–16.3	0.31–16.3	1.29–15.2	0.57–7.33	0.39–9.95
	average	24.59	3.85	5.1	7.96	1.89	4.59
SO_4	range	249–471	250–460	236–442	235–539	81–461	81.1–461
	average	355	316	309	329	330	301

Concentration given in $mg\,l^{-1}$.

the best available practice is to treat the mine drainage. This was the case at Wheal Jane where discharged acidic metal-rich mine water is currently being treated by conventional metal precipitation processes at a current annual cost of £750 000 pa. It is hoped in the future that passive treatment cells comprising anaerobic and aerobic wetland cells can be used (Cambridge 1996). Such a scheme is currently in operation as a pilot plant in the Pelenna coalfield in South Wales (Edwards *et al.* 1997*a, b*). Here a system of four pilot cells has been designed in series principally to remove iron, although sulphate is also removed from the system (Table 9).

Iron retention by hydroxide precipitation is active on the surface of all four beds. Some sulphate reduction does occur in beds 2 and 4; however, gypsum precipitates have also been observed on the surface of the treatment cells. Cell 3 currently has a higher through flow rate than was designed and consequently is performing inadequately due to poor control on the cell discharge. The overall performance appears successful with a reduction from $22\,mg\,l^{-1}$ Fe to 2 or $3\,mg\,l^{-1}$. The reduction of $20\,mg\,l^{-1}$ Fe occurs in an area of $200\,m^2$ and a flow of $0.75\,l\,sec^{-1}$. This is equivalent to about $7\,mg\,m^{-2}\,day^{-1}$.

Selection of appropriate technology

A review on mine waste containment would not be complete without reference to criteria by which suitable containment is applied to a site. Ideally these cover the following aspects (Connelly *et al.* 1995):

- identification of options which would be of potential benefit
- selection of individual options which will achieve an acceptable contaminant load for discharge to receiving waters
- selection of combinations of options which will fully remediate the discharge

- environmental impact assessment of remediation works
- risk assessment of success/failure of a particular option
- assessment of Net Present Value (NPV) costs for the options on the basis of capital and operating costs and long-term interest rates
- combination of the above as cost benefit analysis
- selection of one or more options for further investigation/implementation.

It is often the case that not all of the above are justified. For example, where the level of remediation is determined by receiving water quality objectives the benefits are equivalent for each case and this can be removed from the analysis, or where the scale of the problem does not justify such rigorous analysis.

However, for most, if not all, mining operation, the 'cost' is by far the most important selection criterion in selecting a containment strategy for waste. In a feasibility study, all the costs are assessed including the costs of environmental protection. The approach to the design of any aspect of the mine is generally BATNEEC (Best Available Technology Not Entailing Excessive Cost). The detail will be a function of the planning authority requirements and the philosophy of the mining company.

However, a mine is developed on the basis of a return on investment for the investors. There is therefore a limit as to the environmental costs beyond which the mine will not go ahead. For example, the Crown Butte mine, Montana, USA, found that despite a very rich Cu–Ag–Au deposit, the environmental costs associated with containment of waste and protection of the environment made the deposit uneconomic. Although risk assessments are done, the methods of assessment and monitoring cannot be considered as exhaustive and engineering decisions are made at some stage of investigation, which are commensurate with the level of investment in the mining project itself.

Table 10. *Comparison of cost and efficiency for effluent treatment*

Technique	Salt removed (%)	Water recovery (%)	Cost*	
			Operating	Capital
Electro dialysis	80	80	2.81	4.25
Tubular RO	80	85	2.61	5.03
GYPCIX				
(TDS > 2 g l^{-1})	90	90	1	3.27
(TDS 2–6.5 g l^{-1})	80	80	1.93	3.93
BaCO$_3$	60	50	1.79	2.33
BaS	75	85	1.85	2.43
Lime	50	40	1.65	1.59
Wetland	55	70†	0.2	0.8

Adapted from Bowell *et al.* 1997.
* Cost in MUS$ gallon day.
† Wetland discharges treated water and does not usually recycle water which is often used to dilute untreated overflow effluent.

Conclusions

The recognition that mining of sulphide-bearing material has the potential to generate low-quality drainage for long periods has led to alternative proposals for management and prediction. These proposals involve various options for active and passive treatment of mine waste. Selection of the most appropriate criteria involves an understanding of hydrogeological and geochemical characteristics.

The remediation of the environmental effects of mine drainage include collection and treatment of waters using conventional treatment facilities, passive downstream treatment involving constructed wetlands or a porous reactive wall. Controls on sulphide oxidation include the mantling of sulphides, oxygen-diffusion barriers and use of bactericides. Source and migration controls are the preferred option but source control particularly is very difficult to achieve.

Conventional treatment methods are still the most widely used for metal mines, and among the various techniques in controlling sulphide oxidation, water storage seems to be the most utilized.

It is beneficial to assess the potential for source and migration control and implement them if appropriate, before mines are closed.

Discharge control, particularly active chemical treatment, may involve significant operating costs at a closed site for many years.

Selection of the appropriate system of remediation requires a thorough understanding of the hydraulic system and the hydrogeochemical mechanisms of discharge development.

The authors would like to express their thanks to numerous colleagues in SRK and in the University College of Wales including; D. Bentel, J. Cowan, Dr M. Dey, R. Dorey, A. McCraken, G. Muller, J. Parshley, Professor F. Pooley, R. Stuart and Dr A. Wood. A. P. Edwards is acknowledged for providing more recent information on the performance of the Pelenna wetlands and J. Barta of Getchell Gold for review and discussion. Review and comment of this manuscript by C. Rochelle, D. Banks and two reviewers has greatly improved the manuscript.

References

ACHMED, S. M. 1991. Electrochemical and surface chemical methods for the prevention of atmospheric oxidation of sulfide tailings and acid generation. *Proceedings of the Second International Conference on Abatement of Acidic Drainage*, MEND Secretariat, Ottawa, Ontario. **2**, 305–319.

ADAMS, M. D. & FLEMING, C. A. 1989. The mechanisms of aurocyanide adsorption onto activated carbon. *Metallurgical Transactions*, **20B**, 315–325.

——1993. Influence of surface chemistry and structure on adsorption by activated carbon. *XVIII International Mineral Processing Conference*, Sydney 23–28 May 1993, 1175–1187.

AGRICOLA, G. 1556. De Re Metallica, (trans. HOOVER, H. C. & HOOVER, L. H.) Dover, NY.

ALPERS, C. A. & NORDSTROM, D. K. 1991. Geochemical evolution of extremely acid mine waters at Iron Mountain, California: Are there any lower limits to pH? *Proceedings of the Second International Conference on Abatement of Acidic Drainage*, MEND Secretariat, Ottawa, Ontario. **2**, 321–342.

ANGUS ENVIRONMENTAL. 1991. *Review of Environmental Quality Standards*. Science Services, Inland Water Directorate, Ottawa.

APPELO, C. A. & POSTMA, D. 1993. *Geochemistry, Groundwater and Pollution*. Balkema, Rotterdam.

ARNESEN, R. T. *et al.* 1991. Monitoring water quality during filling of the Løkken mine: Role of sulfate reducing bacteria in metals removal. *2nd Int. Conf. Abatement Acid Drainage*, Ottawa, Canada, 16–18/9/91.

ARNOULD, M, BARRES, M & COME, B (eds) 1993. Geology and Confinement of Toxic Wastes, *Proceedings of 'Geoconfine 93', Montpellier.* Balkema, Rotterdam.

BENTLEY, S. P. (ed.) 1996. *Engineering Geology of Waste Disposal.* Geological Society, London, Engineering Geology Special Publication, **11**.

BERENTSEN, E. J., NANNA, R. F., HAZLITT, S. J. & ESTES, L. D. 1996. Discovery and geology of the Turquoise Ridge gold deposit. *Mining Engineer,* October 1996, 31–35.

BIGHAM, J. M. 1994. Mineralogy of ochre deposits. *In:* JAMBOR, J. L. & BLOWES, D. W. (eds) *Environmental Geochemistry of Mine Waste.* Mineralogical Association of Canada, Ottawa, 103–131.

BISH, D. L. & GUTHERIE, G. D. 1993. Mineralogy of clay and zeolite dusts. *In:* GUTHRIE, G. D. & MOSSMAN, B. T. (eds) *Health effects of mineral dusts,* Mineralogical Society of America, Reviews in Mineralogy, **28**, 139–184.

BLOWES, D. W. & JAMBOR, J. L. 1990. Pore water geochemistry and mineralogy of the vadose zone in sulfide tailings, *Applied Geochemistry,* **5**, 327–346.

——, REARDON, E. J., JAMBOR, J. L. & CHERRY, J. A. 1991. The formation and potential importance of cemented layers in inactive mine tailings. *Geochimica et Cosmochimica Acta,* **55**, 965–978.

——, PTACEK, C. J. & JAMBOR, J. L. 1994. Remediation of low quality drainage in tailings impoundments. *In:* JAMBOR, J. L. & BLOWES, D. W. (eds) *Environmental Geochemistry of Mine Waste.* Mineralogical Association of Canada, Ottawa, 365–379.

BOSMAN, D. J., CLAYTON, J. A., MAREE, J. P. & ADLEM, C. J. L. 1991. Removal of sulfate from mine water. *In: Proceedings of the ARD in Pyrite Environments,* Lisbon, Portugal, 16–19 September 1990, 211–221.

BOWELL, R. J. 1994. Arsenic sorption by Iron oxyhydroxides and oxides. *Applied Geochemistry,* **9**, 279–286.

—— & BRUCE, I. 1995. Geochemistry of ochre and mine waters, Levant mine, Cornwall. *Applied Geochemistry,* **10**, 237–250.

——, WARREN, A., MJENGERA, H. A. & KIMARO, N. 1995. Environmental impact of metal mining on wildlife in the Serengeti national park, Tanzania. *Biogeochemistry,* **28**, 131–160.

——, FUGE, R., CONNELLY, R. J. & SADLER, P. K. J. 1996. Controls on ochre chemistry and precipitation in coal and metal mines. *Minerals, Metals & the Environment,* Prague, September 3–6 1996. Institution of Mining and Metallurgy, 293–322.

——, ELLIS, J., CONNELLY, R. *et al.* 1997. Options for Sulfate control and remediation. *Proceedings of 4th International Symposium on Environmental Geochemistry.* Vail, Colorado. October 1997.

BROADBENT, C. P. & WARNER, N. A. 1985. Review of the current status of energy recovery from slag. *Institute of Mining Engineers conference 'Energy and the Process Industry, London, June 1985',* 23–29.

——, MACHINGANUTA, N. C., VAN DEN KERKHOF, A. R. F. & BONGERS, E. A. M. 1994. Heavy metal immobilization in slags. *Pyrometallurgy for complex materials and wastes.* TMS, USA.

BURBANK, A. L., OLIVER, D. J. & PRISBREY, K. A. 1990. Bacterial cycling in Acid Mine Drainage Design for Closure. GAC-MAC Metting, May 1990, 463–470.

BURGRESS, S. G. & WOOD, L. B. 1961. Pilot plant studies in production of sulfur from sulfate-rich sewage. *Journal of the Science of Food and Agriculture,* **20**, 326–341.

CABRI, L. J., CHRYSSOULIS, S. L., DEVILLERS, J. P. R., LAFLAMME, J. H. G. & BUSECK, P. R. 1989. The nature of 'invisible' gold in arsenopyrite. *Canadian Mineralogist,* **27**, 353–362.

——, ——, CAMPBELL, J. L. & TEESDALE, W. J. 1991. Comparison of *in situ* gold analyses in arsenian pyrite. *Applied Geochemistry,* **6**, 225–230.

CAMBRIDGE, M. J. 1996. Development of treatment strategy for acid mine drainage at Wheal Jane, Cornwall. *Minerals, Metals & the Environment.* Prague, September 3–6 1996. Institution of Mining and Metallurgy, 337–340.

CHAPMAN, J. T. & HOCKLEY, D. E. 1996. Innovations in asssessment of mine waste acid generation characteristics. *Minerals, Metals & the Environment.* Prague, September 3–6 1996. Institution of Mining and Metallurgy, 323.

COLMER, R. M. & HINKLE, M. E. 1947. The role of microorganisms in acid mine drainage. *Science,* **106**, 253–256.

CONNELLY, R. J., HARCOURT, K. J., CHAPMAN, J. & WILLIAMS, D. 1995. Approach to remediation of ferruginous discharge in the South Wales coalfield and application to closure planning. *Minerals Industry International,* May 1995, 43–48.

COSTECH 1990. *Manual for the Assessment of Acid Rock Drainage.* British Columbia Acid Mine Drainage Task Force.

COTTER-HOWELLS, J. & THORNTON, I. 1991 Sources and pathways of environmental lead to children in a Derbyshire mining village *Environmental Geochemistry & Health,* **13**, 127–135

CRAVOTTA, C. A., III. 1994. Secondary iron-sulfate minerals as sources of sulfate and acidity. *In:* ALPERS, C. N. & BLOWES, D. W. (eds) *Environmental Geochemistry of Sulfide Oxidation.* American Chemical Society Symposium Series, **550**, 345–364.

DWAF. 1993. *South African Water Quality Guidelines,* 4 vols. Department of Water Affairs and Forestry, South Africa.

DIEL, B. N. & TOLPPI, J. B. 1991. Kinetics and mechanics of metal leaching from cement matrices. *Metals Behaviour and Treatment,* **5**, 126–131.

ECKENFELDER, W. W. 1989. *Industrial Water Pollution Control.* McGraw Hill, New York.

EDWARDS, P. J., BOLTON, C. P., RANSON, C. M., & SMITH, A. C. 1997a. The river Pelenna minewater treatment project. Proceedings of the CIWEM conference on wetlands for minewater treatment, Sheffield, September 1997.

——, BOWELL, R. J., CONNELLY, R. J., SADLER,P. J. K. & WOOD, A. (1997*b*) Biogeochemical mechanisms of metal and sulfur removal in natural and constructed wetlands *In: Proceedings of 4th International Symposium on Environmental Geochemistry*. Vail, Colorado. October 1997

ENGINEERING AND MINING JOURNAL 1989. Ghana's golden glow. June 1989.

——1996. News articles. June 1996.

EMMET, M. & KHOE, G. 1994. Environmental stability of As-iron hydroxides. *Proceedings EPD Congress 1994*, 153–166.

EPPS, J. 1993. Site Reclamation. *In: United Nations Interregional Workshop on Environmental Management in Developing countries, Konkola*, Zambia 6–10/12/93. **A.7**.

ERICKSON, P. M. & LADWIG, J. 1985. *Control of Acid Formation by inhibition of bacteria*. Final Report, West Virginia Department of Natural Resources.

EVERETT, D. J., DU PLESSIS, J. & GUSSMAN, H. W. 1994. The removal of salt from underground mine waters. *Mining Environmental Management*, March 1994, 12–14.

FETTER, R. 1986. *Applied Hydrogeology*. Prentice-Hall, Englewood Cliffs, NJ.

FLANAGAN, P. J. 1990. *Parameters of water quality: Interpretation and standards*. Environmental Resources Unit, Dublin.

FUGE, R. 1993. Acid Mine Drainage: A UK perspective. oxidation *In: Short Course Notes, Sulfide Oxidation and the Generation of Acid Mine Drainage, Proceedings of the Mineral Deposit Study Group Annual Meeting*, London, 13–18/12/93.

——, PEARCE, F. M., PEARCE, N. G. & PERKINS, W. T. 1993. Geochemistry of Cd in the secondary environment near abandoned metalliferous mines, Wales. *Applied Geochemistry*, Supplement 2, 29–35.

GOULD, W. D., BECHARD, G. & LORTIE, L. 1994. The nature and role of microorganisms in mine drainage. *In: JAMBOR, J. L. & BLOWES, D. W. (eds) Environmental Geochemistry of Mine Waste*. Mineralogical Association of Canada, Ottawa, 185–199.

HARCOURT, K. & WICKHAM, S. 1996. Omai aftermath: A case study. *In: Minerals, Metals & the Environment*. Prague, September 3–6 1996. Institution of Mining and Metallurgy, 435–451.

HARRIS, G. B. & KRAUSE, E. 1993. The disposal of arsenic from metallurgical processes: its status regarding ferric arsenate. *Extractive Metallurgy of Nickel, Cobalt, and Associated Metals*. TMS Annual Meeting, Denver, February 21–25. Abstracts volume.

HEGENBERGER, E., WU, N. L. & PHILIPS, J. 1987. Interaction between iron and activated carbon. *Journal of Physical Chemistry*, **19**, 5067–5071.

HOCKLEY, D., PAUL, M., CHAPMAN, J., JAHNS, S. & WEISE, W. 1997. Relocation of waste rock to the Lichtenberg pit near Ronneburg, Germany. *ICARD Proceedings*, Vancouver, June 1–5, 1997.

HUANG, X. & EVANGELOU, V. P. 1994. Suppression of pyrite oxidation rate by phosphate addition. *In: ALPERS, C. N. & BLOWES, D. W. (eds) Environmental Geochemistry of Sulfide Oxidation*. American Chemical Society Series **550**, 562–573.

JUBY, G. J. G. & PULLES, W. 1990. *Evaluation of Electrodialysis reversal for desalination of brackish mine water*. WRC Report **179/1/90**.

KAGEY, B. T. & WIXSON, B. G. 1983. Health implications of coal development. *In: THORNTON, I. (ed.) Applied Environmental Geochemistry*. Kluwer, Amsterdam, 463–480.

KALIN, M., CAIRNS, J. & MCCREADY, R. G. L. 1991. Ecological engineering for acid drainage treatment of coal wastes. *Resources Conservation and Recycling*, **5**, 265–275.

KLEINMANN, R. L. P. & PACELLI, R. R. 1991. Biogeochemistry of acid mine drainage. *Mining Engineering*, **33**, 300–6.

KOERNER, R, 1990. *Designing with Geosynthetics*, 2nd. edn, Prentice Hall, Englewood Cliffs, NJ.

KONTOPOULOS, A., KOMNITSA, K. & XENDIS, A. 1996. Environmental characterisation of lead smelter slags in Laviron. *In: Minerals, Metals & the Environment*, Prague, September 3–6, 1996. Institution of Mining and Metallurgy, 405–420.

KRAICHEVA, M. E. 1996. Bulgarian copper mining and the environment. *Minerals, Metals & the Environment*, Prague, September 3–6, 1996. Institution of Mining and Metallurgy. 43–54.

KRAUSE, E. & ETTEL, V. A. 1985. Ferric arsenates – are they safe ? *CIM Annual Meeting*, Vancouver, Canada, 1–20.

—— & ——1989. Solubilities and stabilities of ferric arsenate compounds. *Hydrometallurgy*, **22**, 311–57.

KUO, S., HEILMAN, P. E. & BAKER, A. S. 1983. Distribution and forms of copper, cadmium, iron and manganese in soils near a copper smelter. *Soil Science*, **135**, 101–109.

LEVY, D. B., CUSTIS, K. H., CASEY, W. H. & ROCK, P. A. 1997. A comparison of metal attenuation in mine residue and overburden material from an abandoned copper mine. *Applied Geochemistry*, **12**, 203–212.

LOWSON, R. I. 1982. Aqueous oxidation of pyrite by molecular oxygen. *Chem. Rev.*, **82**, 461–497.

MCENAMEY, B. 1988. Adsorption and structure in carbons. *Carbon*, **26**, 267–274.

MCINTIRE, P. E. & EDENBORN, H. M. 1990. Sulfur cycling by bacteria. *Proceedings 1990 Mining and Reclamation Conference*. 409–415.

MACHINGANUTA, N. C. & BROADBENT, C. P. 1994. Incorporation of As into silicate slags. *Trans. I.M.M*, **103**, C1–8.

MALHOTRA, V. M. 1991. Fibre enforced high volume fly ash shotcrete for controlling aggressive leachates from exposed rock surfaces and mine tailings. *In: Proceedings of the Second International Conference on the Abatement of Acidic Drainage*. MEND Secretariat, Ottawa, Ontario, **2**, 27–41.

MANN, A. 1983. Hydrogeochemistry and weathering on the Yilgarn block, Western Australia- ferrolysis and heavy metals in continental brines. *Geochimica et Cosmochimica Acta*, **47**, 181–190.

MARREE, J. P. 1989. *Sulfate removal from industrial effluents*. PhD Thesis, University of the Orange Free State.

—— & BOSMAN, D. J. 1989. Chemical removal of sulfate, calcium and metals from mining and power station effluents. *Water Sewage and Effluent*, **9**, 10–25.

MARREE, J. P. & DU PLESSIS, P. 1994. Neutralization of acid mine water with calcium carbonate. *Water Science and Technology*, **29**, 285–296.

MCGILL, S. L. & COMBA, P. G. 1990. A Review of Existing Cyanide Destruction Practices, USBM, Reno Research Centre. *Presented at Nevada Mining Association – Wildlife/Mining Workshop*. Reno, Nevada, March 29.

MIDDLETON, A. C. & LAWERENCE, A. W. 1977. Kinetics of microbial sulfate reduction. *Journal of Water Pollution*, 1659–1670.

MIRO. 1994. *Final Report, Phase II, RC 59: Arsenate Stability*. MIRO, Litchfield, Staffordshire.

MORSE, J. W. 1983. The kinetics of calcium carbonate dissolution and precipitation. *In*: REEDER, R. J. (ed.) *Reviews in Mineralogy*. Mineralogical Society of America, **11**, 227–264.

MURPHY, W. M. 1997. Are pit lakes susceptible to limnic eruptions? *Tailings and Mine Waste '97*. Balkema, Rotterdam, 543–548.

NICHOLSON, R. V., GILLHAM, R. W., CHERRY, J. A. & REARDON, E. J. 1989. Reduction of acid generation in mine tailings through the use of moisture-retaining cover layers as oxygen barriers. *Canadian Geotechnology Journal*, **26**, 1–8.

NORDSTROM, D. K. 1982. Aqueous pyrite oxidation and the consequent formation of secondary minerals. *In*: *Acid Sulfate Weathering*. Soil Science Society of America, 37–56.

OLESZKIERICZ, J. A. 1986. Anaerobic treatment of high sulfate wastes. *Canadian Journal of Civil Engineering*, August, 423–428.

PANKOW, J. F. & CHERRY, J. A. 1996. *Dense Chlorinated Solvents*. Waterloo Press, London, Ontario.

PRICE, J., SHEVENELL, L., HENRY, C. D. *et al.* 1995. *Water Quality at Inactive and Abandoned Mines in Nevada*. Report of a cooperative project among state agencies. Nevada Bureau of Mines and Geology Open File Report, 95-A.

RICKS, G. & CONNELLY, R. J. 1996. Preparation of an environmental action plan for the Erdenet Cu-Mo mine, Mongolia. *Minerals, Metals & the Environment*, Prague, September 3–6 1996. Institution of Mining and Metallurgy, 141–162.

REYNOLDS, J. S. *et al.* 1991. Sulfate reduction in a constructed wetland. *American Chemical Society*, Atlanta, GA., April 1991, 14–19.

ROBERTSON, R. M. 1987. *Proceedings of the 11th Annual BC Mine Reclamation Symposium*, April 1987, Campbell River, B.C.

ROBINS, R. G., BERG, R. B., DYSINGER, D. K. *et al.* 1997. Chemical, physical and biological interaction at the Berkerley Pit, Butte, Montana. *In*: *Tailings and Mine Waste '97*. Balkema, Rotterdam, 529–541.

RUNNELLS, R. G., DUPON, D. P., JONES, R. L. & CLINE, D. J. 1997. Determination of natural background concentrations at mining sites *Tailings and Mine Waste '97*. Balkema, Rotterdam, 471–478.

SADLER, P. J. K. 1997. Minewater remediation at a French zinc mine: sources of acid mine drainage and contaminant flushing characteristics. *In*: MATHER, J., BANKS, D., DUMPLETON, S. & FERMOR, M. (eds) *Groundwater Contaminants and their Migration*. Geological Society, London, Special Publications, **128**, 101–120.

SCHUILING, R. D. & VAN GAANS, P. 1997. The waste sulfuric acid lake of the TiO$_2$-plant at Armyansk, Crimea, Ukraine. Part 1: Self-sealing as an environmental protection mechanism. *Applied Geochemistry*, **12**, 181–186.

SCHWERTMANN, U. & CORNELL, R. M. 1991. *Iron oxides in the laboratory*. Verlag Chemie, Weinheim, Germany.

SCOTT, K. M. 1987. Solid solution in, and classification of, gossan-derived members of the alunite-jarosite family, Queensland, Australia. *American Mineralogist*, **72**, 178–187.

SIDES, A. 1996. Environmentally acceptable gold mining. *Minerals, Metals & the Environment, Prague, September 3–6 1996*. Institution of Mining and Metallurgy. 109–116.

SMITH, A. & MUDDER, T. 1991 *Chemistry and Treatment of Cyanidation Wastes*. Mining Journal London.

STEFFEN, ROBERTSON and KIRSTEN, (B.C.) Inc, Norecol Environmental Consultants and Gormerly Process Engineering 1989. *Draft Acid Rock Drainage Technical Guide*, vol. 1. British Columbia Acid Mine Drainage Task Force. Victoria, B.C.

STUMM, W. & MORGAN, L. J. 1985. *Aquatic Chemistry*. 3rd edn. Wiley, New York.

SWASH, P. M. & MONHEMIUS, A. J. 1995. Hydrothermal precipitation, characterization and solubility testing in the Fe-Ca-AsO$_4$ system. *In*: *Sudbury '95*.

—— & ——1996. The characteristics of calcium arsenate compounds and their relevance to industrial waste disposal. *Minerals, Metals and the Environment II*, 353–362.

TASSE, N., GERMAIN, M. & BERGERON, M. 1994. Composition of interstitial gases in wood chips deposited on reactive mine tailings. *Environmental Geochemistry of Sulfide Oxidation*. American Chemical Society Symposium Series, **550**, 631–634.

THOMAS, J. 1996. Excursion guide to uranium ISL at Straz pod Ralskem. *Minerals, metals and the environment conference*, Prague September 3–9 1996. IMM-CZS.

THORNHILL, M. & WILLIAMS, K. P. 1995. *The Multigravity Separators as a Trial for Reprocessing of Metalliferious Spoil*. Institute of Chemical Engineers Research Event, Edinburgh, **2**, 598–600.

THORNBER, M. R. 1992. Chemical processes during weathering. *In*: BUTT, R. M. C. & ZEIGERS, H. (eds) *Handbook of Exploration Geochemistry, vol 4: Regolith Exploration Geochemistry in Tropical Terrains*, Elsevier, Amsterdam, 65–99.

——1993. Electrochemical aspects of sulfide oxidation *In*: *Short Course Notes, Sulfide Oxidation and the Generation of Acid Mine Drainage, Proceedings of the Mineral Deposit Study Group Annual Meeting.* London, 13–18/12/93.

TRESCASES, J.-J. 1992. Chemical weathering *In*: BUTT, C. & ZEEGERS, H. (eds) *Handbook of Exploration Geochemistry, vol 4: Regolith Exploration Geochemistry in Tropical Terrains.* Elsevier, Amsterdam.

TRUSLER, G. E. 1988. *Chemical removal of sulfates using barium salts.* MEng Thesis, University of Natal.

TSITSISHVILI, G. V., ANDRONIKASHVILI, T. G., KIROV, G. N. & FILIZOVA, L. D. 1992. *Natural Zeolites.* Series in Inorganic Chemistry, Ellis Horwood, Chichester.

USNAS 1977. *Drinking water and health.* 2 vols. National Academy of Science.

USEPA 1995. *Guidelines for Ecological Risk Assessment.*

VAN GAANS, P. & SCHUILING, R. D. 1997. The waste sulfuric acid lake of the TiO_2 plant at Armyansk, Crimea, Ukraine. Part II: Modelling the chemical environment. *Applied Geochemistry*, **12**, 187–202.

VAN DER MERWE, P. F. & VAN DEVENTER, J. S. J. 1988. The influence of oxygen on the adsorption of metal cyanides on activated carbon, *Chemical Engineering Communications*, **65**, 121–128.

VISSER, W. J. 1993. *Contaminated Land Policies in some countries.* Technical Soil Protection Committee *TCBRO2*. The Hague.

WAYCHUNAS, G. A., REA, B. A., FULLER, C. C., & DAVIS, J. A. 1993. Surface chemistry of ferrihydrite: Part 1: EXAFS studies of the geometry of coprecipitated and adsorbed arsenate. *Geochimica et Cosmochimica Acta*, **57**, 2251–2269.

WIDDEL, F. 1987. New types of acetate oxidizing sulfate reducing *Desulfobacter*-species. *Arch. Microbiol.*, **148**, 286–291.

WILLIAMS, T. M. & SMITH, B. 1994. Report on visit to Globe and Phoenix mine, Zimbabwe. *British Geological Society Overseas Geological Survey Report* **WC/94/65/R**.

WHO 1984. *Guidelines for Drinking Water quality.* 2 vols. WHO, Geneva.

WOOLLACOTT, L. C. & ERIC, R. H. 1994. *Mineral and Metal Extraction: An overview.* SAIMM, Monograph **M8**, Johannesburg.

WOOD, A. 1995. Passive Treatment. *Workshop Notes on Solutions to Problems of Acidic Mine Drainage.* Workshop organized by Steffen Robertson & Kirsten, Cardiff. July 1995.

YOUNGER, P. L. 1995. Hydrogeochemistry of minewaters flowing from abandoned coal workings in County Durham. *Quarterly Journal of Engineering Geology*, **28**, S101–S113.

Toxic metal mobility and retention at industrially contaminated sites

L. JARED WEST,[1] D. I. STEWART,[2] J. R. DUXBURY[1]
& S. RICHARD JOHNSTON[2]

[1] *Department of Earth Sciences, University of Leeds, Leeds LS2 9JT, UK*
[2] *Department of Civil Engineering, University of Leeds LS2 9JT, UK*

Abstract: This paper is concerned with industrial sites contaminated with heavy metals, specifically cadmium (Cd), chromium (Cr), arsenic (As), mercury (Hg), lead (Pb), copper (Cu), nickel (Ni) and zinc (Zn), as these are the heavy metal contaminants that cause most concern. It is intended as a source of information to aid engineers and geologists concerned with risk assessment and remediation at such sites. The risks from heavy metals depend strongly on their chemical form, and therefore literature on the chemical forms of metals released at industrial sites is reviewed. Changes in the chemical forms of heavy metals resulting from environmental interactions are discussed, and methods for determining metal chemistry (leaching tests, chemical extractions, direct methods) are outlined. The paper concludes with a case study where the chemical forms of heavy metals at a former metal smelting and vehicle scrapyard site are investigated using a scanning electron microscope with energy dispersive X-ray spectroscopy and selective sequential chemical extractions. The case study showed that Zn is currently mobile at the site whereas Pb, Cu, or Ni are not, and this highlights the importance of adopting a multidisciplinary approach to investigating toxic metal mobility and retention.

As a result of industrial activity over the last 200 years, the UK has a legacy of industrially contaminated land. This has often been contaminated by poorly disposed of wastes, and the release of solids, liquids or solutions during their production, transport, storage and use. Prior to designing any remediation strategy or engineered containment system it is essential to evaluate the degree to which toxic metals are mobile and bioavailable, in order to determine whether they present potential hazards to humans or other biota. Unfortunately, much current engineering practice for determining how metal-contaminated soils and wastes at contaminated sites are treated/categorized still depends on straightforward comparison of total levels of metals with generic values such as the Interdepartmental Committee on the Redevelopment of Contaminated Land trigger values (ICRCL 1987).

The ICRCL trigger values are intended to be maximum total concentrations for a particular land use that do not increase risk to selected targets (Simms & Beckett 1987). However, the potential toxicity to specified targets (such as humans, plants, etc.) will depend very much on the form in which the metal exists (Solomon & Powrie 1992), as well as the site hydrology. While comparison of total contaminant levels with generic trigger values may be an appropriate first step in risk assessment, it can only indicate that there may be a hazard, not that there definitely is one (Institute of Civil Engineers 1994). Indeed, Tessier *et al.* (1979) state that where metals are in the residual fraction, they are essentially immobile, so it is inappropriate to use total metal levels for risk assessment.

For a health hazard to exist, there must be a contaminant pathway from the soil to the biota. Firstly the contaminant must reach the location of the target biota, and secondly, it must be able to enter the target biota. Both mobility and bioavailability depend strongly on chemical speciation (the chemical form in which the contaminant metal is present). Hazard via aqueous pathways will be low where contaminants are only released slowly from particles, or in non-bioavailable dissolved forms. However, poorly soluble metal compounds are often chronically toxic by inhalation, because particles have a long residence time in the lungs.

Typically, heavy metals in soils and solid wastes are relatively immobile under conditions where they are:

- present as the native metal
- in large particles, such as in slags and crushed rock mine debris

From: METCALFE, R. & ROCHELLE, C. A. (eds) 1999. *Chemical Containment of Waste in the Geosphere.* Geological Society, London, Special Publications, **157**, 241–264. 1-86239-040-1/99/$15.00 © The Geological Society of London 1999.

- in low solubility phases such as silicates, sulphides, phosphates and carbonates
- adsorbed by clay minerals, immobile organic matter, immobile oil phases or hydrated Fe/Mn oxide coatings

Metals will be relatively mobile where they are present as

- 'free-aquo' ions
- complexed with inorganic ligands such as Cl^-
- complexed with soluble organic molecules such as humic and fulvic acids
- bound to colloidal particles of organic matter or colloidal sized oil droplets

However, not all mobile forms are equally toxic: free-aquo ions are generally more toxic than complexed metals (Schnoor 1996).

Clearly, knowledge of the speciation of metals is required in order to assess the risk from existing contamination at industrial sites. In many cases, natural chemical containment will operate in the local geosphere to reduce the migration or bioadsorption of toxic metals to safe levels. However, where metals are sufficiently mobile and bioavailable to present potential risks to humans, resources or ecosystems, appraisal of metal speciation is an essential precursor for the effective selection and design of remediation strategies. For example, the effectiveness of bentonite barriers will depend on the partitioning of mobile metals between 'free-aquo' ions (that can become strongly sorbed) and neutral metal complexes (that are only weakly sorbed).

Factors influencing metal speciation at contaminated sites

The metals present at industrially contaminated sites will either be in the form in which they were released, or a derivative that is closer to equilibrium with the environment. Speciation will therefore depend on (i) the original chemical form, (ii) the form that is most thermodynamically stable and (iii) local factors affecting the extent to which equilibrium is achieved, such as reaction rates (which depend on particle size, bacterial activity, presence of catalysing compounds, etc.), time of residence of the contaminant on the site and hydrological fluxes.

The thermodynamically favoured metal species at a contaminated site depends on the pH and the major element chemistry of the site (Salomons 1995). Generally, low pH conditions favour the mobility of metal cations whereas neutral to high pH conditions favour the solubility of oxyanions such as chromate (CrO_4^{2-}),

and heavy metals strongly complexed to organic matter (Bourg & Loch 1995). The stable forms of metals such as chromium, mercury and arsenic also depend on the redox conditions, as they have several stable oxidation states within the range of redox conditions found in the environment. Both pH and redox conditions have an indirect effect on heavy metal speciation by controlling the availability of sorbants like Fe/Mn oxides (which are subject to dissolution in mildly reducing and in mildly acidic conditions, Chuan *et al.* 1996), the exchange capacity of clay minerals, organic matter availability and the oxidation state of sulphur (metals are often precipitated as sulphides in strongly reducing conditions).

Many of the metals found at contaminated sites will not be in thermodynamic equilibrium with their surroundings. For example, many metal ores (formed at high temperatures and pressures) are not at equilibrium near the surface, but will weather to stable forms. Often this occurs at low rates on the engineering timescale because of physical and chemical kinetic barriers. Similarly, refined metals such as zinc, chromium, etc. are thermodynamically unstable but are highly resistant to corrosion due to the formation of coherent oxide films on their surface, so are likely to persist for long periods. Likewise, metals locked into fused slags may be immobile as a result of the low permeability of the medium (Cheeseman *et al.* 1993). However, metals deposited from solutions (either directly from anthropogenic activity or from leaching of solid wastes) are more likely to be close to equilibrium with their environment.

Speciation and mobility of metal contamination at industrial sites

In the following sections, the original speciation of metals released during various industrial activities are reviewed, and possible changes in speciation in the environment outlined. The spatial location of metallic compounds at specific types of site is not discussed in detail; the reader is referred to the Department of the Environment's Industry Profiles (Department of the Environment 1995*a–o*). The present review is focused on cadmium, arsenic, chromium, mercury, lead, copper, nickel and zinc because they are common at industrially contaminated sites in forms which are considered hazardous. Cadmium, arsenic, chromium, mercury and lead were identified as groundwater contaminants representing threats to human health in a comprehensive US survey (Knox & Canter 1996).

Cadmium and mercury are European Community (EC) list I substances, whereas arsenic, chromium, lead, copper, nickel and zinc are list II substances (European Community, 1980). Copper, nickel and zinc are highly toxic to plants (phytotoxic) and may cause problems in revegetating derelict land (Cairney 1995). Metals such as aluminium and iron can be present in very high levels on many contaminated sites and can damage aquatic ecosystems both as particulates and in dissolved forms. However, their effects are covered extensively in the acid mine drainage literature (e.g. Bowell & Bruce 1995; Gray 1996, Kelly *et al.* 1988), and these metals are not addressed in this paper. Compounds of other heavy metals and metalloids such as tin, silver, selenium, tellurium, thorium, vanadium and beryllium can also be toxic but are not considered separately here because the instances of their contaminating industrial sites to high levels are relatively uncommon.

Metal mobility fields

Before considering toxic metals separately, general trends in their environmental behaviour and mobility are outlined. Schematic diagrams showing general fields of metal mobility in aqueous

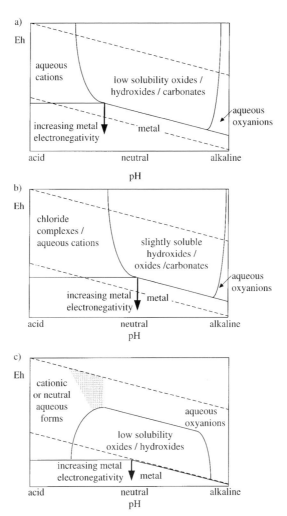

Fig. 1. General fields of metal mobility at contaminated sites. (**a**) Copper, nickel, lead; (**b**) zinc, cadmium, mercury; (**c**) chromium and similar transition metals (i.e. metals with oxyanions that are soluble over a wide pH range). Dashed lines show region of water stability.

systems are provided for guidance. For example, Fig. 1a is a schematic *Eh*–pH diagram for metals such as lead, copper and nickel in the presence of water and carbon dioxide, but in the absence of organic complexing agents. The diagram has been developed from data presented by Pourbaix (1966) by using the HCS geochemical modelling package (Outokampu Research, Finland) to investigate the effects of sulphur, carbonates and chloride ions on the stability fields. Note that the positions of the field boundaries vary greatly between specific metals, for example, the elemental metal stability field overlaps the water stability field (as in Fig. 1a) only for metals that are more electronegative than hydrogen (e.g. copper). Figure 1a shows that lead, copper and nickel form aqueous cations in acidic conditions, oxide/hydroxide or carbonate forms in neutral to alkaline conditions, and anionic forms in extremely alkaline conditions (although in natural systems, anionic species of these metals rarely reach significant concentrations). In the environment, the predominant form of each metal is often a highly insoluble oxide/hydroxide or carbonate, hence these metals are usually immobile (although synthetic organoleads are mobile, and inorganic lead is more mobile in the absence of carbonate). Furthermore, even in *Eh*–pH conditions favouring aqueous cations, these may remain strongly sorbed to solids (as solid surfaces usually carry a negative charge). The main conditions in which these metals become mobile in the subsurface is where they are complexed by dissolved or colloidal organic matter.

Metals such as zinc, cadmium and mercury are often more mobile at industrially contaminated sites than copper, nickel and lead (Fig. 1b shows their general mobility fields). This is because they can form aqueous complexes with chloride, carbonate and hydroxide, even where these ligands are present at the modest levels often found at contaminated sites. Such complexes are often only weakly sorbed to solids.

Chromium, and similar transition metals such as vanadium and molybdenum, have high order oxidation states that form oxyanions and so can be extremely mobile in oxidizing conditions. Such oxyanions are soluble over a wide pH range (Fig. 1c shows general mobility fields), and are relatively weakly sorbed to solids. Chromium, vanadium and molybdenum are less mobile in reducing conditions as their lower oxidation states tend to form low-solubility oxides and hydroxides. Arsenic (a metalloid in Group V of the periodic table) also forms soluble oxyanions over most of the water stability field. However, it differs in that the (+3) oxidation

state also forms soluble mobile oxyanions, and under extremely reducing conditions very toxic compounds of the (−3) oxidation state such as arsine can be produced.

Cadmium

Cadmium is toxic and bioaccumulative in the environment (Department of the Environment 1984). In wastes, cadmium is nearly always present as (+2) oxidation state compounds rather than as native metal. As a dust, cadmium compounds (particularly CdO) are toxic by inhalation. Otherwise, the toxicity of cadmium compounds depends strongly on their solubility. The chloride, sulphate and nitrate are soluble and highly toxic whereas the carbonate, sulphide and hydroxide are less soluble.

Most Cd containing waste is produced as a by-product of refining other metals such as zinc. Other sources of cadmium are wastes from electroplating, pigments and plastic manufacture and spent batteries (Moore & Ramamoorthy 1983). Cases of high Cd body burdens, such as those in the Jintsu Valley, Japan (Fasset 1980), and Shipham, Somerset, UK (Simms & Morgan 1988), are usually associated with the consumption of food grown on areas contaminated with Cd by Zn and Pb mining. Details of the speciation of cadmium contamination are listed in Table 1 (note that the data in Tables 1 to 6 are from a variety of sources and are intended to aid desk studies and not as substitutes for site specific data).

In the environment, the behaviour of Cd(II) is strongly dependent on pH and redox conditions. In neutral to alkaline soils and oxidizing conditions the metal tends to be sorbed, primarily to Fe/Mn oxyhydroxides but also to organic matter. However, in the presence of chloride, complexes such as $CdCl_2^0$, $CdCl_3^-$ and $CdCl_4^{2-}$ are formed that are mobile in neutral conditions (Yong *et al.* 1992). Under mildly reducing or acidic conditions Cd(II) may be relatively mobile because Fe/Mn oxyhydroxides are soluble, whereas under strongly reducing conditions Cd(II) will precipitate if S^{2-} is present.

Mercury

All forms of mercury are acutely toxic to humans and animals, accumulating in the liver, kidney and brain, and some forms may cause abnormalities in foetal development (teratogenic). The main forms of mercury in the environment are the native metal, inorganic salts of the (+1) and (+2) oxidation states and

Table 1. *Speciation of cadmium contamination*

Source	Concentration	Speciation	Contaminated Media
Pb/Zn mining	$<500\,mg\,kg^{-1}$	CdS (with ZnS)	rock waste, tailings, soils
Non-ferrous metal smelting	wastes: up to 5% soils: 1 - 750 mg kg^{-1}	CdS in ZnS, ZnCO$_3$	slag (leaches slowly), soils
Electroplating	–	metal, sulphate, others	solutions, sludges, soils
Domestic waste incineration	$600\,mg\,kg^{-1}$	various	ash
Vehicle breaking (scrapyards)	50–500 mg kg^{-1}	various	dusts and sludges, soils
Cd stabilizer production	various, up to 5%	stearate	–
Paints and pigments	various, up to few %	selenide, nitrate	–
Cd cell manufacture	up to 2%	metal, nitrate, sulphate, oxide/hydroxide	sludge
Phosphoric acid production	4–10 mg kg^{-1}	sulphate in gypsum	soil
Sewage disposal sites	up to 3500 mg kg^{-1}	organic complexes, sulphides	sludges, soils
Domestic waste landfills	–	CdCl$_2$, other mobile complexes	waste, soils
Iron and steel production	$<5\,mg\,kg^{-1}$	oxide	slag (leaches slowly)
Coal ash (e.g. at power stations)	$<14\,mg\,kg^{-1}$	oxide, sulphate (insoluble carbonate formed on weathering)	slag

Sources: Department of the Environment (1984, 1995*m*); Alloway (1995); Simms & Morgan (1988); Coppin *et al.* (1992).

organomercury compounds (Windmoller *et al.* 1996; Department of the Environment 1977). Organomercury compounds are the most potent toxins, and have the longest residence time within organisms (Craig 1986). Unfortunately, organomercury compounds can be produced from the inorganic forms by bacteria in the environment (Schnoor 1996).

Much mercury used by humans is in consumer products (such as batteries) which are eventually disposed of with municipal waste. Also, sewage sludge contains mercury from both domestic and industrial sources (Department of the Environment 1977). As a result, both landfills and sewage sludge disposal sites are likely to be contaminated with mercury, although mercury may be precipitated as the sulphide in landfills. Data on the speciation of mercury contamination are given in Table 2.

The main source of concentrated mercurial wastes in the UK has historically been the chlor-alkali industry, where mercury anodes are used in the electrolysis of brine. Such wastes are confined to a few sites, where the effluents from mercury cells have been placed in lagoons for settlement, or disposed of into canals where mercury accumulated in the bottom sediment (Bridges 1987). In one such disposal lagoon in the UK, the sludge was found to contain about 400 mg kg^{-1} total mercury of which 1 mg kg^{-1} was organomercury (Department of the Environment 1977). According to Windmoller *et al.* (1996) the original speciation of mercury from chlor-alkali plants is the native metal. However, in soils this becomes oxidized to Hg(I) and Hg(II), with Hg(II) bound to humic substances where these are present.

Other sources of mercury contamination include Zn/Cu mining and smelting (generally Hg occurs as a trace element in ZnS), paper pulping, paint manufacturing and armament storage and disposal (mercury fulminate, Hg(CNO)$_2$, is an explosives detonator). Also, organic mercury compounds have been used as fungicides. In developing countries, contamination of river sediments by metallic Hg has resulted from gold mining, which has led to serious poisoning incidents and human deaths (Lacerda *et al.* 1995).

Table 2. *Speciation of mercury contamination*

Source	Speciation	Contaminated media
Chlor-alkali industry	elemental Hg, organo-mercury	sludges from settlement lagoons dredgings from canals/rivers
Hg/Cu/Zn mining, extraction	HgS	mining: tailings, rock waste, soils. smelting: slags, dusts, soils
Gold and silver mining	elemental Hg	river sediments
Paper pulping	organo-mercury (e.g. phenylmercury)	sediments from canals/rivers
Paint manufacture and use, esp. at shipyards	organo-mercury (oleate, phenylmercury)	soils, wastes
Pesticide manufacture	organo-mercury	soils
Weapons dumps	mercury fulminate $Hg(CNO)_2$ mercury thiocyanate mercurous nitrate	soils ($<600\,mg\,kg^{-1}$)
Electrical lighting manufacture/use	elemental Hg, iodide	soils (associated with glass)
Fungicide manufacture (pre-1970)	organo-mercury	soils
Domestic waste landfills	HgS	waste
Sewage disposal	sulphides / organically complexed	sludges, soils

Sources: Department of the Environment (1977, 1995*b*, 1995*m*); Bridges (1987); Windmoller *et al.* (1996); Mitra (1986).

Arsenic

Most forms of arsenic are acute toxins and carcinogens to humans and the higher animals, and can be absorbed into the body by ingestion, inhalation and skin contact. Arsenic exists in the environment in its (-3), $(+3)$ or $(+5)$ oxidation states, forming arsine (e.g. AsH_3), arsenite (AsO_2^-) and arsenate (AsO_4^{3-}) species respectively. The toxic effects of AsO_2^- are due to enzyme blocking, whereas those of AsO_4^{3-} are due to interference with cellular energy transfer mechanisms, with arsenates being less toxic than arsenites because they are eliminated from the body faster (Moore & Ramamoorthy 1983). The toxicity of As compounds is related primarily to solubility and volatility; elemental As is less harmful as it is not strongly absorbed by the gut, whereas arsines are noxious by inhalation (Moore & Ramamoorthy 1983). Arsenite is considerably more mobile than arsenate, because the latter strongly absorbs to and sometimes co-precipitates with iron hydroxides (Schoor 1996).

Arsenical wastes are produced by metal mining, the use of pesticides in tanneries, as a by-product of Cu, Zn, Pb refining, car battery reclamation, timber treatment, fertilizer and glass manufacture, and in smaller quantities by ammonia production, detergent production and the manufacture of xerographic components (Department of the Environment 1980*a, b*; Moore & Ramamoorthy 1983; Sadler *et al.* 1994; Dutre & Vandecasteele 1995; Machinga-

wuta & Broadbent 1994; Porter & Peterson 1977). The levels of arsenic in wastes produced by such operations and an indication of their speciation are given in Table 3.

The impact of arsenic wastes on the environment depends strongly on As leachability. Machingawuta & Broadbent (1994) report a slag from metal smelting that contained tens of per cent of As, but leaching tests showed low leachability ($<10\,mg\,l^{-1}$ in leachates in contact for periods of years). In contrast, arsenical dusts produced by copper extraction from ore are far more leachable ($>1000\,mg\,l^{-1}$ in leachates in contact for periods of days; Dutre & Vandecasteele 1995). Arsenic in pesticides is even more leachable. Arsenical pesticides buried at two sites in Minnesota led to soil and groundwater contamination and subsequent poisoning of humans via drinking water (Bridges 1987).

In neutral pH and oxidized soil environments, inorganic arsenic forms arsenates which are strongly sorbed by iron oxyhydroxide and oxide minerals. Inorganic arsenic is more mobile in acidic or reducing conditions (Bowell 1994), because iron oxyhydroxides and oxides are partially dissolved, and because arsenite rather than arsenate species predominate. However, in strongly reducing conditions where sulphide is present, both arsenite and arsenate precipitate as sulphides. Arsenites may also be relatively mobile in the presence of organic complexing agents (Bowell 1994), and under reducing conditions, bacteria may methylate arsenites and arsenates to form arsines (O'Neill 1995).

Table 3. *Speciation of arsenic contamination*

Source	Concentration	Contaminated Media and Speciation
Metal mining, esp. Cu, Zn, Pb, Sn	up to 3.6%	tailings and rock waste; AsS, FeAsS soils around tailings
Cu, Zn, Pb, Sn refining	0.2–5%	sludges with As trioxide, trisulphide, scorodite (FeAsO$_4$)
	up to 25%	slags
	up to 50%	fly ashes and precipitated dusts with As (III)
	–	soils
Pesticide/herbicide production	–	arsenates, arsenites of sodium, calcium, lead, organoarsenides such as phenyl arsenic and diarseno-diphenyl compounds, probably in soils
Pb recycling	1%	sodium sulphate sludges containing As (V)
Fine chemical works	<25%	pentoxide (fungicide), trioxide (dyeing mordant). trisulphide (pigment), others
Timber treatment	1.5%	sludges with As pentoxide, organoaresenates (aromatic),
	5–10%	ash from burned treated timber, up to 20% of this is soluble
Glass manufacture	<4%	trioxide, trisulphides, in brickwork and flue dusts
Ammonia production (VetroCoke)	12%	potassium ferri-arsenate
Phosphoric acid production	4–7%	trisulphide
Tanneries (historical)	0.1–1%	sulphides, sodium arsenite, calcium arsenite in tannery landfills
Coal ash (e.g. at power stations)	<1%	silicate ash with leachable As anions sorbed to surface sites

Sources: Bowell (1994); Department of the Environment (1980*a, b*; 1995*f*, 1995*h*, 1995*n*, 1995*o*); Moore & Ramamoorthy (1983); O'Neill (1995); Sadler *et al.* (1994); Dutre & Vandecasteele (1995); Machingawuta & Broadbent (1994); Porter & Peterson (1977); Lee & Spears (1995).

Chromium

Chromium has a relatively low toxicity (compared with cadmium and mercury) but the high solubility of its compounds makes it a significant environmental concern. The most common oxidation states of chromium are (+3) and (+6), and these are the ubiquitous oxidation states in environmental systems (Department of the Environment 1978). Cr(VI) compounds are irritants and are adsorbed by ingestion, inhalation and skin contact, and some Cr(VI) compounds are carcinogenic to humans. Cr(III) compounds are soluble in stomach acid and may be poisonous by ingestion (Department of the Environment 1981).

Most chromium-rich solid waste is produced by mining, steel manufacture and chromium smelting, but significant amounts of land have also been contaminated by tanneries and metal plating operations. According to Breeze (1973), the uncontrolled tipping of about 10^6 tonnes of smelter waste near Bolton, Lancashire, UK,

resulted in severe river pollution by chromium. Leaks from industrial processes using chromium solutions have also led to large-scale river and aquifer contamination (McKinley *et al.* 1992). Details of the speciation of chromium contamination are listed in Table 4.

Solutions and sludges from electroplating, timber treatment and printing works contain mainly chromium (+6) species, whereas chromium smelter waste contains both the (+3) and (+6) species (Breeze 1973; Gemmel 1973). Most other chromium wastes contain some (+6) species, the exception being sludges from tanning of hides, which typically contains 3.5% by weight of Cr(III) as the hydrated oxide (Department of the Environment 1978, 1995*f*; Bridges 1987), but only rarely contains Cr (VI) species because reducing agents are added to the waste stream. However, most of the Cr(III) in tannery waste is complexed to dissolved organic matter such as proteins and organic acids from the hides, and are therefore mobile under a wide range of *Eh*–pH conditions (Walsh &

Table 4. *Speciation of chromium contamination*

Source	Concentration	Speciation	Contaminated media
Mining of Cr	<50%	$PbCrO_4$, $FeCr_2O_4$	rock waste, tailings
Smelting of Cr	2–6%, about one third of this is highly leachable	$FeCr_2O_4$, Na and Ca chromate and chromites mixed with blocks of sodium sulphate	greyish yellow waste with up to 2% water soluble Cr
Metal finishing (Electroplating, dipping, pickling, etching)	up to several per cent in solution	chromic acid, chromate solutions	liquids, sludges, soils
Tanning hides	3.5%	Cr(III) hydrated oxides, Cr (III) complexed to organics (e.g. proteins, organic acids)	sludges liquids, soils
Timber treatment compounds (production and use)	up to several per cent	Cr(VI) as sodium dichromate or chromic acid, with Cu, As	liquids, sludges, soils, sawdust
Paints, pigments and dyes, printing works	–	Cr (VI) salts chromium (III) oxide strontium chromate	fine powders, liquids, sludges, soils
Paper pulping	–	Cr(VI)	sludges in landfills, soils
Refractory brick manufacture	–	chromium (III) oxides	brick waste
Corrosion inhibitors/ fungicides (production and use)	–	some Cr(VI)	solid salts, liquids
Coal Ash	<10 000 mg kg^{-1}	–	ash
Sewage disposal sites	<5% <2000 mg kg^{-1}	Cr(III) Cr(III)	sludge soil

Sources: Department of the Environment (1976, 1978, 1995*d*, 1995*f*, 1995*g*); Alloway (1995); Breeze (1973); Jacobs (1992); Gemmel (1973).

O'Halloran 1996). The proteins are likely to degrade quickly in the environment leading to precipitation of the Cr(III), whereas the organic acid complexes are more persistent.

The CrO_4^- anion and related forms are highly mobile in soils, whereas the Cr^{3+} cation is strongly and specifically adsorbed above about pH 5 and its solubility is strongly restricted by formation of $CrOH_3$ above pH 6 (Nriagu & Nieboer 1988). However, according to Wernicke *et al.* (1993), a proportion of Cr^{3+} from soils contaminated by electroplating solutions was mobile, possibly due to organic complexation. Also, conversion of chromium from the (+3) to the (+6) state can occur in oxic conditions in sediments (Walsh & O'Halloran 1996). In soils, reduction of Cr(VI) to Cr(III) by organic matter and Fe/Mn oxyhydroxide can occur, particularly under acidic conditions (Puls *et al.* 1994).

Lead

Lead compounds are both acute and chronic poisons, the principal toxic effects being on the central nervous system, blood and kidneys; children are more susceptible to nervous system damage than adults (Grandjean 1975). In the environment, lead is usually in Pb(II) inorganic salts and organolead compounds (although Pb(IV) occurs as the dioxide). Toxicity depends on solubility, but even poorly soluble compounds can be chronic poisons as inhaled dusts, which dissolve slowly in the lungs (also some lead compounds which are poorly soluble in water can be absorbed by ingestion). Most inorganic lead (II) salts are poorly soluble, the exceptions being the acetate, chlorate, nitrate, nitrite and to a lesser degree the chloride. As a result Pb^{2+} ions are relatively immobile above about pH 5–6 due to precipitation, unless

Table 5. *Speciation of lead contamination*

Activity	Concentration	Speciation	Contaminated media
Mining Pb, Zn etc.	<15% <5000 mg kg^{-1}	PbS, PbSO$_4$, PbCO$_3$	rock waste, tailings soils
Refining of Pb, Zn, Cd	up to several per cent up to several thousand mg kg^{-1}	as above, and silicates	slags, dusts, drosses soils
Petrol station/oil refineries	<1% 0.025%	inorganic tetraethyl lead	sludges, soils
Sports shooting/military uses including weapons disposal	up to several per cent <10%	metallic Pb metallic Pb and others	soils
Paint manufacture and use (esp. at shipyards)	–	White lead (II) carbonate Pb (IV) 'red lead' primer	soils soils
Battery reclamation/ scrapyards	up to several thousand mg kg^{-1}	native metal, oxide, sulphate, carbonates	soils, calcium sulphate sludges
Precious metal recovery	–	oxide, nitrate	soils
Explosives manufacture	–	lead azide Pb(N$_3$)$_2$, lead styphnate (i.e. Benzene, 2,4- dihydroxy-1,3,5-trinitro-lead)	soils
Glassmaking/pottery	–	Pb (IV) oxide 'red lead'	soils
Sewage sludge disposal	up to several thousand mg kg^{-1}	sulphides, organo-complexes	sludge, soils
Pesticide production	–	lead arsenate	soils
Coal ash	up to several thousand mg kg^{-1}	silicates	ash

Sources: Bridges (1987), Yong *et al.* (1992), Davis (1995), Gul (1994), Coppin *et al.* (1992); Department of the Environment (1995*a, b, e*), Weast (1973).

complexed to humic acids (Moore & Rama-moorthy 1983).

The speciation of lead contamination is summarized in Table 5. The main source of lead-rich wastes is mining, where tailings and rock waste containing up to 2% lead (usually as the sulphide, carbonate or sulphate) are commonly produced (Bridges 1987). Lead-rich slags (2–4% lead) and dusts (10–20% lead) are produced by lead smelting, refining and reclamation, and in some cases these have been disposed of on site (Department of the Environment 1995*a*). Scrap steel slags may contain up to several per cent by weight of lead along with other heavy metals. Atmospheric deposition of dust from metal refining and from wind dispersal of mine wastes can produce levels of several thousand mg kg^{-1} in soils within a few hundred metres of the stacks (Coppin *et al.* 1992; Davis 1995).

The lead contamination mentioned above is in relatively immobile forms, which are therefore unlikely to contaminate marine and freshwaters significantly. However, dusts from tailings etc. may still represent a hazard by ingestion/inhalation. Although sludges from the cleaning of leaded petrol storage tanks (e.g. at oil refineries or service stations) contain organoleads such as tetraethyl lead (Yong *et al.* 1992), these convert to Pb(II) carbonates and oxides in soils (Davis 1995). Furthermore, Pb(II) is not very mobile in soils because, in addition to the low solubility of its compounds, it is strongly sorbed by organic matter and Fe/Mn oxides.

Copper, zinc and nickel

The primary hazards from copper, zinc and nickel are their phytotoxicity, although nickel (and to a lesser extent copper) is also toxic to humans. In the environment, these metals usually occur as inorganic salts of their (+2) oxidation states, although Cu also occurs as the native metal and as salts of the (+1) state. The solubilities of Cu(II) and Ni(II) salts are generally low in neutral and alkaline conditions (in the absence of complexing agents). However, zinc may be mobile in neutral conditions as $Zn(OH)^+$, and in alkaline conditions as $Zn(OH)_2^0$ (Moore & Ramamoorthy 1983). Zinc is often mobile in saline conditions due to the formation of neutral and anionic complexes with chloride ions.

The main sources of contamination by Cu, Ni, or Zn are areas where these metals have been

Table 6. *Speciation of copper, zinc and nickel contamination*

Process	Cu	Zn	Ni
Mining	$CuFeS_2$, Cu_5FeS_4, Cu_2S, Cu_3AsS_4, CuS, Cu, cyanides	ZnS, in crushed rock wastes <5%	$(Ni, Fe)_9S_8$, NiAs, $(Cu, Ni)_3S_4$
Refining Cu, Zn, Ni	$CuCl_2$ sludges, silicate slags	sludges with sulphides, carbonates, silicate slags	silicate slags
Steel refining	silicates/ carbonates	silicates/ carbonates	silicates/carbonates
Pesticide production	$CuSO_4$ Cu(II) acetoarsenite	zinc phosphide	–
Engineering works	elemental fragments, oxide dross	metal, primers: ZnO, $ZnCrO_4$, zinc molybdate, zinc phosphate	elemental fragments, oxide dross
Pigments and paints, textile dye	$CuCl_2$, $Cu(NO_3)_2$, CuO, CuS, $CuCO_3$	elemental fragments, ZnO, ZnS	–
Timber treatment	$CuSO_4$, CuO, Cu(II), naphthanate, arsenite	ZnO, $ZnCl_2$, naphthanate, carboxylates	–
Metal finishing (Electroplating, galvanizing, pickling)	elemental fragments, $CuSO_4$, $Cu(OH)_2$, $Cu(CN)_2$, $CuCl_2$, Cu(II) pyrophosphate	elemental fragments, $ZnSO_4$, $ZnCl_2$, $Zn(CN)_2$	elemental fragments, $NiSO_4$, sulphamate, $NiCl_2$
Chemical catalysis, e.g. pharmaceutical works	Cu, $CuCl_2$, acetate, stearate	ZnO, $ZnCl_2$, Zn	elemental, NiO
Scrapyards	elemental fragments	elemental fragments, oxides	elemental fragments
Shipyards	elemental fragments	elemental fragments, oxide, chloride, chromate, epoxy	
Sulphuric acid manufacture	CuS, $CuSO_4$	ZnS, $ZnSO_4$	NiS, $NiSO_4$
Battery and electrical goods manufacture	elemental Cu (finely divided, windings), in alloys with Cd, Cr, Be, $CuCl_2$	elemental Zn, $ZnCl_2$, ZnO, hydroxides	NiO(OH), $NiSO_4$
Gas works		ZnO, cyano complexes in 'spent oxide'	NiO
Domestic waste landfills	elemental fragments, CuS (precipitate)	elemental fragments, $ZnCl_2$ (in leachate)	elemental fragments, oxides, NiS (precipitate)
Sewage farms	organic complexes/ sulphides	organic complexes/ sulphides	organic complexes/ sulphides

Sources: Alloway (1995); Department of the Environment (1976, 1995*c, g, i, j, k, l, m*), Hutchinson (1979, 1982); Moore & Ramamoorthy (1983), Thornton (1979), West & Stewart (1995), Coppin *et al.* (1992).

mined or refined, although copper contamination from plating and chemical works is also widespread. Refining and wind dispersal from spoil heaps releases large quantities of these metals into the atmosphere as dust, leading to soil contamination to levels of several thousand $mg\,kg^{-1}$ up to several km downwind (Coppin *et al.* 1992; Hutchinson 1979; Lavender 1980). The speciation of Cu, Ni and Zn contamination is summarized in Table 6.

Hudson-Edwards *et al.* (1995, 1996) investigated changes in speciation of Zn, Cu, Pb and Cd derived from mine wastes. Mineralogical analysis of mining wastes, mining age alluvial deposits, and recent river sediments showed that the primary sulphide ore minerals undergo weathering to produce carbonate, silicate, phosphate and sulphate species. Ultimately Cu, Cd, Zn and Pb become sorbed to iron and manganese oxyhydroxide phases.

When released into soils, aqueous forms of Cu(II) and Ni(II) are often immobilized by sorption. For Cu(II), the most important sorbant is organic matter (Baker & Senft 1995), whereas for Ni(II), iron oxides are also important (McGrath 1995). However, both Cu(II) and Ni(II) may be mobilized by dissolved or colloidal organics (Thornton 1979; Hutchinson 1982). Zn(II) is often relatively mobile because aqueous species such as $Zn(OH)^+$ and $Zn(OH)_2^0$ (the dominant species above pH 6.7 and pH 8 respectively) are weakly sorbed.

Methods for speciating metals

Methods for investigating heavy metal speciation include leaching tests (batch leaching tests, serial batch and column leaching tests, sequential chemical extractions, characterization leaching tests) and direct methods. A comprehensive review of leaching and extraction tests is beyond the scope of this paper. The following section includes a brief description of each technique, a summary of its advantages and limitations, and review of the application of the technique to industrially contaminated sites. For a detailed review of leaching tests the reader is referred to National Rivers Authority (1994) and van der Sloot *et al.* (1997); a detailed review of chemical extractions is provided by Ure and Davidson (1995).

Leaching and extraction tests are designed to reach one of three end points:

(1) Exhaustion of a solid contaminant phase;
(2) Equilibrium between the contaminant aqueous and solid phase;
(3) Steady kinetically controlled release.

Single stage batch leaching tests are usually designed to reach equilibrium at the end of the test (van der Sloot 1996) and therefore use a low liquid to solid (L/S) ratio (≤ 10), whereas sequential chemical extraction procedures are usually designed to exhaust all of the contaminant phases that are soluble under particular pH/*Eh* conditions and hence use a high L/S ratio (50–100). Leaching tests carried out at a range of L/S ratios and contact times can identify those contaminants that reach solubility control.

Batch leaching tests

A batch leaching test involves placing a test material in a liquid (leachant) under specified conditions. Agitation is carried out to bring the fluid phase into intimate contact with all solid surfaces. Then leachate and solid are separated (commonly by using a 0.45 μm membrane filter). Analysis of the leachate gives some indication of the soluble components in the test material. The composition of the leachate is controlled by the equilibrium pore fluid composition, kinetic factors controlling the rate of achievement of equilibrium and the contact time. The equilibrium composition depends on pH, redox potential, temperature and chemical composition of the solid (i.e. hydrophobicity, buffer capacity, complexation capacity), whereas kinetic controls include surface area, the porosity/permeability of the test material, the liquid to solid ratio, contact time and temperature. Batch leaching tests are generally more reproducible if they are conducted in closed vessels, as this minimizes the effect of atmospheric CO_2 and O_2 on the system chemistry (van der Sloot *et al.* 1997).

A number of agencies have adopted batch leaching tests for regulatory compliance testing of materials. For example, the American Society for Testing Materials (1979, 1985) have proposed two standard batch tests for measuring leachability of waste materials. In the water shake extraction method, a known weight of waste is shaken in water at an L/S ratio of four for 48 hours. This method is intended as a rapid means of evaluating water extractable materials in wastes. In the acid shake extraction method, the protocol is similar except the leachant is an acidic buffer solution. This test is intended to indicate leachability under acidic conditions, such as those found in Municipal Solid Waste landfills. However, it is difficult to determine total availability of a contaminant or its speciation from single stage batch tests because there is no way of identifying the test end point from a single L/S ratio and duration (van der Sloot 1996).

A two-stage serial batch extraction has been proposed in the draft European Standard Leaching test for Granular Waste Materials (British Standards Institute 1996) to evaluate hazards from landfills for waste compliance purposes. This procedure uses an L/S ratio of two for a contact time of six hours in the first stage, and an L/S ratio of eight (to give a cumulative L/S ratio of ten) for a contact time of 18 hours in the second stage. The advantage of a two-stage batch test is that it allows some indication of whether leachate concentration is solubility controlled (contaminant species under solubility control will produce aqueous activities that are invariant with contact time and L/S ratio). This draft European Standard Leaching test for Granular Waste Materials has been adapted as a compliance leaching test for assessment of contaminated land by the National Rivers Authority (now part of the Environment Agency). The NRA test is a single stage extraction using simulated rainwater at an L/S ratio of 10:1, with agitation for 24 hours at room temperature (National Rivers Authority 1994).

Other regulatory authorities have developed 'total availability' batch extractions, which are designed to assess the amount of a contaminant that is leachable. For example, the Dutch Availability test NEN 7341 (van der Sloot 1996) is aimed at assessing the fraction of the total mass potentially available for leaching under environmental conditions. It is designed to exhaust potentially mobile phases (it is implicitly assumed that the silicate matrix and other low solubility mineral phases do not dissolve significantly) so it is performed on very finely ground material tested with an L/S ratio of 100. It is conducted in two stages, with the pH controlled to 7 and 4 respectively. However, some contaminant phases show leachabilities that are sensitive to redox conditions, which are not controlled in these tests (van der Sloot *et al.* 1994).

Serial batch leaching and column leaching tests

During long-term leaching in field situations, phases that initially buffer the pH/*Eh* of the leachate can become exhausted with important consequences for the release rate of contaminants. For example, Baverman *et al.* (1997) showed that copper release from sulphidic mine waste became much more rapid after exhaustion of a pH-buffering calcite phase. In order to investigate such behaviours, serial batch tests and column leaching tests have been devised.

In serial batch tests, the same specimen is repeatedly leached with batches of fresh leachant, in order to crudely simulate a hydrological flux. Column leaching tests (e.g. American Society for Testing Materials 1991) impose a hydrological flux through a column of material in order to better replicate site-specific conditions. Unfortunately, as it is difficult to interpret the tests in fundamental terms, extrapolation of data to longer time periods is problematic. Also, simulation of large scale heterogeneity of wastes and fluid pathways is not possible in small-scale laboratory columns (National Rivers Authority 1994).

Selective sequential chemical extractions

Selective sequential chemical extractions are batch leaching tests where a series of progressively stronger leaching reagents are added to a sample, and the solution removed for analysis after each stage. The aim is to categorize trace metals into fractions that are likely to be released into solution under various conditions. For example, Tessier *et al.* (1979) proposed a multi-stage extraction that located trace metals in each of five fractions:

(1) Exchangeable: metals that can be desorbed by changes in solutional ionic strength;
(2) Bound to carbonates: this fraction is released by a reduction in pH (typically to pH 5);
(3) Bound to iron and manganese oxides: this fraction is soluble under reducing conditions;
(4) Bound to organic matter and sulphides: these fractions are soluble under oxidizing conditions;
(5) Residual: the metals are present within silicate mineral lattices, resistant sulphides and organic matter; these phases do not dissolve except in strong mineral acids.

Sequential extractions have been developed for use on river sediments (e.g. Breward & Peachey 1995; Davidson *et al.* 1994; Hudson-Edwards *et al.* 1995; Tessier *et al.* 1979); and more recently have been applied to soils from contaminated sites (Kubota & Orikasa 1995; Phillips & Chapple 1995; Ramos *et al.* 1994; Tuin & Tels 1990; Yarlagadda *et al.* 1995; Yong *et al.* 1993). However, river sediments have smaller particles and are more homogeneous than specimens from industrially contaminated sites, so serious consideration should be given to sample size when conducting sequential extractions on industrial site materials. Also, highly soluble materials may remain on industrial sites, so water leaching stages should be included.

Tuin & Tels (1990) suggest that there are problems associated with sequential extractions because the reagents are not as selective as sometimes suggested; there is overlap and incompleteness on extraction stages; soil composition and indigenous metal speciation can be altered (particularly with arsenic and chromium; Walsh & O'Halloran 1996). For example it is very difficult to remove all exchangeable metals without dissolving some of the 'bound to carbonates' fraction; it is very difficult to remove all metal sulphides without attacking silicate mineral phases (Tessier et al. 1979). Also, the sequence in which reagents are applied can influence the results. In short, sequential extractions only indicate which elements are soluble in specific leaching agents, rather than giving actual chemical speciation.

Table 7 shows a summary of the findings of various sequential extraction studies on contaminated soils. The data from individual studies in Table 7 are averages of soil from more than one source. No studies of artificially spiked soils have been included, because they exhibit very different behaviour to waste site soils (Tuin & Tels 1990). Despite the wide range of soil and contaminant sources included in Table 7 some clear trends are apparent. For example, extractable copper is mainly associated with the oxidizable phase, which may indicate that it is bound to organic matter (Cu can be strongly bound by soil organic matter; Baker & Senft 1995).

Extractable cadmium is often in the exchangeable fraction, where it is very bioavailable (van der Sloot et al. 1997). Zinc shows a similar pattern, probably because the ions Zn^{2+} and Cd^{2+} have similar complexing behaviour (i.e. both complex strongly to inorganic ligands such as chloride rather than to organic ligands).

Table 7 indicates that lead is often distributed across all the fractions apart from the exchangeable fraction. In a study whose results are not included in Table 7 Phillips & Chapple (1995) investigated soils from a contaminated shipbuilding site, and found that most extractable lead was in the reducible phase, with smaller levels in the oxidizable and carbonate fractions.

Where present, nickel is often mainly in the residual phase and to a lesser extent in the reducible phase, where it is probably bound to Fe/Mn oxides. In contaminated soils from a ceramics factory and a metal plating works, significant chromium was found in the reducible phase, and in the organic/sulphide phase, but most was found in the residual phases (Tuin & Tels 1990). This suggests that metals added as solutions to soils can become part of the residual fraction, perhaps by substitution into silicate minerals. Once in this fraction such metals probably present very little risk.

Characterization leaching tests

Suites of leaching tests for the purpose of fully characterizing solubility behaviour have been developed (van der Sloot et al. 1997). These involve leaching at a range of pH, redox conditions, liquid to solid ratios and contact times. Their results have been used to infer the presence of particular solid phases in industrial waste materials by inverse geochemical equilibrium modelling. For example, by using a wide range of batch leaching tests and appropriate thermodynamic data, Fallman (1997) was able to infer that chromium in a steel arc furnace slag was present in a barium chromate-sulphate solid solution. Unfortunately, the results of inverse modelling are often non unique; several different contaminant mineral phases and solid

Table 7. *Results of sequential extractions on contaminated soils*

Source	Exchangeable	Carbonate	Reducible	Oxidizable	Residual
Tuin & Tels (1990)	Cd		Zn	**Ni**, Cd, Cr, Pb, Zn	**Cu**, Cr, Ni	**Pb**, Cd, Cr, Ni
Ramos et al. (1994)		**Cd**, Zn	**Pb**, **Zn**	**Cu**	Cd, Cu, Pb
Kubota and Orikasa (1995)	Cd		Pb, Zn	Cu, Pb	Cu, Pb, Zn
Yarlagadda et al. (1995)		Pb	**Pb**, Cu	Cu	Cu
Work reported here		**Zn**	Zn, Pb	**Cu**, Pb, Zn	**Ni**, Pb

M Majority of metal.
M Significant proportion of metal.

solutions are typically identified that could give the measured aqueous activities. Also, the leaching tests required are extremely time-intensive and inverse modelling may be problematic where a contaminant is complexed with organic matter, or sorbed to amorphous oxide phases, due to a lack of relevant thermodynamic data.

Direct techniques

In recent years, advanced analytical methods have been used to directly investigate metal speciation in contaminated soils and sediments. Use of the following techniques has been reported in the literature.

Scanning electron microscopy with X-ray energy dispersive spectroscopy (SEM-EDS). Soil specimens can be imaged using a Scanning Electron Microscope and the elemental composition of individual particles can be analysed using X-ray Energy Dispersive Spectroscopy (Watt 1985). Backscatter electron luminescence can be used to identify particles rich in elements of high atomic number, and EDS analysis can then be used to determine the elemental composition of the particles thus identified. Standard 'thick window' EDS can only detect elements above Na in atomic number; however, 'thin window' detectors can detect carbon and oxygen, permitting the identification of carbonates.

X-ray photoelectron spectroscopy (XPS). X-ray Photoelectron Spectroscopy allows the identification of elements heavier than helium present to a depth within particles of approximately 1 nm (a few monolayers) from their core electron binding energies (see Briggs & Seah 1983; Smith 1994). It has a detection limit of about 0.1% by weight (Smith 1994). Minor shifts in the core level electron binding energies from their elemental values provide information on chemical speciation and bonding. The technique is mostly useful for detection and analysis of surface coatings, and its spatial resolution is relatively poor compared with SEM-EDS.

Secondary ion mass spectrometry (SIMS). This technique yields elemental analysis for all elements including hydrogen, by analysing secondary ions displaced from the surface by a beam of primary ions (usually oxygen or argon; Lodding 1988). The depth of the specimen analysed by SIMS is only about 0.5 nm, so it has the potential to analyse very thin coatings and adsorbed layers. A high energy ion beam can be used to etch the surface of the specimen for depth profiling. One advantage of the technique is that the detection limit is very low, at the parts per million level or below (Smith 1994).

Confocal laser raman spectroscopy (LRS). Chemical speciation of elements in soil particles can be analysed by LRS, which yields information on chemical bonding (Banwell & McCash 1994). A portion of any light scattered by the specimen has certain discrete frequencies above and below that of the incident radiation, which are characteristic of electrical polarizability of chemical bonds within the specimen. Chemical compounds within soil particles can be identified by comparison of their vibrational spectra with those of reference compounds. Microscope Raman devices can analyse areas of down to 1 μm^2. As water is a weak Raman scatterer (so there is little interference with the signal), LRS can be used with wet samples that are close to their natural state.

Previous studies using direct techniques

Many of the spectroscopic techniques described above have been used to investigate the speciation of metals in waste materials. For example, Eighmy *et al.* (1995) carried out a comprehensive suite of direct analyses, including SEM-EDS, XPS, SIMS, X-ray powder diffraction (XRPD) and characterization leaching tests in order to investigate the speciation of metals in municipal solid waste precipitator ash. SEM-EDS provided good spatial resolution for abundant elements such as Pb, Zn and Ti, which were present at percentage levels. SIMS was more successful with trace metals, and it was able to provide both depth profile data and some limited chemical speciation data (e.g. it showed that reduced sulphur species were present). XPS gave the speciation of more abundant contaminant metals (Pb, Zn and Ti) near the surface of particles (top 10–30 Angstroms), but would probably be less successful where contaminant levels are relatively low. Equilibrium geochemical modelling based on the measured leaching behaviour identified many of the same solid phases as the spectroscopic techniques.

A few studies have been carried out on contaminated soils using spectroscopic techniques. For example, Puls *et al.* (1994) have carried out 'thick window' EDS and chemical extractions on soils contaminated with chromate from an electroplating works. These techniques showed that chromium was present associated with iron in coatings on silicate minerals. Sequential extractions indicated it was in the reducible phase, which may indicate a mixed chromium–iron hydroxide coating. XPS showed that the chromium, which was originally in a (+6) oxidation state in the plating solution, had been reduced to a (+3) state in the coatings. SIMS depth profiling of the iron hydroxide coatings showed that the degree of chromium contamination fell sharply within the top 10 nm.

Yarlagadda *et al.* (1995) conducted SEM-EDS and sequential extractions on size fractionated soil specimens from sites contaminated with lead and copper by wire manufacturing (acidic pickle liquors) and steel fabrication. They found that Cu and Pb were distributed fairly uniformly across sub 2 mm size fractions. SEM-EDS showed variations between the form of contamination between size fractions. The <75 μm fraction contained a few irregular shaped Cu-enriched particles and some smaller isolated Pb-enriched particles, whereas the 0.85–2.0 mm fraction contained no discrete contaminant particles. Here, lead was detected adhering to larger particles, either as small particles or as coatings. Sequential extractions indicated that Pb was primarily in carbonate and reducible phases, whereas Cu was associated with carbonate, reducible and oxidizable fractions.

LRS has been used extensively for speciating metal precipitates in aqueous systems. For example, Grimes *et al.* (1995) used it to identify lead oxide, phosphate and carbonate species on the surface of lead pipes. However, the application of LRS to speciating metals in contaminated soils is uncommon, although Johnston *et al.* (1997) report a pilot study where it was used to identify chromate and sulphate groups in a chromium waste tip soil.

There have also been studies of artificially spiked soils using direct techniques, but these have not been included in this review as spiked soils are generally very different from waste-site soils (Yarlagadda *et al.* 1995).

Case study of a smelter and scrapyard site

From the previous review, it is apparent that many factors must be considered when evaluating the mobility of toxic metals from contaminated industrial sites. Consequently, any investigation must determine the total level of each contaminant, identify its chemical speciation and its likely leaching rate. An indication of current leaching rate can be obtained from the site hydrology and metal levels in site waters. Establishing solid phase chemical speciation requires the synthesis of data from direct tech-

niques (e.g. SEM-EDS, SIMS, XPS) and appropriate leaching tests. This is illustrated by the following case study of a metal-contaminated scrapyard. This study included total metal analysis of different size fractions, sequential chemical extractions, SEM-EDS and determination of metal levels in ponded surface water.

Sampling

The site under investigation was formerly a vehicle scrapyard and battery store, now derelict and owned by the local authority. It is situated in Batley, West Yorkshire, in a residential area and is bordered by domestic gardens. Before 1950, it was used for unknown metal smelting operations. Desk studies indicate that the site overlies Middle Coal Measures Sandstone and that the soils in the area are stiff brown/grey clays. The site was visited in March 1996, when the weather was clear. However, surface water was present and was running off the site. This water was visibly contaminated with oils and had a pH of 7.3. Site inspection revealed many different types of waste to be present at the surface, including smelter slags, demolition wastes, vehicle components, waste oils, car battery cases and, in one area, domestic refuse along with the remains of various domestic appliances.

Samples were taken using a trowel and hand auger from depths of up to 40 cm (obvious large metallic fragments were excluded). Detailed analyses have been conducted on two samples. Sample LOB2 was a silty sediment comprising brick fragments and weathered slag taken from an area of surface water ponding within the site boundary. Sample LOB4 was sandy topsoil (with brick fragments) from an off-site area of distressed vegetation where contaminated run-off from the site was soaking away.

Chemical analysis

Samples were air dried and passed through sieves of mesh sizes 2 mm, 400, 300 and 63 μm. Samples of the four size fractions (2 mm–400 μm, 400–300 μm, 300–63 μm and <63 μm) and of the entire sub-2 mm fraction were ground in a Tema Agate grinder for 2–4 minutes (which reduces particles size to below 63 μm), pressed into pellets and analysed using a ARL9400 Sequential X-ray Fluorescence Spectrometer (XRF) with a rhodium tube running at 60 kV and 40 mA. Elements with atomic numbers below that of sodium cannot be detected by this method.

The level of contaminant metals in various fractions of each specimen (measured by XRF)

Table 8. *Soil analyses from the scrapyard site, determined by XRF*

		Weight (%)	Metal content of fraction ($mg\,kg^{-1}$)			
			Cu	Ni	Pb	Zn
LOB2	2 mm–400 μm	53	420	240	1230	1530
	400–300 μm	11	600	150	1940	2100
	300–63 μm	27	740	150	2260	2360
	<63 μm	9	1030	170	3050	2730
Weighted average of fractions			580	200	1750	1920
Measured metal content			430	150	1480	2050
LOB4	2 mm–400 μm	76	170	50	570	730
	400–300 μm	6	290	70	710	710
	300–63 μm	11	520	70	1150	1340
	<63 μm	7	350	60	850	1040
Weighted average of fractions			230	50	660	820
Measured metal content			110	60	560	590

are given in Table 8. Contaminant metals were present in all size fractions below 2 mm, but Pb, Zn and Cu were concentrated in the finer fractions. Disagreements between weighed averages of the analyses of separate size fractions and the bulk analysis (e.g. for Cu) probably results from a heterogeneous distribution of contaminants.

Based on the XRF analyses, sample LOB2 had a Pb level which exceeded its ICRCL trigger value for domestic gardens and allotments, and Zn, Cu and Ni levels that exceeded their ICRCL trigger values for areas where plants are grown (ICRCL 1987). Sample LOB4 also had Pb and Zn levels above trigger values.

Induction Coupled Plasma Atomic Emission Spectroscopy on site surface water showed detectable levels of Zn ($300\,\mu g\,l^{-1}$), Ni ($300\,\mu g\,l^{-1}$), Cu ($30\,\mu g\,l^{-1}$), and As ($20\,\mu g\,l^{-1}$). Levels of Zn exceeded its Dutch B value ($250\,\mu g\,l^{-1}$), whereas that of Ni exceeded its Dutch C value ($200\,\mu g\,l^{-1}$; Cairney 1995).

SEM-EDS

Samples of the 300–400 and <63 μm fractions were mounted on aluminium pedestals using carbon based double sided tape, and coated with carbon by vacuum evaporation. These were viewed in both secondary electron (SE) and back-scattered electron (BSE) mode using a Camscan series 4 Scanning Electron Microscope. Heavy metal contaminated particles were identified by their high BSE luminescence. Points on individual particles were analysed under the SEM with a Link AM10,000 Energy Dispersive X-ray Analysis System. The depth of penetration for this method varies with the atomic number,

but the analysis is weighted towards the surface. As with XRF, low atomic number elements are not detectable using 'thick window' EDS.

A BSE micrograph of the <63 μm fraction of specimen LOB2 is shown in Fig. 2. EDS on the bright particle in the centre of the frame indicates that it is a particle of elemental lead, or a lead compound such as lead sulphate (the sulphur signature is not distinguishable when lead is present). Other electron micrographs show metals such as tin associated with small clusters of clay particles.

Figure 3 is a close-up SE micrograph of the surface of a sand sized particle from the 400–300 μm sieve fraction of specimen LOB2, which shows that it is an agglomeration of finer particles. EDS on the silt sized prismatic structures shows that they are quartz grains, whereas the finer, amorphous material gave a clay signature. EDS on the region arrowed showed that copper, iron and halogens (iodine and chlorine) are present on the surface of the agglomeration. Figure 4 shows a BSE image of another region of the same particle. EDS on the bright flake showed a dominant copper peak, with few other signatures (but note that elements lighter than sodium may have been present).

A SE micrograph of the <63 μm fraction of specimen LOB4 (Fig. 5) shows a silt sized clay floc surrounded by fainter uncontaminated angular quartz and iron sulphate particles. EDS detected lead on this floc, but the use of different forms of electron micrograph indicated that it was present as particles within the floc rather than as a continuous coating.

A BSE micrograph of the 400–300 μm fraction of specimen LOB4 (Fig. 6) shows sand sized agglomerations of clay platelets. The brightness

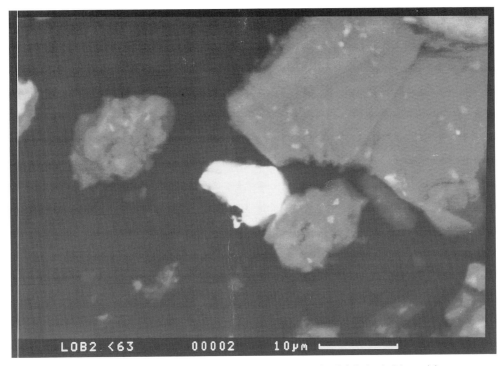

Fig. 2. Backscatter electron image of <63 μm fraction of LOB2, showing bright lead rich particle.

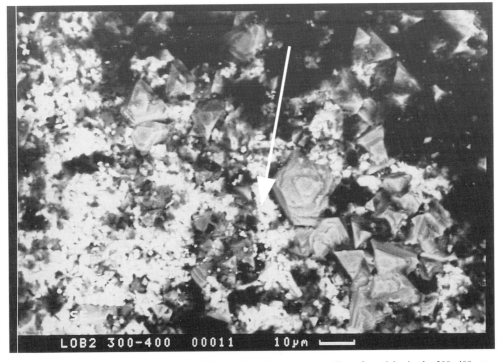

Fig. 3. Close-up secondary electron image of the surface of an agglomeration of particles in the 300–400 μm fraction of LOB2. Arrowed particle was analysed by EDS (see text).

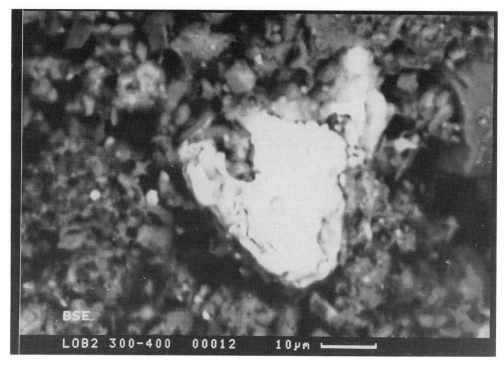

Fig. 4. Close-up backscatter electron image of the surface of an agglomeration in the 300–400 μm fraction of LOB2, showing a bright copper-rich flake.

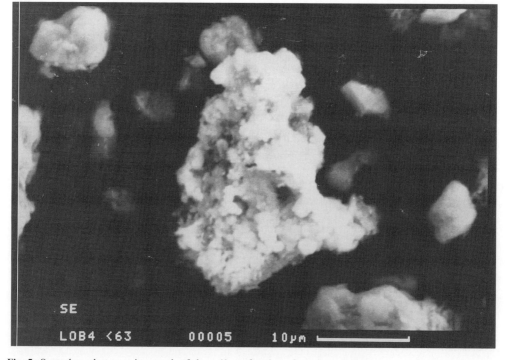

Fig. 5. Secondary electron micrograph of the <63 μm fraction of specimen LOB4.

Fig. 6. Backscatter electron micrograph of a sand-sized agglomeration of clays in 300–400 μm fraction of specimen LOB4. Area indicated by the arrrow is enlarged in Fig. 7.

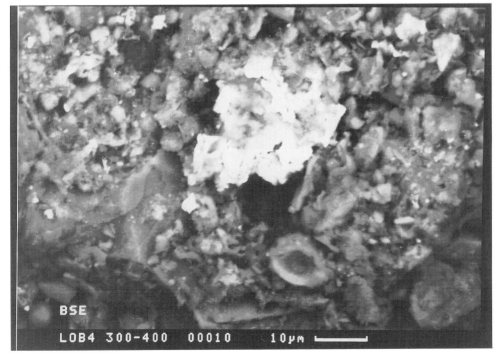

Fig. 7. Close-up of a lead-contaminated clay cluster arrowed in Fig. 6.

of the area arrowed, shown enlarged in Fig. 7 indicates contamination. EDS on this area gave a strong aluminosilicate signature from the clay along with a lead signature.

Sequential extractions

Sequential extractions were carried out following a procedure modified from that proposed by the Commission of European Communities Bureau of Reference (Davidson *et al.* 1994) for stream sediments. The aim was to distinguish between metals which are (i) in highly soluble forms, (ii) in the exchange complex and bound to carbonates, (iii) reducible (bound to Fe/Mn oxides) and (iv) oxidizable (e.g. organically complexed). All extractions were carried out on air dried soil passing the 2 mm sieve. Duplicate extractions were performed to check reproducibility.

Extractions were carried out in 50 ml Nalgene centrifuge tubes with sealable watertight caps. Laboratory vessels were rinsed in dilute nitric acid and distilled water before each use. Analysis of solutions was carried out using a ARL MAXIM simultaneous measurement Induction Coupled Plasma Atomic Emission Spectroscope.

Extraction stages

Step one. 25 ml of deionized water was added to 0.625 g of dry soil in a centrifuge tube, agitated for 1 h, and then centrifuged at 5000 g for 15 min. The supernatant was decanted off and retained. This was repeated with the supernatant retained separately.

Step two. 25 ml of 0.1 M acetic acid was added to the soil residue and the tube was agitated overnight. The supernatant was separated and retained. The soil residue was washed with 10 ml distilled water.

Step three. 30 ml of 0.1 M hydroxylammonium chloride (adjusted to pH 2 with dilute nitric acid) was added to the residue and the procedure in step two repeated.

Step four. 6 ml of 30% hydrogen peroxide was added to the soil residue and the tube was shaken manually until effervescence ceased. Digestion was continued by heating in an 85°C water bath until the contents had evaporated down to a small volume (<2 ml). This was repeated with a further 6 ml of H_2O_2. Then 30 ml of 1 M ammonium acetate (adjusted to pH 2

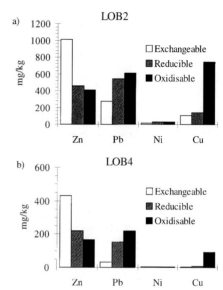

Fig. 8. Sequential extraction data: (**a**) LOB2; (**b**) LOB4.

with dilute nitric acid) was added, the tube was shaken overnight and the supernatant extracted and retained.

The sequential extraction results (Fig. 8) show a similar pattern for both specimens, with Zn mainly in the exchangeable phase, Pb mainly in the reducible and oxidizable phases and Cu in the oxidizable phase. The lack of Ni in the extractions indicates that it is mainly in the residual phase. Water extractions (data not shown) from both specimens contained about $50\,\mu\mathrm{g}\,\mathrm{l}^{-1}$ of Zn but no detectable Cu, Ni or Pb.

Discussion

XRF analysis of the various size fractions showed Pb, Cu and Zn are concentrated in the finer fractions in soils both on and near the site. SEM-EDS analysis revealed that Pb and Cu in the coarser fractions were usually associated with fine particles in large agglomerations. It also revealed that even in the fine fractions, Pb and Cu contamination was not uniform, but was restricted to specific particles. Zn-rich areas were not located despite this metal having the highest concentration in the bulk soils.

Electron micrographs of the off-site soil showed localized Pb-rich areas (perhaps precipitates) on a number of clay flocs. In contrast, the images of the on-site soil showed Pb and Cu separately in small non-silicate particles. This suggests that the on-site soil is contaminated with particulates, while the off-site is contaminated by dissolved metals that have been precipitated.

Sequential extractions indicated that Zn is more soluble than the other metals. It was present in surface water migrating off site and in the off-site soil. SEM-EDS failed to detect discrete zinc-rich areas, which suggests that Zn was thinly dispersed in the exchange complex. The absence of Pb and Cu in the on-site water and in the water extractions suggests there is little leaching of these metals at present, although the presence of Pb precipitates in the off-site soil indicates that soluble forms of Pb have entered these soils. This may have been associated with battery acid spills.

The case study illustrates the importance of applying a range of techniques for the investigation of contaminant metal speciation. In the absence of SEM-EDS data, it would not have been possible to determine that the Pb in the off-site soil was precipitated on clay flocs. Hence the distinction between primary contamination (i.e. lead particulates), and secondary contamination mobilized from the original contamination could not have been made. Conversely, SEM-EDS failed to detect zinc, and without sequential extraction data and surface water analysis the mobility of this metal would not have been apparent.

Summary

Industrial land uses that result in the highest risk from heavy metals are mining of non-ferrous metals, smelting of such metals (particularly Cr), chlorine-alkali manufacture (Hg contamination), timber treatment (As, Cr, Cd contamination), tanning (Cr contamination), metal finishing (Cr, Cd, Cu, Zn, Ni contamination), vehicle breaking (Pb, Cd, Zn, Cu contamination), weapons disposal (Pb, Hg, Cu, Zn, As contamination), paint manufacture and use (Pb, Cd, Hg, Zn contamination) and sites where sewage has been disposed of (contamination by Cd, Hg, Cr, others).

Of the metals considered, cadmium and mercury are the most toxic to humans and are also bioaccumulative; historically, poisoning incidents by these metals has been via the contaminated foodstuffs pathway. Arsenic is less toxic than cadmium or mercury, as it is more easily eliminated from organisms, but is a more common contaminant and is often present at relatively high levels. In the developing world the most common pathway to humans is via drinking water, but in the UK soil ingestion is of more concern. Chromium is extremely mobile

in its ($+6$) oxidation state, hence contamination of drinking water and eco-toxicity are the main concerns, although the very high levels present in smelter waste may also make skin contact and ingestion serious risks. Lead is relatively immobile in the geosphere, and does not usually represent a threat to waters unless in organic forms; the main hazards from contaminated sites are ingestion of contaminated soil by children. Zinc, copper and nickel are not very toxic to humans and usually only present local problems with revegetation of sites.

Of the techniques for determining speciation at contaminated sites, simple batch leaching tests currently offer the simplest method for incorporating mobility into risk assessment. However, single leaching test results cannot be interpreted in fundamental terms. Sequential chemical extractions offer more details on the conditions under which metals are soluble and therefore greater insight on bioavailability. Characterization of leaching behaviour over a range of pH/*Eh* conditions, contact times and L/S ratios may allow identification of a range of possible solid phases that control leachate composition. Spectroscopic techniques such as SEM-EDS give direct information on the contaminant phases present, although where used alone they are likely to miss important dispersed phases present below their detection limits. Unambiguous speciation of contaminant metals in soils requires an integrated approach using both spectroscopic techniques and leaching tests.

As an example of such an integrated approach, a case study is reported in which a soil from a former smelter and scrapyard was investigated. The case study demonstrates the importance of an investigative approach that employs a range of analytical techniques to evaluate toxic metal distribution and mobility. It showed that total element analyses of separated size fractions, characterization leaching tests such as sequential extractions, and detailed electron microscope analysis can provide complementary information on the location and chemistry of toxic metals in soils.

The authors would like to thank Professor B. Yardley, Dr D. Banks, and Dr E. Condliffe for their assistance.

References

ALLOWAY, B. J. 1995. *Heavy Metals in Soils*. Blackie, New York.
AMERICAN SOCIETY FOR TESTING MATERIALS. 1979. *Proposed Methods for Leaching of Waste Materials*. Annual book of ASTM standards, Part 31 (Water), 1258–1261.

——1985. D3987-81. *Shake Extraction of Solid Waste with Water.* Annual book of ASTM standards, Section 11 (Water and Environmental Technology), 22–26.

——1991. D4874-89. *Standard Test Method for Leaching Solid Waste in a Column Apparatus.* Annual book of ASTM standards, Section 11 (Water and Environmental Technology), 158–164.

BAKER, D. E. & SENFT, J. P. 1995. Copper. *In:* ALLOWAY, B. J. (ed.) *Heavy Metals in Soils.* Chapman & Hall, London, 179–202.

BANWELL, C. N. & MCCASH, E. M. 1994. *Fundamentals of Molecular Spectroscopy.* 4th edn, McGraw-Hill, New York.

BAVERMAN, C, MORENO, L. & NERETNIEKS, I. 1997. *Using CHEMFRONTS, a geochemical transport programme to simulate leaching from waste materials.* WASCON '97, The International Conference for the environmental and technical implications of construction with alternative materials, Houthem, The Netherlands.

BOURG, A. C. M & LOCH, J. P. G. 1995. Mobilisation of Heavy Metals as Affected by pH and Redox Conditions. *In:* SALOMONS, W. & FORSTNER, A. U. (eds) *Biogeodynamics of Pollutants in Soils and Sediments; Risk Assessment of Delayed and Non-Linear Responses.* Springer, Berlin, 213–238.

BOWELL, R. J. 1994. Sorption of arsenic by iron oxides and oxyhydroxides in soils. *Applied Geochemistry,* **9**, 279–286.

—— & BRUCE, I. 1995. Geochemistry of iron ochres and mine waters from Levant Mine, Cornwall. *Applied Geochemistry,* **10**, 237–250.

BREEZE, V. G. 1973. Land reclamation and river pollution problems in the Croal Valley caused by waste from chromate manufacture. *Journal of Applied Ecology,* **10**, 513–525.

BREWARD, N. & PEACHEY, D. 1995. The development of a rapid scheme for the elucidation of the chemical speciation of elements in sediments. *Science of the Total Environment,* **29**, 155–162.

BRIDGES, E. M. 1987. *Surveying Derelict Land.* Clarendon, Oxford.

BRIGGS, D. & SEAH, M. P. 1983. *Practical Surface Analysis (by Auger and X-ray Photoelectron Spectroscopy).* Wiley, New York.

BRITISH STANDARDS INSTITUTE. 1996. Draft European Standard PrEN 12457 leaching: compliance test for leaching of granular materials. **96/105207DC**.

CAIRNEY, T. 1995. *The Reuse of Contaminated Land – a Handbook for Risk Assessment.* Wiley, Chichester.

CHEESEMAN, C. R., BUTCHER, E. J., SOLLARS, C. J. & PERRY, R. 1993. Heavy metal leaching from hydroxide, sulphide and silicate stabilised/solidified wastes. *Waste Management,* **13**(8), 545–552.

CHUAN, M. C., SHU, G. Y. & LIU, J. C. 1996. Solubility of heavy metals in a contaminated soil: effects of redox potential and pH. *Water, Air and Soil Pollution,* **90**, 543–566.

COPPIN, N. J., STAFF, M. G. & JOHNSON, M. S. 1992. Environmental impact of old metal mines. *Proceedings of the Mining and Metallurgy Conference: Minerals Metals and the Environment.* Elsevier, London, 104–114.

CRAIG, P. J. 1986. *Organometallic Compounds in the Environment; Principles and Reactions.* Longman.

DAVIDSON, C. M., THOMAS, R. P., MCVEY, S. E., PERALA, R., LITTLEJOHN, D. & URE, A. M. 1994. Evaluation of a sequential extraction procedure for the speciation of heavy metals in sediments. *Analytica Chimica Acta,* **291**, 277–286.

DAVIS, A. & OLSEN, R. L. 1995. The geochemistry of chromium migration and remediation in the subsurface. *Ground Water,* **33**(5), 759–768.

DAVIS, B. E. 1995. Lead. *In:* ALLOWAY, B. J. (ed.) *Heavy Metals in Soils.* Chapman & Hall, London, 206–223.

DEPARTMENT OF THE ENVIRONMENT. 1976. *Waste Management Paper 11. Metal Finishing Wastes.* HMSO, London.

——1977. *Waste Management Paper 12. Mercury Bearing Wastes.* HMSO, London.

——1978. *Waste Management Paper 17. Wastes from Tanning, Leather Dressing and Fellmongering.* HMSO, London.

——1980a. *Waste Management Paper 16. Wood Preserving Wastes.* HMSO, London.

——1980b. *Waste Management Paper 20. Arsenic Bearing Wastes.* HMSO, London.

——1981. *Waste Management Paper 23. Special Wastes: a technical memorandum providing guidance on their definition.* HMSO, London.

——1984. *Waste Management Paper 24. Cadmium Bearing Wastes.* HMSO, London.

——1995a. *Metal manufacturing, refining and finishing: lead works.* Industry Profile.

——1995b. *Chemical works: explosives, propellants and pyrotechnics manufacturing works.* Industry Profile.

——1995c. *Chemical works: Inorganic chemicals manufacturing works.* Industry Profile.

——1995d. *Chemical works: coatings (paints and printing inks) manufacturing works.* Industry Profile.

——1995e. *Waste Recycling, treatment and disposal sites: metal recycling sites.* Industry Profile.

——1995f. *Animal and animal products processing works.* Industry Profile.

——1995g. *Timber treatment works.* Industry Profile.

——1995h. *Profile of miscellaneous industries.* Industry Profile.

——1995i. *Gas works, coke works and other coal carbonisation plants.* Industry Profile.

——1995j. *Engineering Works: shipbuilding, repair, shipbreaking.* Industry Profile.

——1995k. *Chemical Works: pharmaceuticals manufacturing works.* Industry Profile.

——1995l. *Engineering Works: mechanical engineering and ordnance works.* Industry Profile.

——1995m. *Engineering Works: electrical and electronic equipment manufacturing works.* Industry Profile.

——1995n. *Power Stations (excluding nuclear power stations).* Industry Profile.

——1995o. *Chemical works: fine chemicals manufacturing works.* Industry Profile.

DUTRE, V. & VANDECASTEELE, C. 1995. Solidification/stabilisation of arsenic-containing waste: leach tests and behaviour of aresenic in the leachate. *Waste Management,* **15**(1), 55–62.

EIGHMY, T. T., EUSDEN, J. D. JR, KRANOWSKI, J. E., DOMINGO, D. S., STAMPFLI, D., MARTIN, J. R. & ERICKSON, P. 1995. Comprehensive approach toward understanding element speciation and leaching behaviour in municipal solid waste incineration electrostatic precipitator ash. *Environmental Science and Technology*, **29**, 629–646.

EUROPEAN COMMUNITY. 1980. The protection of groundwater against pollution caused by certain dangerous substances. EC directive 80/68/EEC. *Official Journal of the European Communities*, **23**, L20, 43–48.

FALLMAN, A.-M. 1997. *Leaching of chromium from steel slag in laboratory and field tests – a solubility controlled process?* WASCON '97, The International Conference for the environmental and technical implications of construction with alternative materials, Houthem, The Netherlands.

FASSET, D. W. 1980. Cadmium. *In*: WALDRON, H. A (ed.) *Metals in the Environment*. Academic, London, 61–110.

GEMMEL, R. P. 1973. Revegetation of derelict land polluted by a chromate smelter, part I: chemical factors causing substrate toxicity in chromate smelter waste. *Environmental Pollution*, **5**, 181–197.

GRANDJEAN, P. 1975. Lead in Danes; Historical and toxicological studies. *In*: GRIFFIN, T. B. & KNELSON, J. H. (eds) *Lead*. Academic, London, 6–75.

GRAY, N. F. 1996. The use of an objective index for the assessment of the contamination of surface water and groundwater by acid mine drainage. *Journal of the Chartered Institution of Water and Environmental Management*, **10**, 332–340.

GRIMES, S. M., JOHNSTON, S. R. & BATCHELDER, D. N. 1995. Lead Carbonate system; solid-dilute solution exchange reactions in aqueous systems. *Analyst*, **120**, 2741–2746.

GUL, R. 1994. Heavy metal content of the Keban Lead Plant Slag and Movement of Metals in Soil of the Surrounding Region. *Water Pollution Research Journal of Canada*, **29**(4), 531–544.

HUDSON-EDWARDS, K. A., MACKLIN, M. G., CURTIS, C. D. & VAUGHAN, D. J. 1995. Characterisation of Pb- Zn- Cd- and Cu-bearing phases in river sediments by combined geochemical and mineralogical techniques. *In*: *Contaminated Soils. Proceedings of the Third International Conference on the Biogeochemistry of Trace Elements*, Paris.

——, ——, —— & ——1996. Processes of formation and distribution of Pb-, Zn-, Cd- and Cu-bearing minerals in the Tyne Basin, Northeast England: Implications for Metal-Contaminated River Systems. *Environmental Science and Technology*, **30**(1), 72–80.

HUTCHINSON, T. C. 1979. Copper contamination of ecosystems caused by smelting activities. *In*: NRIAGU, J. O. (ed.) *Copper in the Environment*. Wiley, New York, 451–502.

——1982. Nickel. *In*: LEPP, N. W. *Effect of Heavy Metal Pollution on Plants*. Applied Science, London, 171–212.

INSTITUTE OF CIVIL ENGINEERS. 1994. *Contaminated Land – Investigation, Assessment and Remediation*. Thomas Telford, London.

INTERDEPARTMENTAL COMMITTEE ON THE REDEVELOPMENT OF CONTAMINATED LAND (ICRCL). 1987. *Guidance on the Assessment and Redevelopment of Contaminated Land*. Guidance Note 59/83 (2nd edn). Department of the Environment, London.

JACOBS, J. H. 1992. Treatment and stabilisation of hexavalent chromium containing waste material. *Environmental Progress*, **11**(2), 123–126.

JOHNSTON, S. R., WEST, L. J. & STEWART, D. I. 1997. Microscope Spectroscopy on Contaminated Soil. *In*: YONG, R. N. & THOMAS, H. R. (eds) *Geoenvironmental Engineering. Contaminated Ground: Fate of Pollutants and Remediation*. Thomas Telford, London, 91–98.

KELLY, M., ALLISON, W. J., GARMAN, A. R. & SYMON, C. J. 1988. *Mining and the Freshwater Environment*. Elsevier.

KNOX, R. C. & CANTER, W. 1996. Prioritisation of ground water contaminants and sources. *Water, Air and Soil Pollution*, **88**, 205–226.

KUBOTA, T. A. M. & ORIKASA, K. 1995. Distribution of different fractions of cadmium, zinc, lead and copper in unpolluted and polluted soils. *Journal of Water, Air, and Soil Pollution*, **83**(3/4), 187–194.

LACERDA, L. D., MALM, O., GUINARAES, J. R. D., SALOMANS, W. & WILKEN, R. D. 1995. Mercury and the New Gold Rush in the South. *In*: SALOMANS, W. & FORSTNER, A. U. (eds) *Biogeodynamics of Pollutants in Soils and Sediments; Risk Assessment of Delayed and Non-Linear Responses*. Springer, Berlin, 213–238.

LAVENDER, S. J. 1980. *New Land for Old*. Adam Hilger, Bristol.

LEE, S. & SPEARS, D. A. 1995. The long term weathering of PFA and implications for groundwater pollution. *Quarterly Journal of Engineering Geology*, **28**, S1–S15.

LODDING, A. 1988. Secondary ion mass spectrometry. *In*: WINEFORDNER, J. D. & KOLTHOFF, I. M. (eds) *Inorganic Mass Spectrometry*. Monographs on Chemical Analysis No. 95, Wiley, London, 125–172.

MCGRATH, S. P. 1995. Chromium and Nickel. *In*: ALLOWAY, B. J. (ed.) *Heavy Metals in Soils*. Chapman & Hall, London, 152–174.

MCKINLEY, W. S., PRATT, R. C. & MCPHILLIPS, L. C. 1992. Cleaning up chromium. *Civil Engineering*, March 1992, 69–71.

MACHINGAWUTA, N. C. & BROADBENT, C. P. 1994. Incorporation of Arsenic in silicate slags as a disposal option. *Transactions of the Institute of Mining and Metallurgy*, **103**, C1–C8.

MITRA, S. 1986. *Mercury in the Ecosystem*. Transtech, Lancaster, PA.

MOORE, J. W. & RAMAMOORTHY, S. 1983. *Heavy Metals in Natural Waters: Applied Monitoring and Impact Assessment*. Springer-Verlag, New York.

NATIONAL RIVERS AUTHORITY. 1994. *Leaching tests for Assessment of Contaminated Land: Interim NRA Guidance*. R&D Note 301.

NRIAGU, J. O. & NIEBOER, E. (eds) 1988. *Chromium in the Natural and Human Environments*. Wiley, New York.

O'NEILL, P. 1995. Arsenic. *In*: ALLOWAY, B. J. (ed.) *Heavy Metals in Soils.* Chapman & Hall, London, 105–121.

PHILLIPS, I. & CHAPPLE, L. 1995. Assessment of a heavy metals contaminated site using sequential extraction, TCLP and risk assessment techniques. *Journal of Soil Contamination,* **4**(4), 311–325.

POURBAIX, M. 1966. *Atlas of Electrochemical Equilibria in Aqueous Solutions.* Pergamon, London.

PORTER, E. K. & PETERSON, P. J. 1977. Arsenic tolerance in grasses growing on mine waste. *Environmental Pollution,* **14**, 255–265.

PULS, R. W., CLARK, D. A., PAUL, C. J. & VARDY, J. 1994. Transport and Transformation of Hexavalent Chromium Through Soils and into Groundwater. *Journal of Soil Contamination,* **3**(2), 203–224.

RAMOS, L., HERNANDEZ, L. M. & GONZALEZ, M. J. 1994. Sequential fractionation of copper, lead, cadmium and zinc in soils from or near Donana National Park. *Journal of Environmental Quality,* **23**, 50–57.

SADLER, R., OLSZOWY, H., SHAW, G., BILTOFT, R. & CONNELL, D. 1994. Soil and water contamination by arsenic from a tannery waste. *Water, Air and Soil Pollution,* **78**, 189–198.

SALOMONS, W. 1995. Long term strategies for handling contaminated sites and large-scale areas. *In*: SALOMONS, W. & FORSTNER, A. U. (eds) *Biogeodynamics of Pollutants in Soils and Sediments; Risk Assessment of Delayed and Non-Linear Responses.* Springer, Berlin, 1–30.

SCHNOOR, J. L. 1996. *Environmental Modelling: Fate and Transport of Pollutants in Water, Air, and Soil.* Wiley, New York.

SIMMS, D. L. & BECKETT, M. J. 1987. Contaminated land: setting trigger concentrations. *Science of the Total Environment,* **65**, 121–124.

—— & MORGAN, H. 1988. The Shipham report – an investigation into cadmium contamination and its implications for human health –Introduction. *Science of the Total Environment,* **75**, 1–10.

SMITH, G. C. 1994. *Surface Analysis by Electron Spectroscopy.* Plenum, New York.

SOLOMON, C. L. & POWRIE, W. 1992. Assessment of health risks in the redevelopment in contaminated land. *Geotechnique,* **42**(1), 5–12.

TESSIER, A., CAMPBELL, P. G. C. & BISSON, M. 1979. Sequential extraction procedure for speciation of particulate trace metals. *Analytical Chemistry,* **51**(7), 844–850.

THORNTON, I. 1979. Copper in soils and Sediments. *In*: NRIAGU, J. O. (ed.) *Copper in the Environment.* Wiley, New York, 171–216.

TUIN, B. J. W. & TELS, M. 1990. Distribution of six heavy metals in contaminated clay soils before and after extractive cleaning. *Environmental Technology,* **11**, 935–948.

URE, A. M. & DAVIDSON, C. M. (eds) 1995. *Chemical Speciation in the Environment.* Chapman & Hall, London.

VAN DER SLOOT, H. A. 1996. Developments in evaluating environmental impact from utilization of bulk inert wastes using laboratory leaching tests and field evaluation. *Waste Management,* **16**(1–3), 65–81.

——, HEASMAN, L. & QUEVAUVILLER, PH. 1997. *Harmonisation of Leaching/Extraction Tests.* Studies in Environmental Science 70, Elsevier, Amsterdam.

——, HOEDE, D. & COMANS, R. N. J. 1994. The influence of reducing properties on leaching of elements from waste materials and construction materials. *In*: GOOMANS, J. J. J. M., VAN DER SLOOT, H. A., & AALBERS, TH. G. (eds) *Environmental Aspects of Construction with Waste Materials.* Elsevier, Amsterdam.

WALSH, A. R. & O'HALLORAN, J. 1996. Chromium speciation in tannery effluent – II. Speciation in the effluent and in a receiving estuary. *Water Research,* **30**(10), 2401–2412.

WATT, I. M. 1985. *The Principles and Practice of Electron Microscopy.* Cambridge University Press, Cambridge.

WEAST, R. C. 1973. *The Handbook of Chemistry and Physics.* 53rd edn. CRC Press.

WERNICKE, C., WIENBERG, R., GERTH, J., WILICHOWSKI, M., FORSTNER, U., & WERTHER, J. 1993. Methods of treating soil contaminated with chromium. *In*: ARENDT, F. (ed.) *Contaminated Soil '93.* Fourth International KfK/TNO Conference on Contaminated Soil, Berlin, vol. 2, 1467–1468.

WEST, L. J. & STEWART, D. I. 1995. *Electrokinetic processing of sewage sludge contaminated with phytotoxic metals.* Symposium on Natural Hazards and Environmental Geotechniques, Bangkok, Research Papers Vol., 27–36.

WINDMOLLER, C. C., WILKEN, R. D. & JARDIN, W. F. 1996. Mercury speciation in contaminated soils by thermal release analysis. *Water, Air and Soil Pollution,* **89**, 399–416.

YARLAGADDA, P. S., MATSUMOTO, M. R., VANBENSCHOTEN, J. E. & KATHURIA, A. 1995. Characteristics of heavy metals in contaminated soils. *ASCE Journal of Environmental Engineering,* **121**(4), 276–286.

YONG, R. N., GALVEZ-CLOUTIER, R. & PHADUNGCHEWIT, Y. 1993. Selective sequential extraction analysis of heavy-metal retention in soil. *Canadian Geotechnical Journal,* **30**, 821–833.

——, MOHAMED, A. M. O & WARKENTIN, B. P. 1992. *Principles of Contaminant Transport in Soils.* Developments in Geotechnical Engineering 73. Elsevier, Amsterdam.

Evaluation of the containment properties of geological and engineered barriers by pore-water extraction and characterization

SHAUN REEDER & MARK R. CAVE

Analytical and Regional Geochemistry Group, British Geological Survey, Keyworth, Nottingham, NG12 5GG, UK

Abstract: In the construction of underground repositories for the disposal of potentially toxic wastes, it is critical to the development of a safety case that the chemical composition of waters flowing through the proposed containment barrier is fully understood. Because the barrier material is usually chosen to have low permeability, to minimize the risk of long-term transport of toxic material into the biosphere, it is often not possible to obtain sufficient flowing groundwater samples to allow complete hydrogeochemical characterization. In order to study the chemical composition of the water in these formations it is necessary, therefore, to use specialized pore-water extraction techniques.

In this study, the practicalities of three methods for extracting and characterizing pore-waters from potential containment materials are discussed. These are: (i) mechanical squeezing using a hydraulic press; (ii) centrifugation using a heavy liquid displacent; and (iii) aqueous leaching of residual salts from crushed core material.

Three case studies are presented which demonstrate how each of the different pore-water characterization methods may be applied practically, and how the data obtained can prove invaluable in understanding the hydrogeochemistry of sites.

Many types of geological materials have physico-chemical properties that make them potentially suitable for the containment of toxic wastes. For deep disposal, important factors include low permeability, low hydraulic conductivity, suitable physical properties (e.g. plastic rather than brittle deformation behaviour), suitable chemical properties (e.g. reducing conditions) and ease of repository construction. Mudrocks, crystalline rocks and evaporites all display some, if not all, of these relevant properties and have been considered by several countries for the deep disposal of radioactive wastes (Savage 1995).

In demonstrating the long-term safety of a deep underground repository it is necessary to evaluate the site-specific geological and hydrogeological regimes in detail, and to predict the future chemical processes that might control pollutant migration using information about the past movement of pore-waters and solutes. One of the most important features of a rock–water system to consider, therefore, is the spatial variation of groundwater composition. Geochemical data provide essential information on groundwater residence times, flow paths and flow rates that may be used to validate hydrogeological flow models. The data are also essential in understanding radionuclide transport behaviour, including solubility and sorption properties.

Because of their low permeability, flowing groundwaters cannot generally be obtained from clays and mudrocks. Although fracture flows can frequently be induced from crystalline formations, the frequency of features such as open fractures that conduct water flow is normally limited. To evaluate the chemistry within geological and engineered barriers it is necessary, therefore, to characterize matrix pore-waters obtained using specialized extraction techniques.

Methods for pore-water characterization

To effectively eliminate sample handling perturbations, the chemical composition of pore-waters would ideally be determined *in situ*. Downhole geophysical methods for the measurement of parameters such as electrical conductivity and temperature are often used in this manner (Keys 1997). Methods for the *in situ* determination of pH and redox potential are also available (e.g. Grenthe *et al.* 1992). *In situ* techniques for more extensive chemical analysis of major and trace elements are, however, generally either not suitable or insufficiently established for the characterization of pore-waters in the materials used for geological and engineered barriers.

Most pore-water characterization in this field therefore requires the physical removal of samples from their natural environment, normally as

From: METCALFE, R. & ROCHELLE, C. A. (eds) 1999. *Chemical Containment of Waste in the Geosphere.* Geological Society, London, Special Publications, **157**, 265–273. 1-86239-040-1/99/$15.00 © The Geological Society of London 1999.

drillcore, followed by extraction of the pore-water from the rock matrix.

To maintain conditions as close as possible to those *in situ*, it is essential that sample perturbation is minimized. The most important factors to consider are: choosing a method of sampling that restricts oxidation, contamination and other modifications; if drilling fluid is used, using a suitable tracer to enable later studies on the extent of any contamination; handling samples, both during collection and extraction, in a manner that limits oxidation and other modifications; and adequately preserving the samples to prevent desiccation, evaporation and oxidation.

A number of pore-water or solute extraction methods are reported in the literature. Vacuum filtration (Bufflap & Allen 1995) involves the application of a vacuum to the sample causing pore-water to be drawn-out from the pore spaces. Dialysis or diffusion methods involve equilibration of pore-water in the sample with deionized water (Brandl & Hanselmann 1991) or a solid phase material which takes up the chemical constituents of the pore-water (Zhang *et al.* 1998). Both methods are generally restricted to relatively high moisture content, unconsolidated sediments and have limited potential to the characterization of barrier materials. Distillation of water vapour and other volatile solvents may be used to determine stable isotope contents (Moreau-Le Golvan *et al.* 1997), but pore-water solutes are left behind in the rock matrix so cannot be characterized.

There are three methods, however, which are particularly applicable to the extraction of pore-waters or residual solutes from barrier materials. These are: mechanical squeezing using a hydraulic press; centrifugation using a heavy liquid displacent; and aqueous leaching of residual salts from crushed core material. Details of these methods and examples of their application are discussed in the following sections.

Mechanical squeezing

Squeezing of pore-fluids from geological materials is a fundamentally simple technique, analogous to natural consolidation although at a greatly accelerated rate.

The technique is very flexible and for high porosity, high permeability materials such as sediments, sand, peat and chalks, low-pressure or whole core squeezing is effective, simple to operate and inexpensive.

Importantly, heavier duty squeezing apparatus operated by hydraulic or mechanical action (Kruikov 1947; Manheim 1966) is often the only practical way of physically extracting pore-fluid for geochemical analysis from low permeability clays and mudstones.

Entwisle & Reeder (1993) described a new squeezing apparatus that used commercially available hydraulic pumps of maximum output stress 70 MPa and the capability to maintain constant pressure over a number of weeks. Using this method, pore-waters have been obtained from mudstones with moisture contents as low as 7% (with respect to dry sample weight), although tests may take days, and in some cases weeks, to complete. Except for very low porosity samples, the volumes of pore-water obtained by squeezing are often enough for the analysis of many different chemical species. Typically, the percentage recovery is in the range 15 to 60%. The pore-water extracted is also suitable for the analysis of stable and radiogenic isotopes. Stiff, cemented and highly indurated rocks, and materials of moisture content lower than approximately 7%, tend not to deform sufficiently to allow the displacement of pore-water and are, therefore, unsuitable for extraction of pore-water by squeezing.

So long as sample collection, handling and preservation methods are satisfactory, pore-waters obtained by squeezing tend to be representative of *in situ* conditions when compared to chemical compositions of relevant groundwaters or pore-waters obtained by other methods (Bath 1993; Bath *et al.* 1996; Cottour 1997).

Despite this, fractionation effects and other artifacts are frequently observed and the composition of the extracted pore-water tends to change with increasing stress. In general, the concentration of divalent and trivalent species tends to increase with increasing pore-water extraction and the concentration of monovalent species decreases, although these effects can be highly variable. Overall, the total solute concentration tends to be reduced with increasing extraction. Further study is required to understand fully these fractionation effects, but a number of factors probably contribute, including mineral dissolution and re-equilibration, cation exchange and ion exclusion or filtration. Although these effects complicate the interpretation of results, they may also be used to improve understanding of pore-fluid interactions in clays and mudrocks.

Centrifugation

Centrifugation is a fairly rapid, simple and relatively inexpensive technique. Two methods are commonly used. Edmunds & Bath (1976) described a free drainage technique in which sample is placed in a centrifuge bucket above a porous plate support through which pore-water is collected into a detachable cup during centrifugation. Kinniburgh & Miles (1983) described a displacement technique in which an inert, dense, immiscible liquid (1,1,2-trichloro-1,2,2-trifluoroethane) is added to the sample. On centrifugation, the immiscible liquid penetrates the sample causing progressive displacement of the pore-water from larger cracks and interconnected pores. The displaced pore-water is then withdrawn from the surface of the displacant using a disposable pipette.

Centrifugation is the preferred technique for the extraction of pore-water from chalks, sandstones, sands and other materials of relatively high permeability. The method is generally not suitable for very low permeability materials, such as clays. The porosity of samples is often not an issue, so long as the material is relatively permeable, and pore-water may be extracted from sandstones and chalks with moisture contents as low as 1%. The likelihood and efficiency of extraction, particularly from relatively lower porosity and decreasingly permeable material, is increased substantially when using the displacement technique.

So long as sample collection, handling and preservation methods are satisfactory, porewaters obtained by centrifugation methods tend to be representative of *in situ* conditions when compared to the chemical compositions of corresponding groundwaters (Bath *et al.* 1988). This is not unsurprising since the method, although *ex situ*, is relatively simple and non-intrusive. Fractionation effects similar to those reported for mechanical squeezing may occur, but to a much smaller extent.

Aqueous leaching

The process of leaching when applied to the measurement of pore-water compositions is essentially very simple. A leaching solution, normally deionized water, is applied to the sample in such a manner that it accesses the pore space within the rock material, mixing with the pore-water or dissolving residual solutes of the pore-water if the sample has been dried first. The leaching solution is then separated from the rock matrix and analysed for the chemical components extracted from the pore-water.

Leaching has two clear practical advantages over other methods: (i) it is simple to carry out without the requirement for expensive equipment; and (ii) it is applicable to most rock types, regardless of moisture content. There are,

however, considerable problems arising from solubilization of rock solids and effects caused by changing rock–water equilibria that make interpretation of leaching data very difficult. Additionally, uncertainties in sample moisture contents used to correct residual solute data can result in uncertainty in the calculated *in situ* pore-water compositions. Because of these interpretation difficulties, leaching tends to be used as a last resort when other methods cannot be applied because the sample is highly indurated or has low permeability or porosity.

Only relatively unreactive species such as chloride tend to give rise to data that are representative of *in situ* pore-water compositions, although even these data must be used with some caution. If data are required for other determinands, consideration should be given to methods which either try to minimize leaching artifacts or which attempt to correct for them. The correction method for clay materials described by Bradbury & Baeyens (1997), Pearson & Scholtis (1995) and Pearson *et al.* (1998) uses a combination of aqueous leaching, leaching with a nickel ethylenediamine cation displacement solution and geochemical modelling. Another approach described by Cave & Reeder (1995) uses chemometric data processing to correct for leaching artifacts. Subsequent papers by Cave & Harmon (1997) and Cave & Wragg (1997) have developed the chemometric mixture resolution approach to obtain fully quantitative results for leachate data.

Despite the drawbacks of leaching, it is still a useful and in some instances the only method that can be used for pore-water characterization. It is important, however, that its limitations are recognized.

Site investigation 1: Pliocene mudrocks at Orciatico, Tuscany, Italy

In several countries, mudrocks are a potential host rock for the disposal of certain types of radioactive wastes because of their low hydraulic conductivity and suitable mechanical properties. Although laboratory studies provide much useful information on shorter-term migration processes of solutes, including radionuclides, there are limitations when considering long-term performance assessment. Studies of natural sites with some similar features to a possible waste repository and/or its surrounding rocks (natural analogues) can allow insight into processes occurring over thousands or millions of years.

One such natural analogue site is near the village of Orciatico in Tuscany, Italy. Here the emplacement of a laccolith intrusion, approximately 1 km across, about 4 Ma ago provided a heat source that acted upon the surrounding Pliocene marine clays. The time for which this heat acted was the same order of magnitude as that expected for the heat caused by high activity nuclear waste (Leoni *et al.* 1986). Samples from three boreholes (I5, I1 and I2) were collected for pore-water characterization at increasing distance from the exposed contact with the intrusion (40, 150 and approximately 1000 m respectively).

Since the moisture contents of the clay sequences were in the order 11 to 27% with respect to dry weight, hydraulic squeezing was the most appropriate extraction technique. Immediately after collection, samples were preserved in tightly fitting plastic tubes capped with wax. Samples were squeezed using the method and apparatus described by Entwisle & Reeder (1993). Between 20 and 94 g of pore-water was obtained from original sample masses of between 577 and 813 g; only one clay core, of moisture content 10.7%, failed to yield any sample. Squeezing was carried out at pressures up to 70 MPa for a maximum 551 h. The extracted pore-waters were analysed for a range of major and trace inorganic constituents and for stable $\delta^{18}O$ and δ^2H isotope compositions.

The pore-water compositions obtained are all more dilute than seawater, indicating that local meteoric water has infiltrated the clays to a depth of at least 50 m displacing the original saline water.

Most analyte concentrations decrease with depth, giving rise to concentration gradients that appear to represent diffusion profiles. Chloride (Fig. 1a) and other conservative analytes (e.g. sodium and stable isotopes) give rise to a series of approximately parallel concentration profiles for each borehole. These appear to show a relationship to topography, with analyte concentrations increasing with distance from the contact and decreasing altitude. This may reflect gradual evaporation of surface water migrating downhill from the contact towards borehole I2, resulting in an increase in solute concentration.

Sulphate (Fig. 1b) and other more reactive species (e.g. Ca, Mg, K, Al) show a single concentration profile with no differentiation between data from different boreholes. These data may reflect solubilization resulting from pyrite oxidation and carbonate and silicate dissolution.

Two of the deeper samples from borehole I1 show significantly different pore-water chemistry to their neighbours. This may reflect relatively permeable regions that are conducting fluids from different parts of the system. Alternatively, the samples may contain pyrite-rich horizons which oxidize after sampling resulting in a reduction in pH and an increase in carbonate dissolution. There is no evidence for changes in the pore-water chemistry resulting from intrusion of the laccolith, although rainwater infiltration has probably removed all traces of any change that did occur.

The results from this study are thought to provide evidence for diffusion of solutes into the clay formation, from which it may be possible to determine transport properties of the bulk rock. This in turn will enable a better understanding of the migration of contaminants, including radionuclides, within mudrocks.

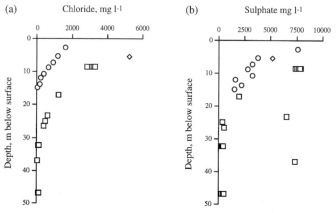

Fig. 1. Chloride (**a**) and sulphate (**b**) squeezed pore-water concentration profiles for three boreholes in Pliocene mudrocks at Orciatico, Tuscany, Italy. Boreholes I5, I1 and I2 (represented by open circles, open squares and an open diamond respectively) are at increasing distance from the exposed contact with the intrusion (40, 150 and approximately 1000 m). Adapted from Rochelle *et al.* (1998).

Site investigation 2: Vraconian (Lower Cretaceous) mudrocks at Marcoule, Southern France

As described in the previous study, argillaceous materials potentially have highly suitable containment properties, and are under consideration for the deep disposal of low and intermediate radioactive wastes in a number of countries. One such investigation is being carried out by the French radioactive waste management agency, ANDRA, in a Vraconian (Lower Cretaceous) mudrock formation near Bagnol-sur-Cèze in southern France.

Unlike the Orciatico study, the clay formation has a very low moisture content, generally no greater than 10% with respect to dry weight and typically less than 5%. Hydrochemical characterization is therefore only possible by studying residual clay pore-waters since flowing groundwaters cannot be collected. Since the limit of the squeezing technique is about 7 to 8% moisture content, squeezing tests had to be supplemented by aqueous leaching.

Samples collected from deep boreholes (up to 1500 m) were preserved by sealing in aluminized bags previously flushed with nitrogen gas and then vacuum evacuated. The extraction of pore-waters by mechanical squeezing was attempted using the method and apparatus described by Entwisle & Reeder (1993). Squeezing was carried out at pressures up to 70 MPa for a maximum of 427 h. Of twelve samples tested, with moisture contents in the range 3.1 to 9.3% with respect to dry sample weight, only those five with moisture

contents greater than 6.4% successfully yielded pore-water. Between 1.77 and 6.96 g of pore-water was obtained from original sample masses of between 536 and 592 g.

The aqueous leaching method described by Cave & Reeder (1995) was used to supplement the squeezed pore-water data. This involved adding 5 g of milled sample to 30 ml of deionized water and shaking at room temperature for 24 h. The resulting supernatant leachate was centrifuged for 0.25 h at 13 500 rpm and filtered to 0.45 μm. The aqueous leachate data were corrected to an equivalent concentration in the pore-water using the sample moisture content.

The pore-water chloride data from two boreholes, MAR203 and MAR403, are plotted in Figs 2a and 2b respectively. Differences in chloride content are observed between aqueous leachate data and pore-water data obtained by squeezing, the reasons for which are not fully understood but may reflect problems in the correction of aqueous leachate data using sample moisture contents. Nevertheless, both boreholes show chloride concentrations decreasing from the middle of the formation to the edges (samples for borehole MAR403 are only from the lower part of the clay formation) and suggest that diffusion of solutes has occurred.

Assuming that the formation is derived from deposition of marine sediments, the initial pore-water composition would approximate to the salinity of seawater. Pure diffusion computer modelling of these data have enabled ANDRA to predict that the chloride profiles could have been derived from seawater diffusing out of the formation over a period of 100 Ma

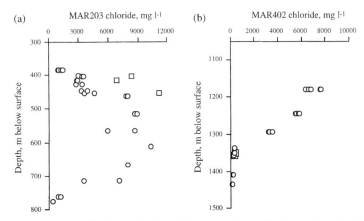

Fig. 2. Estimated pore-water chloride profiles for a Vraconian (Lower Cretaceous) mudrock formation in Southern France. Borehole MAR203 (**a**) profiles the whole formation, whilst MAR402 (**b**) covers the middle to base of the target formation. Data derived from both aqueous leaching of residual solutes (open circles) and mechanical squeezing (open squares) are shown. Adapted from Cave & Reeder (1997).

(Harmand 1997). This timescale for diffusion can be used to support the case for the formation to be potentially acceptable as a repository site.

Site investigation 3: Permo-Triassic red beds and Ordovician volcanics at Sellafield, Cumbria, England

Although argillaceous rocks have been studied extensively, other geological formations may potentially have suitable containment properties. Until April 1997, a site near Sellafield, northwest England was being assessed by the United Kingdom Radioactive Waste Executive, UK Nirex, for its suitability to host a radioactive waste repository. The site consists of a sequence of relatively permeable sedimentary rocks, comprising mainly Permo-Triassic red beds. These rocks overlie the potential host formation, a sequence of low permeability Ordovician volcanic basement rocks, the Borrowdale Volcanic Group (BVG).

Many groundwaters were sampled from deep (up to c. 2 km) boreholes. However, the sampling zones were relatively far apart (typically several tens of metres) owing to the relatively low frequency of fractures or sufficiently permeable matrices from which groundwater could be induced to flow into the boreholes. To supplement the data from sampled groundwaters and provide continuous geochemical profiles down the boreholes, rock core samples were collected to enable characterization of the pore-water composition (Cave et al. 1994).

Unlike the argillaceous lithologies described in the two previous studies, sandstones and volcanic rocks are not particularly suitable for pore-water extraction by squeezing. The sandstones, of moisture content 2.66 to 5.66% with respect to wet weight, were tested using the centrifugation method described by Kinniburgh & Miles (1983) using 11,2-trichloro-1,2,2-trifluoroethane as a displacent. However, the only method available for testing the very low porosity basement volcanics (moisture content 0.79 to 1.96%) was aqueous leaching. The method used was the same as described in Site Investigation 2. For comparison, aqueous leaching was also carried out on the sandstones.

The extracted pore-waters and aqueous leachates were analysed for a range of major and trace inorganic constituents. Because of the relatively high levels of drilling fluid contamination observed in these samples, a simple mathematical correction was applied to the data by subtracting the contamination for each determinand using the ratio of LiCl tracer originally contained in the drilling mud to that in the raw sample. In addition, the multivariate statistical analysis described by Cave & Reeder (1995) was used to try and eliminate some of the problems associated with solute enhancement experienced in the aqueous leaching technique.

For chloride, excellent agreement is obtained between pore-water data obtained by both centrifugation extraction (Fig. 3a) and recalculation of aqueous leaching residual solute data (Fig. 3b) compared to discrete flowing groundwaters. Good agreement is also achieved for Br, Na and, to a lesser extent, Mg, Ca, Sr and SO_4. Other determinands (e.g. Si, K and HCO_3) give rise to compositions that are significantly enhanced compared to expected values.

Although further work is required to understand the mechanisms of the leaching process and to provide an insight into the elevation of some solutes, the pore-water data provide a more comprehensive coverage than would be possible using conventional groundwater data and show fine detail that has proved invaluable in interpreting the hydrogeochemical regime at Sellafield. For example, the chloride data have been used to provide evidence for a vertical saline transition between two distinct groundwater regimes occuring within the Sherwood Sandstone Group (Bath et al. 1996).

Discussion

The previous sections have shown that pore-water extraction and solute leaching procedures can provide useful information concerning the general chemical characteristics of pore-waters from low-permeability media. In particular, it is possible to establish levels of total salinity, determine whether systems are chloride dominated and provide useful information concerning chemical gradients.

The theoretical prediction of whether a particular pollutant will be soluble (and hence potentially mobile) or insoluble (and hence potentially immobile) using pore-water chemistry data is more problematical. This limitation arises because it is impossible to derive reliable estimates of in situ concentrations for several key constituents using the techniques described. In particular, reliable data for in situ pH and redox conditions (Eh) cannot be acquired. This is particularly true when pore-waters cannot be extracted directly and compositions must instead be estimated from leachate data. Although there are limitations in the use of pore-water data to

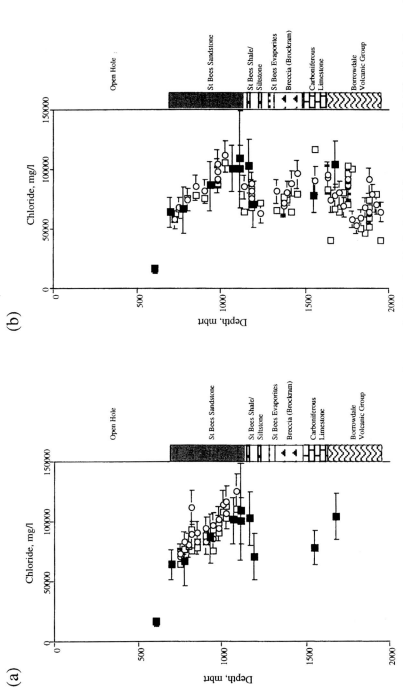

Fig. 3. Pore-water chloride profiles obtained by centrifugation extraction (**a**) and aqueous leaching of residual solutes (**b**) from a deep Sellafield borehole. Pore-water data reconstructed by simple mathematical correction (open circles) and multivariate statistical analysis (open squares) are compared to groundwater data obtained from hydraulic tests (closed squares). Depths are m below rotary table (mbrt). The 2σ error bars are based on a combination of analytical and sampling errors. Adapted from Bath *et al.* (1996).

predict the relative importance of the chemical and physical containment properties of a given medium, pore-water data may still be used to estimate the overall efficiency with which a low-permeability medium is likely to contain waste in two main ways.

Firstly, the chemical data obtained from pore-waters or leachates may be used to constrain the physical hydrogeological properties of low-permeability media used in waste containment. Such data can provide evidence for compositional gradients that may be compared to the outputs of theoretical models for diffusive and/or advective transport of water and solutes. These models may be constructed to study natural situations that are analogous to waste disposal scenarios or compositional gradients in low-permeability media surrounding actual waste disposal sites. They may also be applied to compositional gradients established within low-permeability media in laboratory experiments.

Secondly, data can be acquired for actual or potential pollutants, including heavy metals, within the pore-waters of a low-permeability medium. The concentration gradients shown by these pollutants can be used to establish the efficiency of the containment medium and evaluate whether it might be used to contain waste effectively in the future. As above, pollutants may be evaluated both at natural analogues and in media surrounding actual waste disposal sites. It may also be possible to conduct experiments in which blocks of containment media are placed in contact with aqueous solutions containing pollutants. The migration of these pollutants through the media might then be evaluated by obtaining and analysing pore-waters and/or leachates.

Laboratory experiments have significant potential for producing chemical data for pore-waters and/or leachates that can be used to help evaluate the relative importance of physical and chemical containment properties of low-permeability media. To date, there has been little application of experiments in this way. However, it is worthwhile noting that the chemical and physical properties of low-permeability media tend to be coupled (e.g. Horseman *et al.* 1996; Metcalfe *et al.* 1999).

Conclusions

This paper has demonstrated the value of pore-water characterization techniques in providing hydrogeological data for the evaluation of containment properties of geological and engineered barriers. The characterization of pore-waters is particularly important since flowing groundwaters cannot generally be obtained from the low permeability materials used for containment barriers.

As demonstrated by the case studies, all three of the commonly used extraction methods have the capability to provide unique geochemical data. Pore-water squeezing is the most generally applied method, particularly in the study of argillaceous formations. Centrifugation is less useful for investigating low permeability rocks, but can provide useful data for more permeable lithologies such as sandstones. Aqueous leaching is often used as a last resort, but can provide useful data for conservative species such as chloride.

Although the choice of method is often dependent on the rock type, a combination of several extraction and characterization techniques may frequently be used to provide data that aid understanding of hydrogeological features, particularly when pore-water compositions can be compared with groundwater data.

It should be noted, however, that there are many uncertainties associated with pore-water extraction and characterization, and that more work is required to develop extraction procedures and data correction methods to validate the data obtained. In particular, further studies are required to understand and evaluate differences in the relative compositions of pore-waters in more (fractures or flow zones) and less (rock matrix) permeable parts of rock formations.

The authors wish to acknowledge ANDRA, Nirex and the European Commission DG XII for funding aspects of the work presented in this paper. The paper is published by permission of the Director of the British Geological Survey (NERC).

References

BATH, A. H. 1993. Clays as chemical and hydraulic barriers in waste disposal: evidence from pore-waters. *In*: MANNING, D. A. C., HALL, P. L. & HUGHES, C. R. (eds) *Geochemistry of Clay–Pore Fluid Interactions*, Chapman & Hall, London, 316–330.

——, ENTWISLE, D. C., ROSS, C. A. M. *et al.* 1988. Geochemistry of pore-waters in mudrock sequences – evidence for groundwater and solute movements. *In*: *International Association of Hydrogeologists Symposium on Hydrogeology and Safety of Radioactive and Industrial Hazardous Waste Disposal*, Orléans, France, 7–10 June 1988, **1**, 87–97.

——, McCARTNEY, R. A., RICHARDS, H. G., METCALFE, R. & CRAWFORD, M. B. 1996. Groundwater chemistry of the Sellafield area: a preliminary interpretation. *Quarterly Journal of Engineering Geology*, **29**, S39–S57.

BRADBURY, M. H. & BAEYENS, B. 1997. *Derivation of in situ, Opalinus Clay Porewater Compositions from Experimental and Geomechanical Modelling Studies*. Paul Scherrer Institut Technical Report **97–14**.

BRANDL, H. & HANSELMANN, K. W. 1991. Evaluation and application of dialysis porewater samplers for microbiological studies at sediment-water interfaces. *Aquatic Sciences*, **53**(1), 55–73.

BUFFLAP, S. E. & ALLEN, H. E. 1995. Sediment pore water collection methods for trace metal analysis: a review. *Water Research*, **29**(1), 165–177.

CAVE, M. R. & HARMON, K. A. 1997. Determination of trace metal distributions in the iron oxide phases of red bed sandstones by chemometric analysis of whole rock and selective leachate data. *Analyst*, **122**, 774–781.

——— & REEDER, S. 1995. Reconstruction of *in situ* porewater compositions obtained by aqueous leaching of drill core: An evaluation using multivariate statistical analysis. *Analyst*, **120**, 1341–1351.

——— & ———1997. *Methodologies of Pore-Water Extraction from Low Permeability Argillaceous Rock Samples: Results and Discussion from the Gard Site*. Atlas des Posters Journées Scientifiques CNRS/ANDRA, Bagnols-sur-Cèze, 20 et 21 Octobre 1997, 55.

——— & WRAGG, J. 1997. Measurement of trace element distributions in soils and sediments using sequential leach data and a non-specific extraction system with chemometric data processing. *Analyst*, **122**, 1211–1221.

———, REEDER, S. & METCALFE, R. 1994. Chemical characterization of core pore-waters for deep borehole investigations at Sellafield, Cumbria. *Mineralogical Magazine*, **58A**, 158–159.

COTTOUR, P. 1997. *Projet Mont Terri: Prélèvement des Fluides Interstitiels* in situ *et sur Échantillons Carottès (Expérience Water Sampling)*. Atlas des Posters Journèes Scientifiques CNRS/ANDRA, Bar-le-Duc, 27 et 28 Octobre 1997, 59.

EDMUNDS, W. M. & BATH, A. H. 1976. Centrifuge extraction and chemical analysis of interstitial waters. *Environmental Science and Technology*, **10**, 467–472.

ENTWISLE, D. C. & REEDER, S. 1993. New apparatus for pore fluid extraction from mudrocks for geochemical analysis. *In*: MANNING, D. A. C., HALL, P. L. & HUGHES, C. R. (eds) *Geochemistry of Clay-Pore Fluid Interactions*. Chapman & Hall, London, 365–388.

GRENTHE, I., STUMM, W., LAAKSUHARJU, M., NILSSON, A-C. & WIKBERG, P. 1992. Redox potentials and redox reactions in deep groundwater systems. *Chemical Geology*, **98**, 131–150.

HARMAND, B. 1997. *Interprétation des Profils de Chlorures et de Bromures dans la Couche Silteuse de Marcoule par un Modèle en Diffusion Pure*.

Atlas des Posters Journées Scientifiques CNRS/ANDRA, Bagnols-sur-Cèze, 20 et 21 Octobre 1997, 58–59.

HORSEMAN, S. T., HIGGO, J. J. W., ALEXANDER, J. & HARRINGTON, J. F. 1996. *Water, Gas and Solute Movement Through Argillaceous Media*. Nuclear Energy Agency (NEA) Organisation for Economic Co-operation and Development (OECD), Paris.

KEYS, W. S. 1997. *A Practical Guide to Borehole Geophysics in Environmental Investigations*. CRC, Boca Raton, FL.

KINNIBURGH, D. G. & MILES, D. L. 1983. Extraction and analysis of interstitial water from soils and mudrocks. *Environmental Science and Technology*, **17**(6), 362–368.

KRUIKOV, P. A. 1947. [Recent methods for the physico-chemical analysis of soils: Methods for separating soil solution]. *In*: *Rukovodstvo dlya polevykh I laboratornykh issledovanii pochv*, Moscow Izdat, Akad, Nauk, SSSR, 3–15 [in Russian].

LEONI, L., POLIZZANO, C. & SARTORI, F. 1986. Nuclear waste repositories in clays: the Orciatico metamorphic aureole analogy. *Applied Clay Science*, **1**, 385–408.

MANHEIM, F. T. 1966. *A hydraulic squeezer for obtaining interstitial water from consolidated sediments*. US Geological Survey Professional Paper **550-C**, C256–C261.

METCALFE, R. 1999. The chemical containment properties of argillaceous media: a review. *This volume*.

MOREAU-LE GOLVAN, Y., MICHELOT, J-L. & BOISSON, J-Y. 1997. Stable isotope contents of porewater in a clayey formation (Tournemire, France): assessment of the extraction technique and preliminary results. *Applied Geochemistry*, **12**, 739–745.

PEARSON, F. J. & SCHOLTIS, A. 1995. Controls on the chemistry of pore water in a marl of very low permeability. *In*: KARAKA, Y. K. & CHUDAEV, O. (eds) *Water–Rock Interaction*. Balkema, Rotterdam, 35–38.

———, WABER, H. N. & SCHOLTIS, A. 1998. Modelling the chemical evolution of porewater in the Palfris Marl, Wellingberg, Central Switzerland. *In*: MCKINLEY, I. G. & MCCOMBIE, C. (eds) *The Scientific Basis for Nuclear Waste Management XXI*. Materials Research Society, 789–796.

ROCHELLE, C. A., ENTWISLE, D. C. & REEDER, S. 1998. Porewater chemistry of the Orciatico clays. *Mineralogical Magazine*, **62A**(1–3), 1281–1282.

SAVAGE, D. 1995. *The Scientific and Regulatory Basis for the Geological Disposal of Radioactive Waste*. Wiley, Chichester.

ZHANG, H., DAVISON, W., KNIGHT, B. & MCGRATH, S. 1998. *In situ* measurements of solution concentrations and fluxes of trace metals in soils using DGT. *Environmental Science and Technology*, **32**, 704–710.

A probabilistic risk assessment methodology for landfills, with particular reference to the representation of chemical containment

B. R. PLIMMER, A. B. PRINGLE & S. J. MONCASTER

Golder Associates (UK) Ltd, Landmere Lane, Edwalton, Nottingham, NG12 4DG, UK

Abstract: In Waste Management Paper 26B, the Department of the Environment advocate the use of risk assessment in the design of waste disposal sites, such that the potential impact of the facility on specified receptors may be identified and minimized. In this paper the tool currently used by the regulatory authorities to assess the risk to groundwater (LandSim) is examined, paying particular attention to the treatment of chemical containment. A case study is presented demonstrating two different approaches to the modelling of the potential attenuation of ammonia afforded by the geosphere.

LandSim – a computer-based probabilistic risk assessment model

One of the concerns of the Environment Agency in the UK is the risk posed to groundwater by the disposal of waste in landfill sites. Groundwater itself may be seen as both a pathway (for example, where there is discharge to a surface water course) and a receptor (particularly where there is groundwater abstraction).

'LandSim' is a computer based probabilistic model for assessing the extent of leachate containment within waste disposal sites and the risk of groundwater contamination. The software was produced by Golder Associates (UK) Ltd under contract to the Department of the Environment and the (then) National Rivers Authority (now the Environment Agency). The aim was to provide a user friendly tool for the quick and consistent assessment of licence applications to aid in the decision making process. LandSim has therefore been designed to use data generally available within the application documents. A LandSim assessment does not, on its own, constitute a risk assessment, and the use of LandSim as a *tool* should not be interpreted as use as a *decision maker*. The decision as to whether or not to license a waste disposal facility should never be based purely on the results of a LandSim simulation.

The approach taken within the model is to describe the physical and chemical processes involved using a series of mathematical equations. The uncertainty in the equation parameters is represented through the definition of a range of parameter values describing the probability or chance of each value occurring. In order to incorporate the parameter distributions in the calculation process, LandSim uses a Monte Carlo simulation technique. For further details on the LandSim model, the reader is referred to Hall (1997) or Golder Associates (1996).

Leachate attenuation

Of the chemical compounds in a typical leachate from household, commercial and industrial waste, ammonia has one of the highest concentrations relative to the specified drinking water and surface water standards. In this sense, ammonia is one of the most toxic species in landfill leachate. For this reason the potential for leakage and transport of ammonia away from the landfill is usually studied.

The chemical attenuation of ammonia in leachate is incorporated in LandSim in three ways (excluding the effects of dilution, which may be considered a physical process).

(1) Chemical reaction (accounted for in the initial concentration range) and the wash through of contaminants by clean infiltrating water may cause a decline in species concentration with time. Knox (1995) has described the wash through process in terms of an exponential decay, based on the rate of infiltration, the field capacity of the waste and the thickness of the waste. This exponential decay may optionally be included in a LandSim assessment.

(2) Retardation within the geosphere may be simulated through the inclusion of a contaminant specific retardation factor.

From: METCALFE, R. & ROCHELLE, C. A. (eds) 1999. *Chemical Containment of Waste in the Geosphere.* Geological Society, London, Special Publications, **157**, 275–280. 1-86239-040-1/99/$15.00 © The Geological Society of London 1999.

(3) As an alternative to the above, the attenuation of ammonia may be assessed by considering the process of cation exchange in the liner and geosphere.

A schematic diagram illustrating the interaction of these three processes within the Land-Sim model is shown in Fig. 1. The attenuation afforded by these processes may be simulated probabilistically at a borehole or surface water body downgradient of the landfill.

Leachate attenuation by retardation

The retardation factor describes the extent to which the movement of a species is retarded compared to bulk water flow. The factor is used as a multiplier on the unretarded travel time, and is defined in LandSim using a partition coefficient, K_d, for each species ($K_d = 0$ for unretarded species):

$$R_f = 1 + K_d \times (P/n) \qquad (1)$$

where R_f = retardation factor for contaminant (dimensionless), K_d = partition coefficient for contaminant (ml g^{-1}), P = bulk density of material (g cm^{-3}), n = effective porosity of material (fraction).

The partition coefficient is specific to a given contaminant under given geological and chemical conditions, and care must be taken in its selection. The calculated retardation in LandSim is independent of the initial range of contaminant

concentrations selected, and is therefore particularly useful where there is great uncertainty over leachate composition. LandSim incorporates retardation in both the unsaturated zone and the aquifer, but not within the liner.

Attenuation of leachate through cation exchange

For ammonia, retardation is caused primarily through the sorption of ions onto soil particles, a process known as cation exchange. Therefore, in the case of ammonia, LandSim allows an alternative examination of the extent to which cation exchange affords attenuation. In order to avoid including the process twice, this option is not available in the same simulation as the calculation of a retarded travel time. The total available cation exchange capacity in both the clay liner (if present) and the unsaturated zone is calculated and is compared with the cation loading in the leaking leachate, allowing an estimation of the likely time before the cation exchange capacity is exhausted. It is assumed that following this time ammonia will reach the water table.

$$CEC_{avail} = V \cdot P \cdot CEC \cdot re \qquad (2)$$

where CEC_{avail} is the available cation exchange capacity in the liner/unsaturated zone (meq), V is the volume of liner/unsaturated zone (m^3), P is the bulk density of liner/unsaturated

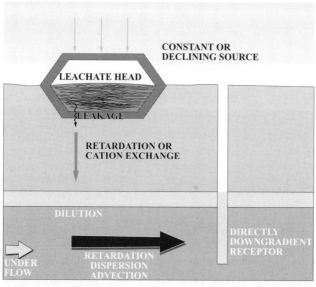

Fig. 1. Schematic illustration of LandSim model.

zone ($kg\,m^{-3}$), CEC is the cation exchange capacity of liner/unsaturated zone ($meq\,kg^{-1}$) and re is the reaction efficiency (dimensionless).

$$LR = Q \cdot C \qquad (3)$$

where LR is the loading rate of contaminant ($meq\,a^{-1}$), Q is the leachate leakage rate ($l\,a^{-1}$) and C is the concentration of contaminant ($meq\,l^{-1}$) giving, for a constant leachate concentration

$$t = CEC_{avail}/LR \qquad (4)$$

and for a declining leachate concentration

$$t = -1/\lambda \ln \{1 - [(CEC_{avail} \cdot \lambda)/(Q \cdot C)]\} \qquad (5)$$

where t is the time to exhaust the cation exchange capacity (a), λ is a parameter describing the decline of the source concentration.

The use of a reaction efficiency in the calculation is designed to include the effects of, for example, other cations competing for exchange sites. Unlike the alternative examination of contaminant retardation, the calculation of the time taken for the cation exchange capacity of the liner and soil to become exhausted is highly dependent upon leachate concentration. Where the range of values used to describe the initial ammonia concentration spans an order of magnitude or more (as do the LandSim default values, based on a study reported by Robinson (1995) encompassing various ages and types of wastes), the resultant time to exhaustion is likely to reflect this. Within LandSim, attenuation due to cation exchange is considered negligible when leachate leakage is through defects in a membrane liner (with a small cross-sectional area of flow), in the unsaturated zone when flow is through fractures, and in the aquifer, where the movement of water may be too fast to allow efficient cation exchange. Such an approach results in a conservative assessment of leachate attenuation, which is in line with the intended use of the software.

When considering the migration of ammonia, the type and perceived accuracy of the data available will determine which of the two approaches is most appropriate.

Although it may be argued that the methods used in LandSim are simplified, it may be seen through the use of a case study that they are justified in the context of a first-pass decision making process.

Case study: Burntstump landfill

Burntstump landfill is located approximately 8 km north of Nottingham. The first recorded waste disposal occurred in two natural valleys in 1974, although it is believed that there was uncontrolled tipping prior to this date. Phase 3, with which this case study is concerned, was filled to a depth of approximately 6 m with domestic waste and builders' rubble in three discrete layers. A thin topsoil capping layer is thought to have been placed in about 1981, and there is no record of any lining system being installed.

Burntstump landfill is underlain by the Sherwood Sandstone, a major aquifer generally about 150 to 200 m thick. Groundwater flows through both matrix and fissures and the water table lies around 47 m beneath the waste in Phase 3 (annual fluctuation of groundwater level is approximately 1 m).

The definition of a leachate plume at Burntstump has been achieved through repeated coring and pore-water analysis over a 12 year period and through bulk groundwater sampling at 2 to 3 year intervals. Pore-water chemistry depth profiles in the matrix have been constructed to identify the position of the pollution plume relative to the waste, and the front currently lies within the unsaturated zone. The average and current migration rates of ammoniacal nitrogen in the unsaturated zone matrix have been used to estimate a total retarded matrix travel time to the water table of about 90 years (compared to a predicted 30 years for the conservative species chloride).

LandSim has been used to model the passage of chloride and ammoniacal nitrogen through the unsaturated zone matrix at Burntstump. Input parameter distributions have been determined using reported data (Lewin et al. 1994; Thornton et al. 1995), both site specific and from analogue sites. Default ranges specified within LandSim have been used for some parameters in the absence of more appropriate data.

LandSim was run initially to confirm that the simulated travel time for chloride matches that predicted using the results of pore-water analysis. At Burntstump, limited flow has been reported in fissures in the unsaturated zone (Lewin et al. 1994). Although LandSim permits an examination of dual porosity flow in the unsaturated zone, the model has been primarily designed for Chalk aquifers in which it is assumed the matrix is fully saturated, and the methodology may not be appropriate in the Sherwood Sandstones. Therefore, it was assumed in this instance (somewhat arbitrarily) that only around 40% of the total quoted infiltration moved within the matrix, leaving the remaining 60% for matrix flow. This value was input to LandSim using the porous medium option. The results of the chloride unsaturated zone matrix travel time showed a close match to the field predictions of around 30 years.

The partition coefficient at Burntstump has been quoted as $0.073–0.106 \, ml \, g^{-1}$ (Thornton et al. 1995). The modelled unsaturated zone travel time using these values is given in Fig. 2, and shows the predicted actual time of around 90 years to fall well within the range at the 60th percentile. It would generally be expected that when using the best estimates of parameter values, the actual result would fall at around the 30th to 70th percentile. Assuming the input value of K_d is appropriate, a similar match would be expected for other retarded species.

Figure 3 presents the predicted time for the cation exchange capacity of the unsaturated zone to become exhausted, using reported cation exchange capacity values for a site close to Burntstump of between 10 and $60 \, meq \, kg^{-1}$ (most likely value of $48 \, meq \, kg^{-1}$), and a reaction efficiency of 8 to 10% (Young et al. 1995). Like the retarded result, the predicted actual time of around 90 years falls well within the range at the 68th percentile. The two results, as would be expected, are similar, with the cation exchange capacity approach indicating a greater degree of conservatism, and reflecting the large range in initial ammoniacal nitrogen concentration.

If the program were to have been used in the assessment of a licence application, it is likely that results falling at the 50th percentile would not have been considered to have allowed a sufficient margin for inherent uncertainties; it is far more usual to consider the 90th or 95th percentile. The two methods compare well at these percentiles, with the retarded travel time predicted at about 80 years and cation exchange capacity modelled to last almost 30 years (90th percentile). Again, the cation exchange capacity method reflects conservatism.

Strengths and weaknesses

In common with all assessment methods, the validity of the results calculated through Land-Sim is highly dependent on the accuracy of the input parameters. Although the inclusion of a range to incorporate parameter uncertainty gives the probabilistic method a considerable advantage over traditional 'worst case' scenarios, the use of a parameter distribution must not be considered a substitute for data collection. Similarly, the expression of too wide a range, indicating that the user has little or no specific data, while probably encompassing all eventualities may, nonetheless, be meaningless.

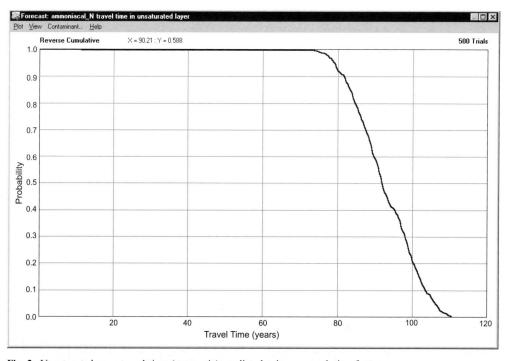

Fig. 2. Unsaturated zone travel time (ammonia) predicted using a retardation factor.

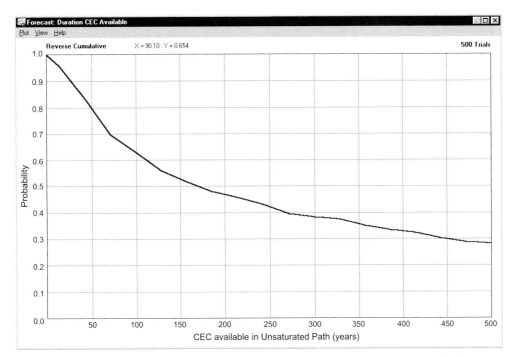

Fig. 3. Unsaturated zone travel time (ammonia) estimated using predicted time to exhaustion of cation exchange capacity.

In the same vein, user friendliness should not be mistaken for absolute simplicity. It is essential that LandSim is used only by those familiar with the concepts of hydrogeology and contaminant transport processes, since it is vital that the correct parameters are incorporated and their effects upon the results recognised.

LandSim has not been designed to make decisions. The impact of the calculated results must be carefully considered, and they cannot be taken in isolation. The choice of confidence level is not trivial.

LandSim is particularly useful when comparing landfill designs, and allows optimization of the containment system at the design stage to permit maximum use of the degree of chemical containment afforded by the engineered and natural zones.

event may be used with this likelihood to make an informed decision. LandSim has been set up to provide a consistent method for Regulators to apply the risk assessment, but is intended purely as a tool in assisting the decision making process. It is vital that the carefully considered, site-specific input parameters are used.

The authors would like to thank all colleagues who have been involved in the development of 'LandSim' and who have assisted in the production of this paper. The development of the 'LandSim' program was carried out for the NRA (now Environment Agency) and Department of the Environment under the contract 'Risk Assessment Methodology for Landfill Engineering' EPG 1/7/04/CLO 172. Particular thanks go to J. Gronow and B. Harris of the Environ-ment Agency for their assistance. The views expressed in this paper are solely those of the authors and are not necessarily those of either the Department of the Environment or the Environment Agency.

Conclusion

Probabilistic risk assessment methodology allows a decision to be made based on the likelihood of certain events occurring, and allows movement away from the concept of a worst case scenario. The importance of the effects of the

References

GOLDER ASSOCIATES. 1996. *LandSim: Landfill Performance Simulation by Monte Carlo Method.* Report to the DOE CWM 094/96 under Contract EPG 1/7/04 CLO172.

HALL, D. H. 1997. Risk and Performance Assessment of Landfills: The LandSim Model. *In: Proceedings Sardinia 97, Sixth International Landfill Symposium, Cagliari, Italy.*

LEWIN, K., YOUNG, C. P., BRADSHAW, K., FLEET, M. & BLAKEY, N. C. 1994. *Landfill Monitoring Investigations at Burntstump Landfill, Sherwood Sandstone, Notts: 1978–1993.* Report CWM 035/94.

KNOX, K. 1995. *Leachate Recirculation and its Role in Sustainable Development.* 5th Joint INCPEN/ IWM/BPF Seminar.

ROBINSON, H. D. 1995. *A review of the composition of leachates from domestic wastes in landfill sites.* Department of the Environment, CWM/072/95.

THORNTON, S. F., LERNER, D. N. & TELLAM, J. H. 1995. *Laboratory Studies of Landfill Leachate – Triassic Sandstone Interactions.* Final Report to the DoE No CWM 035A/94.

YOUNG, C. P., FLEET, M., LEWIN, K., BLAKEY, N. C. & BRADSHAW, K. 1995. *Landfill Monitoring Investigations at Gorsethorpe Landfill, Sherwood Sandstone, Notts: 1978–1992.* Report CWM/034/94.

Index